T0299312

CHEMICALS USED FOR ILLEGAL PURPOSES

Disclaimer

The author and publisher take no responsibility for the actions of anyone as a result of using material in this book. People who use the book to make or prepare drugs, fireworks, explosives, toxins, tear agents, or any of the other products described in the book take full responsibility for their actions. We have taken great pains NOT to give all of the information necessary to make these products. Retention times, some equipment, and cooking temperatures are purposefully left out. We realize that a very good chemist will have the knowledge to fill in the blanks, but amateurs are at great risk mixing these chemicals. This is especially true in the explosives formulas, where proper temperatures can be very critical during a reaction, and an explosion can occur if the temperature is not correct. Any attempt to sue or bring about any form of legal action against the author, or publisher, as a result of injury, death, or property damage caused by the negative intentions of a person or persons who used information in this manual is a direct violation of freedom of speech laws and information right-to-know.

Information contained in the book was taken from public sources and is therefore bound to freedom-of-speech and information protection laws as discussed in the U.S. Constitution. The information in the book is greatly abbreviated. Although we expect that, if used with proper equipment and technique, most of the formulas will produce the product expected, we are also aware of formulas included herein that do not work. These were included because they were in the public domain and could be expected to be tried by persons who were intending some mischief. We did not make any attempt to ensure that a safe, reliable product could be produced using these formulas, as making these products is not the intention of this book.

The book is intended for the use of emergency responders encountering chemicals at a home, abandoned, or at a facility, so that a responder might determine the intention of the person using those chemicals. There is no other purpose. The information is for reference purposes only, and the author and publisher made the book possible to inform, enlighten, and educate persons encountering stashes of chemicals in use or abandoned.

CHEMICALS USED FOR ILLEGAL PURPOSES

A Guide for First Responders to Identify Explosives, Recreational Drugs, and Poisons

ROBERT TURKINGTON

WILEY

A JOHN WILEY & SONS, INC., PUBLICATION

Library of Congress Cataloging-in-Publication Data:

Turkington, Robert.
    Chemicals used for illegal purposes : a guide for first responders to identify explosives, recreational drugs, and poisons / Robert Turkington.
        p. cm.
    Includes bibliographical references and index.
    ISBN 978-0-470-18780-7 (cloth)
    1. Hazardous substances–Identification–Handbooks, manuals, etc.   2. First responders–Handbooks, manuals, etc.   I. Title.

RC87.3.T87 2010
363.17–dc22

                                                                    2008047086

Printed in the United States of America

10 9 8 7 6 5 4 3 2 1

CONTENTS

This is a list of about 900 chemicals that are used to make illegal items. Although not emphasized, equipment is also listed whenever it provides a significant clue as to what activities a person may be engaged in. Wherever possible, each entry includes appearance, odor, hazards (including PELs, TLVs, MACs, or other estimates of hazard), chemical incompatibilities, physical characteristics such as vapor pressure and flash point, a list of the illegal products that can be made using the chemical, regulations (emphasizing DOT, but including EPA, CDTA, FDA, OSHA, DTCC, ATF, and the Australian Group where appropriate). When possible we list a spot test that can confirm that the product is as labeled.

Amphetamines are strongly emphasized in this book because they represent a large proportion of the illegal operations encountered. Two synthesis techniques are most commonly encountered. Four main types of reactions are used to make amphetamines: oxidation, reduction, substitution, and addition. Supplementary reactions used to make hydriodic acid and sodium metal, or to recimize l-methamphetamine are also included.

This section is the battle of the chemists. Although there is definitely activity here, the numbers of people making these drugs at present is very small, perhaps hundreds in the entire country. Most, but not all of these syntheses require a very good chemist and very decent laboratory facilities. Despite this, DEA laboratory personnel have spent considerable time in this area, and the Web is full of formulas to make these drugs.

Designer drugs mimic amphetamine, illegal drugs, and chemical weapon incapacitating agents because they are nothing more than modifications of those drugs. Before 1987, chemists and dealers were hoping to sell these modifications before the DEA changed the law to include the modification. As a result, the DEA revised the law in 1987 to use the effects of the drug, not the formula, as the method of determining if it is illegal. This did not stop the sale of designer drugs. [The law now calls them *controlled substances analogs* (CSAs)]. During the period of experimentation, it was found that many of the CSAs were more potent than the original drugs. One modification of fentanyl is 1000 times as potent as heroin, which means that a person using $600 worth of chemicals can make a kilogram of the drug, which would be the equivalent of a ton of heroin worth millions of dollars. The benefit of that is obvious. Also, many drugs that started out as

CSAs are now considered to be standard drugs. Examples include Cat (methcatanone) and Ecstasy. Little, if any, research has been done on the toxicology or effects of most of these drugs. In this section we often do not even try to give formulas, because, for example, there are over 1000 variations of fentanyl alone. Some of these would have no effect, whereas others might cause death. Most are very unlikely ever to be seen, or are likely to be tried only once. Instead, we include only key formulas or chemicals, which would always be expected to be there. Our only purpose in this book is to avoid a designer drug lab being passed off as just another meth lab and turned over to a cleanup company. The DEA has chemists who would understand the modifications if the chemicals were brought to their attention. Many of these chemicals are associated with plants. In each "cook" we begin where possible with making the drug from scratch, then describe how to extract the desired drug from the plant, then how to refine it for sale. Of these chemicals, probably cocaine, marijuana, and PCP are the most important.

These include drugs that may or may not be illegal, depending on how much the person has, the intent, and where the person is located.

In most of the United States, making ammunition is not illegal. This section is included because many of the chemicals used to make rockets, explosives, and fireworks are also used in the manufacture of ammunition.

In this section we describe how to make compounds that deflagrate and detonate. The range of compounds considered ranges from very sophisticated to bombs, which are crudely improvised. There are over 100 formulas in this section. The first seven processes are methods to create bomb-making ingredients that cannot be purchased.

The distinction between an explosion and a detonation is that in an *explosion*, the heat release rate and the number of molecules per unit volume increase with time more or less uniformly, whereas a *detonation* is propagated by an advancing shock front behind which exothermic reactions take place and thus is (spatially) nonuniform.

HMX, RDX, TNT, tetryl, HNS, picric acid, TATB, PETN, and nitroglycerin are considered to be high explosives. Astrolite, a liquid explosive, is considered to be the most powerful nonnuclear explosive: about twice as powerful as TNT. It is also easier to handle.

Pyrotechnics is a difficult and precise chemical artform. Trying to put the various parts into formulations would have been like trying to explain how an artist chooses colors. Since this book is not designed to go into that much detail, we have listed chemicals in various processes without giving the thousands of possible formulas. Note that depending on where the specific chemical is used, it can be a colorizer, an oxidizer, a fuel, or a retarder. The same chemicals are used over and over for many things. Most mixtures are made using a mortar and pestle.

Pyrotechnic mixtures are designed first to enhance burn rates, also often to retard or delay burn rates. Various oxidizers and fuels are mixed to provide mixtures that will ignite effectively and safely. They are designed to provide specific temperatures or conditions within a hot chemical soup to enhance color formation, spitzing strobes, flashing, and auditory reports. Pyrotechnics must also be designed to be propelled into the air, and to act as far away from their ignition source as possible.

Unlike other portions of this section, possible ingredients, not formulations, are provided. You will have to look through the various list of ingredients for the chemicals that you have encountered. Many are given in more than one list. It is remarkable how few chemicals show up in both explosives and fireworks.

Incendiary Devices (Arson)   310

Arson is more about technique than materials. A successful arsonist knows where to start a fire in a structure. Most of the time arson is very straightforward. In real life, Situations 0.1 through 0.5 are the most likely way to start fires. There are very few or perhaps no professional arsonists. Situations 1 through 9 are published in underground sources and therefore are included here. Incendiary devices are more closely related to terrorist activity. Identification of the crime is not a problem; it may even be a benefit. Arson has been a more common weapon than chemical weapons over the last 20 years of terrorist activity. In this book we do not address timers, which are a major component in setting off these devices as well as explosives.

5. CHEMICAL WEAPONS   313

In terms of casualties, chemical weapons were almost as effective as rifles in World War I. Certainly, the cost per casualty was lower using chemicals. The chemical weapons had one other advantage: terror. All the major powers in World War II had chemical weapons, yet none (with the possible exception of Japan) dared to use them, knowing that the enemy also had them. The present-day terrorist would not necessarily have the same qualms.

Data published by the FBI are pre-9-11; even so, the numbers are probably roughly realistic for today. Fully 70% of the terrorist events were bombings. Incendiary devices account for another 4% of the events. Chemical weapons are listed as 1%. The data are difficult to justify, as the same document mentioned 12 chemical releases (with injuries) in movie theaters, and two biological events, which amounted to more than the total number of chemical events. Depending on the resource quoted, butyric acid, a chemical agent used by anti-abortion groups, is the most common terrorist agent. Finally, events where cyanide was used to poison commercial products have been classified as criminal, not as terrorism. The Unibomber is not considered to be a terrorist. These numbers also do not assign relative importance to events. For instance, the anthrax events in 2001 might be recorded as three events or a thousand events, depending on how the hoaxes were tallied. Either would hide the actual effects of those three events to persons using only the data.

We greatly emphasize chemical weapons in this book, as an emergency responder encountering a facility making chemical weapons would be very likely to be injured. Unlike explosives, where a mortar and pestle are all that might be required, with chemical weapons the apparatus would often determine what is being made.

Nerve Agents   313

Vesicants   330

Blood Agents   338

Toxins are a very broad category. We have tried as best we could to stay with homespun poisons and poisons that have been seen in underground literature, on the Web, or have been tried either historically or recently. We have avoided regurgitating the many lists of such compounds that have been published.

Certain things, such as exotic animals, are not included. We find the possession of a ringed octopus, a cobra, or a black widow farm as not very likely.

Mushrooms are a readily available source of very effective poisons. However, identification of mushrooms is well beyond the scope of this book. Mushroom identification is difficult even for mushroom experts. If mushrooms are encountered, you should find an expert. Death cap mushrooms are determined in the field by amateurs by holding them over black paper and tapping them to knock out the spores, which are white on the black paper. If you do not see white spores, the test is NOT negative. Many other mushrooms are poisonous.

We have, at the urging of our reviewers, included toxins derived from bacteria (such as botulism). We have separated toxins from chemical weapons, as these all seem more personal, homicide-type chemicals. We feel that these are more appropriate to murdering one's mother-in-law than really being a method of mass murder.

6. CONFIRMATION TESTS   389

A confirmation test is only an indicator that a product is labeled correctly. If the product is what it is supposed to be, the test will be positive. However, there are many products that could also test positive. For instance, the alcohol test will pick out all alcohols, glycols, polyols, and glycerides. This means that although the product is probably labeled correctly, there is the possibility that the product is something else. If the product is labeled, appears as it is supposed to appear, and the confirmation test is positive, the responder has three positive indications. In some cases we may give two confirmation tests; for instance, for barium sulfate, we may give a barium test and a sulfate test. If both are positive, the identification is positive.

It is best to think of confirmation tests only as confidence boosters.

PREFACE

In a diverse, affluent, and educated society, with unlimited access to the Internet and relatively uncensored literature, it is not surprising to find that some people will use this information to try to make unavailable or illegal compounds in their own homes. Often, these compounds are being manufactured: for sale (e.g., illegal drugs), for curiosity (e.g., chemical weapons), by "mad chemists," or for political reasons (e.g., explosives). The chemicals that find their way into "at home" manufacturing facilities can be both exotic and dangerous.

Emergency responders might encounter such homemade manufacturing facilities as the result of odor complaints from neighbors, accidental explosions, fires, medical responses, or complaints from friends, neighbors, or family. Since safety is often neglected in "at home" manufacturing facilities, the responders may find themselves walking into an unknown and possibly very dangerous environment with no warning.

It is impossible to anticipate what hazards might be encountered. This manual provides a quick source of information regarding many of these unusual chemicals, and in some cases apparatus, which would indicate an extremely hazardous situation. It is our intention to help both the responder and the safety professional in assessing a potentially hazardous situation as quickly as possible. This manual is also designed to help in the safe breakdown and/or cleanup of "at home" manufacturing facilities.

At present, most of the information regarding these facilities is either in highly technical journal literature or in crude underground sources, which although accessible, would require some effort by users to obtain timely background information for a specific event. It would be difficult for a busy safety professional to locate and apply the information quickly in the detailed form necessary for a person making a "fight-or-flight" decision in the field.

I have tried to encompass any chemicals that might be found in "at home" manufacturing facilities. However, the research indicates that providing a total list of such chemicals or apparatus is impossible. The "at home" chemist is creative, and as any organic chemist will attest, there are an infinite number of ways to make a compound in a laboratory. I do feel that most of the very hazardous situations, such as chemicals and apparatus used to manufacture high explosives and chemical weapons, have been covered with due diligence.

An understanding of the fundamentals of "at home" manufacturing is very important to anyone who must deal with these unique situations. This manual is a resource for responders, cleanup companies, regulators, and safety and industrial hygiene consultants who back up the field personnel.

The manual should help the reader better understand the hazards of the setups and chemicals they encounter. An emphasis is placed on understanding the possibility of exposure and the risk of such exposure. There is information on how to measure employee exposure as well as how to ascertain the type or specific chemicals that have been encountered.

Although this manual is designed primarily as a reference for persons encountering an unusual collection of chemicals or observing unusual cooking apparatus, it has been used in classes where students were challenged to determine what type of chemical weapon was being manufactured. The section on methamphetamines has been used to train fire and law enforcement personnel on the hazards of dealing with clandestine methamphetamine laboratories.

By the very nature of this manual, the references are not always reliable. It is common enough to find an "at home" manufacturer who is following instructions on the Web from a very ill-informed person or from an underground manual in which processes are either questionable or just plain dangerous. In such cases I have tried to help responders understand that they are dealing with a flawed or unpredictable procedure. Where possible we have used sources such as the DEA, FDA, IRA, ATF, and chemical journal articles as our references. Therefore, neither I nor John Wiley & Sons, Inc. can assume responsibility for the content; nor can it be assumed that all acceptable safety measures are contained in this publication, or that additional measures may not be required under particular or exceptional conditions or circumstances.

This manual will be revised periodically. Contributions and comments from readers are invited.

ROBERT TURKINGTON

1

INTRODUCTION

Chemicals Used for Illegal Purposes is a reference book for police and other responders who might encounter chemicals upon entering a home, abandoned alongside the road, or in a facility. The book contains an alphabetical list of chemicals that can be used to make or are used in:

Ammunition	Designer drugs	Other
Amphetamines	Explosives	Recreational drugs
Arson	Fireworks	Tear agents
Chemical weapons	Hobbies (various)	Toxics
Date rape drugs	Illegal drugs	

The purpose of the book is to identify quickly what a person may be doing with any set of given chemicals. Although the majority of these products will be illegal, some, such as recreational drugs and fireworks, may be legal in some areas; hobbies generally include chemicals (although hazardous) that are legal. (We have included these at the request of people who have been embarrassed by overreacting to legal activities involving chemicals.) The inclusion of hobby chemicals may also be a help when determining the nature of an illegal chemical disposal along the side of a road.

A person finding a stockpile of chemicals can first look them up in the alphabetical list. Under each chemical will be a list of products that can be made using that chemical. A typical list entry follows. See the Glossary for definitions of all the acronyms used in the list.

ACETONE
CH_3COCH_3
67-64-1
UN 1090

Appearance: Colorless viscosity 1 liquid. Acetone is available in 1-quart or 1-gallon rectangular metal containers that can be purchased at any hardware store. It can also be purchased at any chemical supply house in 4-liter bottles. Industrially, acetone is sold in 5-gallon metal cans.

Acetone has so many uses that it is NOT a good chemical to key off for the identification of activity.

Odor: Nail polish remover.

Chemicals Used for Illegal Purposes: A Guide for First Responders to Identify Explosives, Recreational Drugs, and Poisons, By Robert Turkington
Copyright © 2010 John Wiley & Sons, Inc.

Hazards: The primary hazard is the high vapor pressure (180 mmHg @ 30°C). The vapors are extremely flammable and intoxicating. Flash point –4° F; LEL 2.5%; UEL 12.8%; Group D atmosphere; LD_{50} 9.75 g/kg; PEL 1000 ppm.

Incompatibilities: Incompatibility Group 6-B. Nitric acid, chromium trioxide, chloroform (in the presence of sodium hydroxide, potassium hydroxide, or ammonia), and acetaldehyde.

Used in

Amphetamines:
Synthesis 3
Synthesis 6
Synthesis 17
Synthesis 18, Suplementary 2
Solvent, methamphetamine hydrochloride extraction (Syntheses 3 and 6)

Most methamphetamine syntheses end up by making a methamphetamine base product. In Synthesis 6, acetone and ether are mixed and chilled and then added to the cook. In Syntheses 3 and 6, methamphetamine · HCl crystals (final product) are obtainable directly without making a base. In these methods the polar solvent acetone is used to extract (really, just wash) the product. Methamphetamine does not dissolve in acetone. *Note*: There is no need to "salt" methamphetamine hydrochloride, as it is already salted. Since acetone will not dissolve methamphetamine hydrochloride, it is used to cleanse methamphetamine hydrochloride crystals. Several washings produce a whiter final product. Acetone may be seen as part of a two-phase (solid and liquid) mixture. Acetone may be used to clean substandard product by the end user. Acetone is added to benzene to make P-2-P (Synthesis 17). It is mixed with calcium hypochlorite to make chloroform.

Chemical Weapons:
Facility 4, N1-1 and 2
Facility 4, N2-1
Facility 4, N3
Facility 5
Facility 14

Explosives:
Acetone peroxide; astrolite, Supplementaries 2 and 3; ADN; DANP; DINA; DNAN; DNAT; DNFA-P; β-HMX; HNBP; HNTCAB; PETN; picryl chloride; plastic explosives, Variation 6; TATB; tetrazide; TNA; TND; TNEN; TNTPB

Illegal Drugs:
Cook 1, Variation 1
Cook 2, Variation 7
Cook 4, Supplementary 2
Cook 5, Variation 1
Cook 6, Variation 2
Cook 9

Tear Agents:
Chloroacetone

Toxins:
Ricin

Regulations: CDTA (List II) 150 kg; RQ 2270 kg; Waste Codes D001 and U002; Hazard Class 3; PG II; ERG 127.

Confirmed by: Carbonyl Test.

Finding one chemical will usually not determine what a person is doing. This is especially true of chemicals such as acetone or sodium hydroxide, which are used for just about everything. It will usually take a group of four or five chemicals to locate a formula. A typical formula will look like this:

Styphnic acid
(Secondary explosive)
TOXIC
Flammable; deflagrates; may detonate
1.
Resorcinol
16% Nitric acid
Sodium nitrite
60% Nitric acid

In this case resorcinol is the key chemical in identification because it is the most unique component of styphnic acid. However, there are many chemicals which on their own will determine a product. An example of this is:

α-BUTYROLACTONE
$C_4H_6O_2$
96-48-0

Appearance: Hygroscopic, colorless, oily liquid with medium solubility in water. This is a common solvent and reagent in chemistry, and is used as a floor stripper, an aroma compound, a stain remover, and a paint stripper.

Odor: Weak characteristic odor.

Hazards: Even before being reacted, this can be a substitute for GHB.

Used in

Date Rape:
GHB, Variation 1

We have tried to provide as much information about every chemical as we could. However, for some of the more obscure chemicals, very little information is available. Since so many of the formulas come from the underground, some of the chemicals named may not exist. We have used the name given in the formula, since that is what you may expect to find at the site; however, we do try to explain what the real chemical may be.

The chemicals we found most difficult to locate were those (1) used to make high explosives [we feel that many of these may be obscure synonyms (the same author may use three or four synonyms for a chemical in the same document) or intermediates (we have identified some as intermediates, such as the Grignard reagents)]; (2) used to make chemical weapons (we feel that once listed, the chemicals tend to disappear from commerce;

computer searches often found them on restricted lists, but not for sale, nor was there information as to their chemical characteristics); (3) used to make designer drugs [once again we often found these listed but not for sale, nor was safety data information available; we have marked such chemicals with a (w)].

We emphasize the common names, not the official names of a chemical. This is a responder's help book, NOT a chemistry book. In every case we try to cross-reference the real name as well as the product that a chemical can be purchased in commercially. This means that the chemical may appear three or four times in the alphabetical list, referring you back to the common name. If we can, we try to present what the chemical looks and smells like, its primary hazards, and, if possible, what it is incompatible with, and regulations that apply.

Since hobbies constitute such a broad area, only hobbies that might be confused with illegal operations are included. This may seem an odd list. For example, making root beer is included in hobbies because it uses chemicals that might also be used to make Ecstasy. Model rocketry is included because it uses many chemicals that might be used to make either fireworks or explosives. Photography is included because it includes the use of many exotic and hazardous chemicals.

The book will not be of very much use when encountering a "mad chemist" who simply has been taking chemicals home for years. This book is directed more toward kitchen, garage, or bedroom setups used to make legal or illegal products.

We are interested in telling the responder how something is made, not how to make it. The formulas are generally all that we provide. However, there are certain cases where we provide more information. Examples include the manufacture of chemical weapons, where some of the apparatus is very dangerous. On the next page is an illustrated expanded reaction (99% nitric acid), which is included because a responder who sees this setup should recognize that it is a very DANGEROUS situation. Opening the apparatus pictured could result in injury or death.

These formulas are from many sources; and all are readily available; some are on very sketchy parts of the Web, some at Amazon.com, some from vanity (self-published) publications, and some from totally undocumented publications. We have included some formulas that *we know will not work*, as we assume that someone will try them. For the same reason, we have included formulas which if attempted by an amateur would probably result in injury or death. Some of the references were written by people who, judging from what they wrote are, at the least, pretty scary.

Process 2
Making 99% Nitric Acid

Variation 1

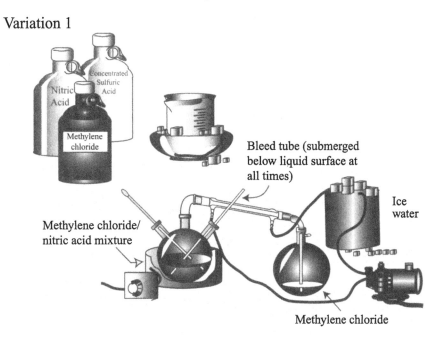

Bleed tube (submerged below liquid surface at all times)

Ice water

Methylene chloride/ nitric acid mixture

Methylene chloride

Variation 2

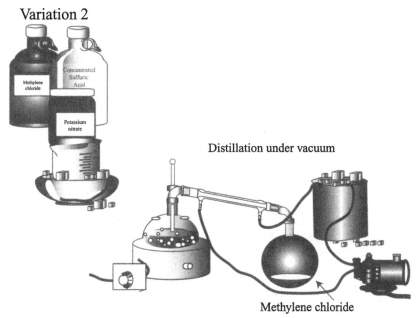

Distillation under vacuum

Methylene chloride

2

ALPHABETICAL LIST

The chemicals, plants, and products listed below are to either use or make illegal compounds or are exotic and hazardous compounds used at home for other reasons. We have also made a considerable effort to list products containing these chemicals that are available commercially. Once a compound has been identified, the information provided here should help in determining its hazard, proper handling procedures, appropriate protective clothing and equipment, and the relevant waste stream.

We were not always able to find all of this information for every chemical. Many chemicals have never been assigned UN numbers, Emergency Response Guides (ERGs), and Packing Groups (PGs), nor do they have regulations specific to them. We were not able to find some chemicals even in the most exotic chemical catalogs or on the Web. These chemicals we listed by name only. Such chemicals are often used to make chemical weapons, and may be banned or intermediates that the authors assume others will be able to make on their own. More important, each chemical is always presented with a list of all the illegal compounds that use it. This is the primary purpose of this section and of the book.

Information on the chemicals listed is provided where possible in the following format:

NAME OF COMPOUND

We try to use the name of the chemical given by the resource, despite the fact that we know of more common synonyms. We also list the more common synonyms alphabetically, referring back to the name used here. Since our resources range from high school dropouts strutting on the Web to seasoned highly educated chemists publishing polished papers, there is often no logic to the synonym that we choose.

CHEMICAL FORMULA

Whenever possible we give the formula in the most polished manner possible. For instance, we show toluene as $C_6H_5CH_3$ rather than as C_7H_8.

CAS NUMBER

We found that although the Chemical Abstracts Service (CAS) registry number is a good way to weed out synonyms, there were some problems using them. We had cases where the resource did not give a CAS number and we had to list several that might be attached to the particular compound. A particularly cumbersome example of that is benzene hexachloride (actually, hexachlorocyclohexane), which has many CAS numbers associated with different isomers of the compound as named. Wherever possible we do list CAS numbers, as in the field there are literally thousands of possible synonyms, which we would not be able to cover here. If a chemical is found that is not in this list,

Chemicals Used for Illegal Purposes: A Guide for First Responders to Identify Explosives, Recreational Drugs, and Poisons, By Robert Turkington
Copyright © 2010 John Wiley & Sons, Inc.

a responder can use a CAS number off the label in the CAS number index to determine if that compound appears in this book.

UN NUMBER

Where possible we give UN (NA) numbers. Occasionally, we ran across UN numbers not covered in the DOT guide.

APPEARANCE

State of matter, viscosity of liquids, color most likely to be seen in clandestine activities, sometimes how to make the compound, and suggestions as to where the compound might be purchased. Whenever possible we give the solubility in water as: miscible, very soluble, soluble, slightly soluble, insoluble, or reacts. We assume that this will help in the possible breakdown of a facility.

ODOR

You should never smell anything on purpose; however, the odor may be obvious, due to the presence of large amounts of the compound. There is a list of common chemical odors at the end of the book. Odors are very subjective and a responder should not trust odor too much as a guide.

HAZARDS

We found that initial buzz words in the literature we used as reference sources varied considerably from document to document. An example is the word harmful. In one document, drinking oleum was described as harmful. Certainly that is a correct statement, but does it really describe the hazard? (I have discussed with the responders what happened to a man who committed suicide by drinking concentrated sulfuric acid. By the time the emergency response team arrived, the man was dissolving from the inside, and very dead.) Other documents we looked at used the word harmful as a default; it was their way of saying "this chemical is not particularly hazardous, but you should never ingest a chemical." A person using any reference information buzz words as a guide should look for the author's specific definitions for those words. We are therefore including our hierarchy here.

Lethal

An extreme inhalation hazard. There is a serious chance that this can kill you more quickly than you can react.

Dangerous

Describing things that are toxic but should be considered in other terms, such as the concentrated sulfuric acid discussed above. Ingestion is only one hazard (and generally unlikely), along with skin contact, reactivity, and immediate effects. There is a serious chance that this can kill or maim you before you can react.

Poisonous!!!

Generally, this is an ingestion hazard, sometimes a skin contact or inhalation hazard; the LD_{50} is very low, perhaps only a couple of grains or less per kilogram.

Toxic

Very toxic by ingestion, sometimes skin contact, or inhalation. The LD_{50} is near but below the g/kg range.

Harmful

We have used this in the default manner described above. LD_{50}'s are generally above the g/kg range.

Very Low Toxicity

We could find no reference to state the obvious.

Inert

Rocks, certain salts, etc.

Nonhazardous Directive 67/548/EEC

The Europeans are not worried about this chemical.

GRAS

The FDA has designated some chemicals, which are not necessarily food but which can be used in or with food, as Generally Regarded as Safe (GRAS). Although we do not suggest that you give this chemical to the kids to play with, it is probably not going to hurt you unless you are very careless.

Generally, we will try to back up these words with LD_{50}'s. We use LD_{50}(rat) oral exclusively, because although we do have other animal data (rabbits, dogs, mice, and even human), we do not necessarily believe that rats are like humans, but the rat oral data is the most common, and we wanted to stay with one so that the reader could use it as a comparison. For each of the chemicals that provided other types of LD% data, the data was quite variable from animal to animal. Less available were OSHA PELs and ACGIH TLVs. Whereas the LD_{50}'s provide information of absolute toxicity, the PELs indicate relative levels of inhalation hazard.

When a compound is considered to be an inhalation hazard, we try to include vapor pressure, flash point, flammable range (LEL and UEL), and Atmosphere Group. Where appropriate, we discuss if the compound can form explosive peroxides, whether it can polymerize spontaneously, and if the compound is a carcinogen, mutagen, or teratogen.

A brief description of the hazards assigned to the compound by the DOT is also presented here. If the hazard is flammable, we try to include vapor pressure, flash point, and flammable range (LEL and UEL). If the hazard is corrosive, we may give the pH of an aqueous solution.

Incompatibilities

We try to list as many incompatible chemicals and a specific list of related compounds that should not be stored or lab-packed with this compound, designating some as reacting and others as simply causing deterioration of the compound. When available, we include the Incompatibility Group.

Used in

Amphetamines
Date rape
Designer drugs
Illegal drugs
Recreational drugs
Ammunition
Explosives
Fireworks
Incendiary devices
Chemical weapons
Tear agents
Other
Pathogens
Hobbies

Regulations

DOT Hazard Classes, Packing Groups (PGs); specific DOT shipping requirements; Emergency Response Guide numbers (ERGs); EPA Reportable Quantities (RQs); Waste Codes; other regulations that could affect how a compound should be handled, including chemicals controlled under the Chemical Diversion and Trafficking Act (CDTA), ATF regulations on rocket fuels, Australian Group schedules on chemical weapons precursors, and FDA's Chemicals Generally Regarded as Safe. Any data not listed was deemed unimportant (i.e., the vapor pressure of nonsubliming solids) or information is not available. Some of the more exotic chemicals do not have very much information available or have not been assigned UN numbers or Waste Codes.

Confirming Test

This is not an identification, only a test that would indicate that a chemical is actually as stated on the label.

Toxicity and Vapor Pressure

LD_{50} is related to ingestion. Since most industrial poisoning is by inhalation, to better understand the hazard of a compound, both toxicity and vapor pressure must be taken into consideration. A compound with high toxicity and low vapor pressure may be far less dangerous than a compound with low toxicity and high vapor pressure.

Common vapor pressures for use as a reference point:

0.24 mmHg @ 30°C	Sodium cyanide
1 mmHg @ 40°C	Phenol
30 mmHg @ 30°C	Water
37 mmHg @ 30°C	Toluene
400 mmHg @ 39.5°C	Acetone
740 mmHg @ 20°C	Acetaldehyde
760 mmHg	Air pressure

Vapor pressure is unfortunately NOT standardized in much of the literature and may be presented in any of several units and at any temperature. Since temperature and vapor pressure are not linear, knowing the vapor pressure at one temperature does not necessarily give an indication of the vapor pressure at another temperature. Even in the

use of such standard reference books as the *NIOSH Pocket Guide to Chemical Hazards*, we find that we cannot compare vapor pressures since they are listed at various temperatures. It is possible to graph vapor pressure against flash point, which factors out the variables in units and allows relative flash points to give an indication of relative vapor pressure. (Some flash points—for example, chlorinated hydrocarbons—do not fit on the curve for most organics.)

Any material that has a flash point below 100°F has sufficient vapor pressure to cause it to be an inhalation hazard. Vapor pressures between 10 and 20 mmHg at 30°C correspond roughly to a flash point near 100°F. For reference, water has a vapor pressure of 30 mmHg at 100°F, and acetone has a vapor pressure of 275 mmHg at 100°F. To convert units of vapor pressure, we provide the following information:

$°C = 5/9 × (F°–32)$
760 mmHg = 1 atm
1.867 mmHg = 1 inch of water
0.350 mmHg = 1 lb/sq ft

Vapor pressure should always be considered along with toxicity when considering how dangerous a material is. For instance, consider acetone and sodium hydroxide. Comparing their relative toxicity by PEL, you come up with the following:

TABLE 1. LD_{50}'s Presented as Standard Measurements

		LD_{50} Toxicity Ratings by Various Routes			
Toxicity Rating	Common Term	LD_{50} Single Oral Dose (Rat)	Inhalation, 4 Hr LC_{50} (Rat)	LD_{50} Skin (Rabbit)	Probable LD_{50} (Man)
1	Extremely toxic	<1 mg/kg	<10 ppm	<5 mg/kg	a taste
2	Highly toxic	1–50 mg/kg	10–100 ppm	5–43 mg/kg	1tsp
3	Moderately toxic	50–500 mg/kg	100–1000 ppm	44–340 mg/kg	1 oz
4	Slightly toxic	0.5–5 g/kg	1000–10,000 ppm	350 mg–2.81 g/kg	1 pt
5	Practically nontoxic	5–15 g/kg	10,000–100,000 ppm	2.81–22.59 g/kg	1 qt
6	Relatively harmless	>15 g/kg	>100,000 ppm	>22.6 g/kg	> 1 qt

Acetone   1800 mg/m³ (750 ppm)
Sodium hydroxide   2 mg/m³

This would indicate that OSHA is willing to accept roughly 1000 times more acetone than sodium hydroxid in a worker's body. Sodium hydroxide is obviously more toxic. However, the vapor pressure of sodium hydroxide is listed at 0 mmHg at 30°C, whereas acetone has a vapor pressure of 180 mmHg at 30°C. If both were spilled into a small space, the vapor pressure would be far more critical than the toxicity. Generally, high vapor pressures indicate a high probability that a material can reach you. A material with low vapor pressure and high toxicity may be much safer to handle than a material with high vapor pressure and low toxicity. Always keep in mind the route of entry when considering these two properties. The most common route of entry for a hazardous material when entering an uncontrolled facility will almost always be inhalation. In the case of sodium hydroxide, skin or eye contact would be a primary consideration when considering the danger of the material.

How to Use this Section:

Certain combinations of chemicals are strong indicators of certain types of compounds. Examples: iodine and red phosphorus are indicators of methamphetamine; 98% sulfuric acid and concentrated nitric acid are very strong indicators of explosives; and ammonium perchlorate, potassium chlorate, and potassium nitrate are strong indicators of fireworks. Compounds containing phosphorus found with compounds containing fluorine are strong indicators of nerve agents.

We recommend, if possible, making a list of chemicals. Look up the chemicals in this section of this book. If none of the chemicals listed appears here, it is a distinct possibility that the activity is not included in the book. If the chemicals do appear here, check off each chemical on the list below. Look for chemicals with very few listings under "Used in," as these will be the most helpful in the determination of what a person is making. Some chemicals, such as hydrochloric acid and sodium hydroxide, are used too generally to be of much help in looking up a formula. Chemicals with very few entries are best for determining what is being made. Once there is a preponderance of evidence, or a chemical or chemicals with only one use are encountered, look up the formula suggested.

A

ACCROIDES (Red) Gum
Also called *accroides resin* or *red gum.*

Appearance: Burgundy-colored resin, exuded from the bark of the Australian grass tree.

Used in

Fireworks:
Binder, sparks

Regulations: There are no specific DOT, EPA, FDA, or ATF regulations.

ACETALDEHYDE
CH₃CHO
5-07-0
UN 1089

Appearance: Colorless viscosity 1 liquid that is soluble in water. This reagent is only available from laboratory supply houses. It will be in 1-quart or 1-pint laboratory brown glass bottles. Acetaldehyde should be kept in a refrigerator. Since the refrigerator is not likely to be spark-proof, this is a dangerous practice.

Odor: The breath of an alcoholic.

Hazards: The primary hazard is the high vapor pressure (740 mmHg @ 20°C). The vapors of this extremely flammable irritating and intoxicating liquid will fill a room immediately upon opening a container.

Flash point –38° F. The wide flammable range of 4 to 57% means that if the container is open for any length of time, an explosion is possible. Group C atmosphere. In air, it can form explosive peroxides. Under certain conditions acetaldehyde can polymerize spontaneously. LC_{50} 20,000 ppm @ 30 min; LD_{50} 1.93 g/kg; PEL 200 ppm.

Incompatibilities: Incompatibility Group 4-A. Nitric acid, chromium trioxide, sodium hydroxide, ammonia, acetic anhydrides, alcohols, and ketones. Polymerizes with acetic acid.

Used in

Amphetamines:
Synthesis 21, Variations 1 and 2

Acetaldehyde is a reagent used to make the intermediate methylaminoethane, which is reacted with a Grignard reagent to make methamphetamine. This is a rarely used method that requires neither P-2-P nor ephedrine.

Regulations: RQ 454 kg; Waste Codes D001 and U001; Hazard Class 3; PG I; ERG 129; forbidden on passenger planes; cannot be sent by U.S. mail.

Confirmed by: Aldehyde Test.

ACETANILIDE
$CH_3CONHC_8H_5$
103-84-4

Appearance: Colorless lustrous crystals, flakes, leaflets, scales, or white powder. Slightly soluble in water.

Odor: None.

Hazards: Used for relief of headaches. Is considered a slight health hazard; however, causes irritation of the mucous membranes upon inhalation. Ingestion of several grams can cause cold extremities, paleness, feeble rapid pulse, and cyanosis. NFPA Health 3. Flash point 174°C.

Incompatibilities: Nitrous ether, alkalis (aniline liberated), alkali bromide, and iodides.

Used in

Hobbies:
Photography

ACETIC ACID
CH_3COOH
64-19-7
UN 2789

Appearance: Colorless viscosity 2 refractive liquid. In storage, clear glass chemical reagent bottle with an orange lid.

Odor: Strong vinegar.

Hazards: The primary hazard is corrosivity. Skin-contact and extreme eye-contact hazard. Lung and eye irritant in the atmosphere. LD_{50}(rat) 3310 mg/kg; PEL 5 ppm. Acetic acid is also a flammable liquid with a flash point of 99°F @ 30°C.

Incompatibilities: Incompatibility Group 6-B. Nitric acid, chromium trioxide, potassium hydroxide, sodium hydroxide, perchloric acid, and sulfuric acid.

Used in

Amphetamines:
Synthesis 0.7, Variation 2
Synthesis 1, Supplementary 5
Synthesis 14
Synthesis 15
Synthesis 16

Reagent to make P-2-P from phenylacetic acid in a tube furnace with phenylacetic acid and thorium oxide (Synthesis 14). Rarely used as a reagent to make α-phenylacetoacetonitrile, an intermediate in the production of P-2-P from benzyl cyanide (Syntheses 15 and 16). To make methcatanone, see Synthesis 0.7, Variation 2.

Chemical Weapons:
Facility 8, Variation 3
Cyanogen chloride

Explosives:
ADN, ADNBF, DANP, β-HMX, lead-TNP, NDTT, picaramic acid, PNT, RDX, tetracene tetrazene, TNA

Hobbies:
Photography; entomology

Illegal Drugs:
Cook 5, Supplementary 4

Regulations: RQ 2270 kg; Waste Codes D001 and D002; Hazard Class 8; PG II; ERG 132.

Confirmed by: Acetic acid indicator tube.

ACETIC ANHYDRIDE
$(CH_3CO)_2O$
108-24-7
UN 1715

Appearance: Colorless, viscous, refractive liquid. Most likely in clear glass bottles or jugs with orange lids.

Odor: Strong vinegar.

Hazards: The primary hazard is corrosivity. Reacts explosively with water, especially in the presence of mineral acids. Skin-contact and extreme eye-contact hazard. Lung and eye irritant in the atmosphere. LD_{50} 1.78 g/kg; PEL 10 ppm; flash point 120°F.

Incompatibilities: Incompatibility Group 4-A. Nitric acid, chromium trioxide, sodium hydroxide, ammonia, acetaldehyde, perchloric acid, and sulfuric acid.

Used in

Amphetamines:
Synthesis 11

Chemical Weapons:
Facility 15

Explosives:
DANP, DINA, DNAT, β-HMX, HNIW, NTDN, MNA, MPG, PVN, RDX, tetracene, tetracine, TNM

Illegal Drugs:
Cook 8, Variations 1 and 2
Cook 9

Regulations: CDTA (List II) 1023 kg; RQ 454 kg; Waste Codes D001and U002; Hazard Class 8; PG II; ERG 137.

Confirmed by: Acetic acid indicator tube.

ACETONE
CH_3COCH_3
67-64-1
UN 1090

Appearance: Colorless viscosity 1 liquid. Acetone is available in 1-quart or 1-gallon rectangular metal containers that can be purchased at any hardware store. It can also be purchased at any chemical supply house in 4-liter bottles. Industrially, acetone is sold in 5-gallon metal cans. Acetone has so many uses that it is NOT a good chemical to key off for identification of activity.

Odor: Nail polish remover.

Hazards: The primary hazard is the high vapor pressure (180 mmHg @ 30°C). The vapors are extremely flammable and intoxicating. Flash point –4°F; LEL 2.5%; UEL 12.8%; Group D atmosphere; LD_{50} 9.75 g/kg; PEL 1000 ppm.

Incompatibilities: Incompatibility Group 6-B. Nitric acid, chromium trioxide, chloroform (in the presence of sodium hydroxide, potassium hydroxide, or ammonia), and acetaldehyde.

Used in

Amphetamines:
Synthesis 3
Synthesis 6
Synthesis 17
Synthesis 18, Supplementary 2

Solvent, methamphetamine hydrochloride extraction (Syntheses 3 and 6) most methamphetamine syntheses end up by making a methamphetamine base product. In Synthesis 6, acetone and ether are mixed and chilled and then added to the cook. In Syntheses 3 and 6, methamphetamine·HCl crystals (final product) are obtainable directly without making a base. In these methods the polar solvent acetone is used to extract (really, just wash) the product. Methamphetamine does not dissolve in acetone. *Note:* There is no need to "salt" methamphetamine hydrochloride, as it is already salted. Since acetone will not dissolve methamphetamine hydrochloride, it is used to cleanse methamphetamine hydrochloride crystals. Several washings produce a whiter final product. Acetone may be seen as part of a two-phase (solid and liquid) mixture. Acetone may be used to clean substandard product by the end user. Acetone is added to benzene to make P-2-P (Synthesis 17). It is mixed with calcium hypochlorite to make chloroform.

Chemical Weapons:
Facility 4, N1-1 and 2
Facility 4, N2-1
Facility 4, N3
Facility 5
Facility 14

Designer Drugs:
Group 1, Variation 1

Explosives:
Acetone peroxide; Astrolite, Supplementaries 2 and 3; ADN; DANP; DINA; DNAN; DNAT; DNFA-P; β-HMX, HNBP; HNTCAB; PETN; picryl chloride; plastic explosives, Variation 6; TATB; tetrazide; TNA; TND; TNEN; TNTPB

Illegal Drugs:
Cook 1, Variation 1
Cook 2, Variation 7
Cook 4, Supplementary 2
Cook 5, Variation 1
Cook 6, Variation 2
Cook 9

Tear Agents:
Chloroacetone

Toxins:
Ricin

Regulations: CDTA (List II) 150 kg; RQ 2270 kg; Waste Codes D001 and U002; Hazard Class 3, PG II; ERG 127.

Confirmed by: Carbonyl Test.

ACETONITRILE
CH_3CN
75-05-8
UN 1648

Appearance: Colorless viscosity 1 liquid that is soluble in water. Most likely to be seen in brown 1-quart bottles from a laboratory supply house. There is no street source for this compound.

Odor: None.

Hazards: Flammable liquid. Flash point 42°F. NFPA Health 2. The resulting fire produces hydrogen cyanide gas. Acetonitrile (methyl cyanide) is toxic. PEL 40 ppm.

Incompatibilities: Oxidizers and magnesium metal.

Used in

Amphetamines:
Synthesis 18, Variations 1 and 2
Synthesis 20, Supplementary 7

Used to make Ecstasy and amphetamine.

Chemical Weapons:
Facility 2, Variation 2

Explosives:
ADN; DATBA; DDD; hexaditon; HNIW, KDN, TNA,TNEN, TND

Illegal Drugs:
Cook 2, Variation 7

Regulations: Hazard Class 3; ERG 131.

Confirmed by: Hydrogen cyanide indicator tube.

Monitored by: Combustible gas indicator.

ACETYL CHLORIDE
CH_3COCl
75-36-5
UN 1717

Appearance: Colorless fuming liquid. Hydrolyzes readily to form hydrogen chloride gas and acetic acid.

Odor: Irritant.

Hazards: Corrosive liquid. Dangerous when heated to decomposition, emits highly toxic fumes of phosgene. Violently water reactive. Flash point 40°F. NFPA Health 3.

Incompatibilities: Water, ethyl alcohol, dimethyl sulfoxide.

Used in

Explosives:
ADBN

Regulations: Hazard Class 8; ERG 132.

ACETYLENE
C_2H_2
74-86-2
UN 1001

OUTLET   CONNECTION

Appearance: Colorless gas. Seen most likely as a tank of acetylene gas used for welding or cutting. Acetylene tanks are stout and have flat tops, left-handed threads, and no pressure relief valve. Calcium carbide (bangsite), a rock-like material, could be used to generate this gas with the addition of water.

Odor: Mild metallic garlic.

Hazards: A flammable nontoxic gas with a flammable range of LEL 2% to UEL 100%, making ignition of a gas leak almost certain in contact with flame. The resulting fire is very intense, and fireballs appear often. Burns with a dark flame and considerable black spider webs in the smoke. Group A atmosphere. Acetylene is an anesthetic.

Incompatibilities: Iodine, chlorine, oxygen, bromine, silver, silver nitrate, mercury.

Used in

Amphetamines:
Synthesis 1, Supplementary 1

Chemical Weapons:
Facility 5

Explosives:
DINA, McGyver

Hobbies:
Welding; auto repair and customizing

Regulations: Waste Code D001; Hazard Class 2.1; PG I; ERG 116; forbidden on passenger planes; cannot be sent by U.S. mail.

Confirmed by: GasTec Polytec IV indicator tube stripe 6 turns brown.

Monitored with: Combustible gas indicator.

ACETYL OXIDE
See Acetic anhydride.

ACID GASES
Chemical Weapon
Facility 15
Acid smoke

ACROLEIN
$CH_2:CH·CHO$
107-02-8
UN 1092
Chemical Weapon

Appearance: Colorless-to-yellowish, very refractive, flammable liquid. Soluble in water. Formulas are found in tear agents.

Odor: Very pungent; highly irritating to eyes and mucosa.

Hazards: Flammable flash point minus −15°F. Unstable, polymerizes to disacryl, which is a solid. Very toxic inhalation hazard, IDLH 5 ppm. PEL 5 ppb. NFPA Health 3.

Incompatibilities: Oxidizers, concentrated sulfuric acid.

Used in

Explosives:
TBA

Tear Agents:
Acrolein

Regulations: Inhibited; Hazard Class 6.1, minor 3; PG I; ERG 131P.

Monitored by: Draeger aldehyde tube.

ADAMANTANE
$C_{10}H_{16}$
281-23-2

Appearance: White-to-off-white powder that comes in a sealed tube, sold in small quantities. Melting point 209°C. Not soluble in water.

Odor: None.

Hazards: None listed. Very stable.

Incompatibilities: Strong oxidizers.

Used in

Explosives:
TNA

Regulations: There are no specific DOT, EPA, FDA, or ATF regulations.

AGAR
9002-18-0

A preserving and gelling agent derived from red seaweed. Used in beverages and baked goods. Used in laboratories to grow bacteria and fungus. Agar is a complex carbohydrate that most species of bacteria cannot digest, so it serves as a good medium and support for holding water and nutrients that "bugs" can digest. Sometimes used with a marker dye to identify colonies. Agar by itself is usually not sufficient to grow various pathogens.

Odor: Characteristic.

Hazards: CAUTION—Agar plates that are not completely clear or which have growth on them may contain pathogens. Agar plates or dishes should be covered and sealed prior to handling.

Used in

Pathogens: Below are listed various specialized agars which are associated with growing various pathogens.

Blood agar	Anthrax
Bile esculin agar	Streptococci
Chocolate agar	Anthrax
Enriched media	*E. coli*
Hektoen enteric	*Salmonella* and *Shigella*
Onoz	*Salmonella* and *Shigella*
Peptone	*E. coli*
Phenylethyl alcohol agar	*Staphylococcus*
Soy agar	General purpose
Trytptic	General purpose
Thayer–Martin agar	Gonorrhea
Yeast extract agar	*E. coli*

Illegal Drugs:
Cook 5, Variation 2
Cook 5, Supplementary 1

Regulations: Unknown or untested agar plates should be disposed of as biological waste.

ALCOHOL (Nonspecific)
Alcohols can be specified in a formulation; however, often the alcohol is just a carrier solvent or an extraction solvent, in which case other alcohols, or even other polar solvents including water, could be substituted. In many cases a specific alcohol is not really indicative of the product.

ALCOHOL LAMP
Alcohol lamps have many uses, both here and in other places. We have listed them here only when the specific formula called for the use of an alcohol lamp. Its use usually implies temperature control. Although none of the explosive formulas listed alcohol lamps, many had upper temperature requirements.

Used in

Recreational Drugs:
Laughing gas 2

Tear Agents:
Tear gas

ALKALI METALS
See Sodium, Potassium, *or* Lithium. Mostly interchangeable.

ALLYL BENZENE
$C_6H_5CH_2CHCH_2$
300-57-2

Appearance: Colorless viscosity 1 liquid that is not very soluble in water. Found in brown glass bottles from a laboratory supply house. There are industrial sources for this material. There are several ways to manufacture this compound, discussed in this book. (*See* Amphetamines Synthesis 18, for details of manufacture.)

Odor: Aromatic.

Hazards: This is an alkene. It is flammable, flash point 33°C, and an irritant.

Incompatibilities: Strong oxidizers.

Used in

Amphetamines:
Synthesis 18
Synthesis 18, Supplementary 5
Synthesis 2, Supplementary 5

Regulations: Hazard Class 3. Should be disposed of by a licensed hauler.

Confirmed by: Determination with iodine.

ALLYL BROMIDE
$CH_3CH_3CH_2Br$
106-95-6
UN 1099

Appearance: Colorless-to-slightly yellow viscosity 1 liquid. Found in brown glass bottles from a laboratory supply house.

Odor: Irritating.

Hazards: Inhalation hazard; PEL 0.1 ppm. Causes headaches and dizziness. Dangerous when exposed to heat. Produces carbonyl bromide and hydrogen bromide. Flammable, flash point 30°F. There is some evidence that allyl bromide is a mutagen.

Incompatibilities: Can react vigorously with oxidizing materials, including peroxides. Incompatible with strong bases.

Used in

Amphetamines:
Synthesis 18, Supplementary 2

Regulations: Hazard Class 3; ERG 131.

ALLYL CHLORIDE
CH_2CHCH_2Cl
107-05-1
UN 1100

Appearance: Colorless viscosity 1 liquid in brown glass bottles from a laboratory supply house.

Odor: Irritating.

Hazards: Flammable. Flash point −25°F. Irritating to the eyes, nose, and throat, on contact of the liquid with the skin. Absorbed through the skin. Acute exposure causes marked inflammation of the lungs. Allyl chloride boils at 45°C. DO NOT CONSIDER ANY EXPOSURE AS INSIGNIFICANT.

Incompatibilities: Reacts vigorously with oxidizing materials.

Used in

Amphetamines:
Synthesis 13
Synthesis 18

Recreational Drugs:
Rush

Regulations: MCA Warning; Red label; ERG 131; Hazard Class 3; not acceptable in passenger transportation.

Confirmed by: Beilstein Test.

4-ALLYL-1,2-METHYLENE DIOXYBENZENE
See Safrole.

ALMOND OIL
See Benzaldehyde.

ALUMINA
See Aluminum oxide.

ALUMINUM
Al
7429-90-5

Appearance: Gray or metallic chips, foil, or powder. Aluminum is most likely to be seen in the form of sandwich wrap. However, aluminum rivets, aluminum can pop-top rings, beer cans, and aluminum nails can also be used.

Odor: None.

Hazards: Aluminum is considered by OSHA to be a nuisance dust. The primary hazard associated with this metal is possible mixing with an incompatible compound while packing up the waste. The reactions noted below depend on the condition and form of the aluminum. Aluminum as a fine bright shiny metal powder is the most dangerous form.

Incompatibilities: Incompatibility Group 2-A. Chloroform, chromium trioxide, iodine, fine aluminum metal mixed with iron oxide, palladium oxide, lead oxide, and probably other metal oxides; becomes an extremely flammable solid. In the presence of oxygen, aluminum can ignite with the release of large amounts of heat.

Used in

Amphetamines:
Synthesis 6
Synthesis 18, Supplementatry 3

Aluminum foil is usually chopped up to be placed in a reaction vessel. When the reaction is finished, the color of the remaining liquid is red orange. Catalyst (Synthesis 6) to substitute hydrogen and methylamine for the carbonyl group on P-2-P to make methamphetamine (Synthesis 18, Supplementary 3).

Ammunition:
Aluminum is an enhancer for some forms of gun powder.

Chemical Weapons:
Facility 15

Explosives:
Binary explosives, Variation 7
McGyver

Aluminum foil with either acid or sodium hydroxide in a plastic pressure pop bottle:
Plastic explosives, Variation 7
Oxidizer–fuel mixtures

TNA, TND
Fine aluminum mixed with nitrates can create high explosives.

Fireworks:
Flame arrester, sparks, salutes, glitter

Hobbies:
Rocket fuel, Variation 3
Rocket fuel, Variation 5
Rocket fuel, Variation 8

Incendiary Devices:
Situation 5
Situation 11

Illegal Drugs:
Cook 2
Cook 5, Variation 1
Cook 7

Recreational Drugs:
Nitrous oxide

Regulations: None specific.

Confirmed by: Aluminum Metal Test.

ALUMINUM CHLORIDE
See Aluminum trichloride.

ALUMINUM OXIDE
Al_2O_3
1344-28-1

Appearance: Alumina is a white amorphous powder. Readily absorbs water from the air. Alumina is insoluble in water and in acid. Used as an abrasive, a refractory, a filler for paints, and for dental cements.

Odor: None.

Hazards: This is a food additive. Nuisance dust.

Used in

Illegal Drugs:
Cook 2, Variation 3
Cook 2, Supplementary 1

In the manufacture of illegal drugs, alumina is packed in a column to be used as a drying agent.

Regulations: None specific.

Confirmed by: Aluminum Test.

ALUMINUM SILICATE
See Bentonite.

ALUMINUM SULFATE
$Al_2(SO_4)_3$
10043-01-3
The commercial product is also known as *cake alum*.

Appearance: White, lustrous crystals, pieces, granules, or powder.

Odor: None.

Hazards: Mild irritant. Can become caustic in water.

Used in

Hobbies:
Photography

Regulations: EPA's PC 1415.

Confirmed by: Aluminum Test; Sulfate Test.

ALUMINUM TRICHLORIDE
(Anhydrous)
$AlCl_3$
7446-70-0
UN 1726

Appearance: White or slightly yellow powder. Aluminum chloride cannot be obtained except at laboratory supply houses. It is most likely to be found in brown jars, which will be sealed with tape or wax. The container will probably hiss when the lid is opened.

Odor: Irritating, due to offgassing of hydrogen chloride gas.

Hazards: Aluminum chloride is very water reactive. The hydrate, which can also be labeled aluminum chloride, is not water reactive. Either salt forms hydrochloric acid when added to water. Aluminum chloride is self-reactive and can explode when a container that has been stored a long time is opened.

Incompatibilities: Incompatibility Group 2-A. Water, nitromethane (in the presence of another organic material).

Used in

Amphetamines:
Synthesis 13

Chemical Weapons:
Facility 1, Variation 1, Supplementary 1
Facility 1, Cyclosarin, Variations 1 and 2
Facility 1, Soman
Facility 2
Facility 5

This is part of the Abuzov reaction, which is almost essential in the manufacture of nerve agents. It is used to make sarin, cyclosarin, and tabun. Other chemical weapons in whose manufacture aluminum trichloride is used include vinyl arsine, smoke agents, diphenyl chloroarsine, lewisite, and phenyl dichloroarsine.

Explosives:
AS-20, PCB, TNA, TNEN

Illegal Drugs:
Cook 9

Tear Agents:
Mace

Regulations: Waste Code D002; PG II; Hazard Class 8; ERG 137.

Confirmed by: PolyTec IV indicator tube.

AMIDOL
2,4-Diaminophenol·HCl
$(NH)_2C_6H_4ClH$
UN 2927 (liquid)
UN 2928 (solid)

Appearance: Colorless crystals soluble in water.

Odor: None found.

Hazards: Toxic.

Incompatibilities: Oxidizers.

Used in

Hobbies:
Photography, developer

Regulations: Poison Solid (Liquid) n.o.s.; ERG 154.

AMINOANTHRAQUINONE (w)
$C_{14}H_9O_2N$
82-45-1

Appearance: Ruby-red crystals insoluble in water.

Used in

Fireworks

Regulations: None specific in DOT, EPA, FDA, AFT, or DEA.

3-(2-AMINOBUTYL)INDOLE (w)
No reference found on the Web or in specialty catalogs. May have a more common synonym or be an intermediate.

Used in

Designer Drugs:
Group 4 chemicals that could be expected to be used to replace P-2-P.

AMINOGUANIDINE·HCl (w)
$CH_6N_4·HCl$

Appearance: Forms large prisms; very soluble in water.

Hazards: Has found use as an antiaging compound in pharmacology.

Used in

Explosives:
Tetracene, tetrazene

2-AMINO-2-METHYLPROPANE
See tert-Butylamine.

2-AMINO-3-PHENYL-*TRANS*-DECALIN
(w)
No reference found on the Web or in specialty catalogs. May have a more common synonym or be an intermediate.

Used in

Designer Drugs:
Group 4 chemicals to replace P-2-P.

2-AMINOTETRALINE (w)
$NH_2C_{10}H_{12}$
No reference found on the Web or in specialty catalogs.

Used in

Designer Drugs:
Group 4 chemicals to replace P-2-P.

5-AMINOTETRAZOLE (w)
CH_3N_5
5378-49-4
UN 1325

Appearance: White crystals that dissolve in water. Used as a polarizer for film and as a gas propellant for airbags.

Odor: Weak characteristic.

Hazards: Explosive flash point 56°C. Has potential explosive property at room temperatures.

Used in

Explosives:
CNTA, SATP

Regulations: Flammable Solid n.o.s.; ERG 133.

5-AMINOTETRAZOLE (Monohydrate) (w)
$CH_3N_5·H_2O$
4418-61-5

Used in

Explosives:
CNTA

3-AMINO-1,2,3-TRIAZOLE (w)

Used in

Explosives:
DNAT

AMMONIA (Anhydrous, Gas)
NH_3
7664-41-7
UN 1005

Appearance: Colorless gas. Aqueous solutions (*see* Ammonium hydroxide). Ammonia is found in laboratory bottles (small gas cylinders). Anhydrous ammonia can be taken from fertilizer storage areas, blueprint azilide machines, or refrigeration systems. Ammonia gas can also be made by distilling any ammonium salt with sodium hydroxide; *see* Explosives, Process 6.

Odor: Ammonia!!!

Hazards: Ammonia is considered to be an irritating gas. In its chemical reactions it is corrosive. The primary hazard associated with ammonia is severe eye and lung irritation. A high inhalation dose will close off the esophagus, which prevents breathing until the person is removed from the area. Very high exposure can dehydrate the skin, causing a health hazard even if the person is wearing an SCBA. Ammonia can form an explosive atmosphere in confined areas. LEL 16%; UEL 25%; Group D atmosphere; LD_{50} 0.35 g/kg; LC_{50} 4837 ppm @ 1 hour; PEL 25 ppm.

Incompatibilities: Incompatibility Groups 3-B and 4-A. Acetaldehyde, chromium trioxide, iodine, nitric acid, mercury, calcium hypochlorite, hydrogen fluoride.

Used in

Amphetamines:
Synthesis 0.5
Synthesis 2, Variation 1
Synthesis 7
Synthesis 18

Look for blue or green valving and piping on those containers that were not designed to hold ammonia. Ammonia is also found in Igloo® thermos bottles being used as dewars or in plastic pipes used as homemade carriers. Ammonia might be found in improper containers which become unstable after a few weeks or in propane or oxygen cylinders. Five gallons of ammonia released simultaneously could easily incapacitate or kill a person. Anhydrous ammonia could be made in the same way that methylamine gas is made from a methylamine solution (*see* Synthesis 7). Ammonia is a reagent; used in ammonia alkali metal method to hydrogenate ephedrine (Synthesis 2); used to make MDA or amphetamine (Synthesis 18).

Chemical Weapons:
Facility 5, Supplementary 1

Ammonia is an industrial chemical weapon. It is an ingredient used to make cyanide from scratch.

Explosives:
ADN; ammonia picrate; astrolite, Supplementaries 1, 2, and 3; AN, DPT; mercury nitride; NENA; sodium azide; silver nitride; TATB

Illegal Drugs:
Cook 1, Variation 1
Cook 6, Variation 1

Regulations: RQ 45.4 kg; Hazard Class 2.3, minor 8; Poison, Inhalation Hazard; Zone D; ERG 125; forbidden on passenger planes; cannot be sent by U.S. mail.

Monitored by: Ammonia indicator tube.

AMMONIUM ACETATE (Anhydrous)
CH_3COONH_4
631-61-8
UN 9079

Appearance: White deliquescent crystals (looks like soft snow). Most likely found in square white plastic 2.5-kg jars with screw-top lids. This is not an over-the-counter item.

Odor: Reminiscent of vinegar.

Hazards: Mildly corrosive (you would not want it on your skin long). Strong irritant to the eyes. Inhalation hazard.

Incompatibilities: Hypochlorites (household bleach).

Used in

Amphetamines:
Synthesis 15

Regulations: RQ 2270 kg; ERG 171.

Confirmed by: Acetic Acid Gas Test.

Monitored by: Ammonia indicator tube.

AMMONIUM BIFLUORIDE
$NH_4F \cdot HF$
1341-49-7
UN 1727

Appearance: White crystals that etch glass readily and are freely soluble in water. Will be found in plastic containers.

Odor: Not found.

Hazards: A toxic corrosive liquid with devastating effects on the human body, made even more dangerous because it does not cause pain on contact. Considerable damage can occur before the victim notices the effects of contact with this compound.

Incompatibilities: Potassium, sodium, lithium, sodium hydroxide, and ammonia.

Used in

Chemical Weapons:
Facility 1
Facility 1, Soman

Often used to make G-agents, sarin/soman, sarin-isopropyl, cyclosarin, etc. Especially if phosphorus compounds are present, look for phosphorus trichloride and phosphorus oxychloride, two very nasty NMAPs. Used to fluoridate Agent DC, to make either Agent DF or di-di.

Regulations: Waste code D002; Hazard Class 8; regulated under Chemical Weapons Convention, TSCA 12(b) and SARA 311/312; Australian Group chemical, not scheduled. Requires export permits from the DTCC.

Confirmed by: Fluoride Test.

AMMONIUM BISULFIDE
See Ammonium sulfide.

AMMONIUM BROMIDE
$(NH_4)Br$
12124-97-9

Appearance: Colorless crystals that are slightly hygroscopic.

Odor: None.

Hazards: Mildly irritating; prolonged exposure may cause rashes.

Incompatibility: Bromine trifluoride, iodine heptafluoride, potassium, and acid salts. Reacts with acids to form toxic hydrogen bromide and bases to form ammonia.

Used in

Hobbies:
Photography

Regulations: Transportation is not regulated.

Confirmed by: Bromide Test.

AMMONIUM CARBONATE
$(NH_4)_2CO_3$
506-87-6

Appearance: Colorless, crystalline plates (crystal ammonia).

Odor: Ammonia.

Hazards: Slight inhalation and irritant hazard.

Incompatibilities: Sodium hypochlorite produces chlorine gas.

Used in

Explosives:
ADN, KDN

Regulations: None specific.

Confirmed by: Carbonate Test.

AMMONIUM CHLORIDE
NH_4Cl
12125-12-9
UN 9085

Appearance: White granular crystals. Most likely found in square white plastic 2.5-kg jars with screw-top lids. This is not an over-the-counter item. Ammonium chloride pulls water out of the air and is used to desiccate other chemicals.

Odor: None.

Hazards: No significant hazard; LD_{50} 1.65 g/kg.

Incompatibilities: Ammonium chloride will liberate chlorine gas in the presence of ammonium nitrate, especially if heated. It is incompatible with sodium chlorate.

Used in

Amphetamine:
Synthesis 19

Explosives:
TADA

Hobbies:
Photography

Illegal Drugs:
Cook 3, Variation 1

Regulations: RQ 2270 kg; ERG 171.

Confirmed by: Chloride Test.

AMMONIUM CITRATE
$(NH_4)_2C_6H_6O_7$
3012-65-5

Appearance: White granules or powder that is soluble in water.

Odor: Slight ammonia.

Hazards: Citrates are used in food and cosmetics.

Incompatibilities: Oxidizers.

Used in

Hobbies:
Photography

Regulations: None specific to this compound.

AMMONIUM DICHROMATE
$(NH_4)_2Cr_2O_7$
7789-09-5
UN 1439

Appearance: Bright orange-red crystals that are soluble in water.

Odor: None.

Hazards: Strong oxidizer. Contact with other material may cause a fire. Corrosive. Causes severe burns to every area of the respiratory system, liver, kidneys, eyes, skin, and blood. May cause allergenic reactions. Causes cancer. This salt is ignitable and will burn.

Incompatibilities: Reducing agents, combustibles, organic materials, strong acids, alcohols, ethylene glycol, carbide, hydrazine, potassium chlorate, sodium nitrite, and water. Breaks down with luminosity when heated.

Used in

Hobbies:
Photography

Regulations: Should be handled as hazardous waste and sent to an RCRA-approved incinerator; Hazard Class 5.1; PG II; RQ 12 kg; ERG 141.

Confirmed by: Chromium Test.

AMMONIUM FLUORIDE
NH_4F
12125-01-8
UN 2505

Appearance: White, small, deliquescent crystals or granular powder. Corrodes glass. Freely soluble in cold water, decomposed by hot water into ammonia and ammonium bifluoride. Will be in plastic containers.

Odor: None.

Hazards: Corrosive. Causes severe irritation and possibly, burns to the skin. NFPA Health 3; Poisonous! Inhalation hazard; PEL 2.5 mg/m^3 (as fluorine).

Incompatibilities: Potassium chlorate, sodium nitrite, chlorine trifluoride, and calcium. Reacts with acids to liberate hydrofluoric acid and bases to liberate ammonia.

Used in

Chemical Weapons:
Facility 1, Variation 1

Often used to make G-agents, sarin/soman, sarin-isopropyl, cyclosarin, etc. Especially if phosphorus compounds are present, look for phosphorus trichloride and phosphorus oxychloride, two very nasty hydrolyzing materials. Used to fluoridate Agent DC, to make either Agent DF or di-di.

Explosives:
NPF

Regulations: Waste Code D002; Hazard Class 8; PG III; ERG 154.

Confirmed by: Fluoride Test.

AMMONIUM HEXACHLORO-PALLADATE(IV)
$(NH_4)_2PdCl_6$
19168-23-1

Appearance: Red-brown cubic crystals that are very slightly soluble in water.

Odor: None.

Hazards: Severe skin irritant.

Incompatibilities: Hypochlorites, strong bases.

Used in

Hobbies:
Photography

Regulations: None specific.

AMMONIUM HYDROXIDE
NH_4OH
Aqueous ammonia
1336-21-6
UN 2672, UN 1005, UN 3318

Appearance: Usually found as a clear liquid. Found in many household cleaning products, usually in plastic bottles. (Cleaning products that include many other ingredients tend to cloud up the product and make it unusable.) Also available from laboratory supply houses in a much stronger solution, in 1-gallon or 4-liter clear glass jugs with green lids.

Odor: Strong ammonia.

Hazards: Although the pH will be close to 11, this is a corrosive and reacts as a base in chemical reactions. This is an inhalation hazard, and if exposed to a large amount as escaping gas, will close the epiglottis, not allowing the victim to breath.

Incompatibilities: Acids, organic acids, hypochlorites (reacts to form chlorine gas).

Used in

Explosives:
Ammonium nitrate; ammonium perchlorate; ammonium picramate; ammonium triiodide; AN; A-NPNT; astrolite, Supplementaries 2 and 3; EDT; nitrostarch; sodium picramate.

Illegal Drugs:
Cook 1, Variations 1 and 2
Cook 2, Variations 1 and 2
Cook 2, Supplementaries 1 and 3
Cook 3, Variation 1
Cook 5, Variation 1
Cook 6, Variation 2
Cook 6, Supplementary 1

Regulations: With more than 10%, but not more than 35% ammonia, ERG 154; with more than 50% ammonia, ERG 125.

Confirmed by: pH test, of 11.

AMMONIUM IODIDE
NH_4I
12027-06-4

Appearance: White tetragonal crystals that are soluble in water.

Odor: None.

Hazards: May cause irritation to the skin.

Incompatibilities: Bromine trifluoride, iodine heptafluoride, and potassium; hypochlorites; strong bases.

Used in

Hobbies:
Photography

Regulations: None specific.

Confirmed by: Iodine Test.

AMMONIUM METAVANADATE
See Ammonium vanadate.

AMMONIUM MOLYBDATE
$(NH_4)Mo_7O_4\cdot 4H_2O$
12054-85-2

Appearance: Colorless or slightly greenish crystals that are soluble in water. A standard laboratory chemical that will be found in brown glass bottles or in plastic jars.

Odor: None.

Hazards: May be harmful if swallowed. Avoid breathing dust. NFPA Health 2.

Incompatibilities: Hypochlorites; strong bases; strong acids.

Used in

Fireworks:
Coating

Hobbies:
Photography, ceramics

Regulations: Not specifically regulated.

AMMONIUM NITRATE
NH_4NO_2
6484-52-2
UN 1942

Appearance: Usually seen as very white to yellow-tinged prills. May be found as a laboratory chemical, in which case it will be in a square white plastic container or a large brown glass jar. Most cost-efficient as a fertilizer, which would come in brown paper bags. (This tends not to be very pure.) Some fertilizers are mixtures, which would change the appearance of the material. Very pure ammonium nitrate can be purchased from drugstores in the form of cold packs.

Odor: None.

Hazards: Ammonium nitrate can be very irritating to the skin. It is an oxidizer. Pure ammonium nitrate can be explosive if tightly contained during a fire. A shipboard fire of ammonium nitrate caused the largest nonnuclear explosion that ever occurred in the continental United States.

Incompatibilities: Strong acids can liberate NO_x. Ammonium nitrate mixed with organic compounds, aluminum, hydrazine, or methyl nitrate are secondary explosives. Hypochlorites or strong bases.

Used in

Amphetamines:
Synthesis 2, Variation 2

Explosives:
ANFO, astrolite G, astrolite A-1-5, binary explosives (β-HMX, RDX, TNTC),

nitrogen trichloride, smokeless powder, tovan

Hobbies:
Rocket fuel, Variation 5; gardening

Other:
Deicing

Recreational Drugs:
Nitrous oxide 1 and 2

Regulations: Pure ammonium nitrate has no special requirements; ERG 140.

Confirmed by: Ammonia Gas Test.

AMMONIUM PERCHLORATE
NH_4ClO_4
7790-98-9
UN 1442

Appearance: White crystalline material that is soluble in water.

Odor: None.

Hazards: Oxidizer; heat decomposes ammonium perchlorate at 130°F; explodes at 350°F.

Incompatibilities: Any oxidizable material that is mixed with any perchlorate becomes a low-order primary explosive. Ferroscene may also form explosive compounds with ammonium perchlorate. Hypochlorites and strong bases.

Used in

Fireworks:
Strobe formulation

Hobbies:
Rocket fuel, Variation 1
Rocket fuel, Variation 3
Rocket fuel, Variation 4
Rocket fuel, Variation 5
Rocket fuel, Variation 8
Rocket fuel, Variation 10

Regulations: ATF: proper storage is in shipping containers, with no more than 750 lb per room, with no special precautions.

Propellants are not considered to be explosives. ERG 143.

Confirmed by: Perchlorate Test.

AMMONIUM PERSULFATE
$(NH_4)_2S_2O_8$
7727-54-0
UN 1444

Appearance: Colorless crystals or white granular powder. Decomposes at 120°C, accelerated by moisture.

Odor: None.

Hazards: Powerful oxidizer.

Incompatibilities: Organic materials, aluminum, reducing agents, sodium peroxide, hypochlorites, strong bases.

Used in

Hobbies:
Photography, retarder

Regulations: Hazard Class 5.1; ERG 140.

AMMONIUM PHOSPHATE
$(NH_4)_2HPO_4$
13011-54-6

Appearance: Colorless crystals or white crystalline powder.

Odor: None.

Hazards: NFPA Health 0.

Incompatibilities: Hypochlorites; strong bases.

Used in

Fireworks:
Coating

Regulations: None specific.

AMMONIUM PICRAMATE (w)
$C_6H_2ONH_4\cdot NH_2(NO_2)_2$

Appearance: Reddish-brown crystalline powder that is soluble in water. Can be made; *See* Explosives, Ammonium picramate.

Odor: None.

Hazards: Oxidizer, explosive, shock sensitive.

Incompatibilities: Heat, deflagrates. Hypochlorites; strong bases.

Used in

Explosives:
Diazodinitrophenol, picaramic acid

AMMONIUM SALTS
There are literally thousands of ammonium salts, the characteristics of which are that of the anion. When heated, or mixed with sodium hydroxide, the ammonium is driven off as ammonia.

Incompatibilities: Hypochlorites; strong bases.

Used in

Explosives:
Process 6

Confirmed by: Ammonia indicator tube.

AMMONIUM SULFATE
Na_4SO_4
7783-20-2

Appearance: Brownish-gray-to-white crystals.

Odor: None.

Hazards: Slight irritant.

Incompatibilities: Sodium nitrite. Hypochlorites; strong bases.

Used in

Explosives:
ADN

Hobbies:
Photography

Toxins:
Clostridium botullinum

Regulations: None specific.

Confirmed by: Sulfate Test.

AMMONIUM SULFIDE
NH_4S
12135-76-1
UN 2683
The name *ammonium sulfide* is usually applied to ammonium hydrogen sulfide, NH_4HS.

Appearance: A porcelain-like mass (yellow crystals) that gives off the odor of hydrogen sulfide (rotten eggs). Freely soluble in water. This is the old skunk perfume of the practical joker.

Odor: Rotten eggs.

Hazards: Toxic vapor pressure 450 mmHg @ 20°C; corrosive, flammable, flash point 22°C; avoid skin contact.

Incompatibilities: Acids, hypochlorites. Strong bases, copper, brass, bronze, zinc aluminum.

Used in

Explosives:
Ammonium picramate

Designer Drugs:
Group 3, Variation 1

Illegal Drugs:
Cook 5, Variation 1

Regulations: Hazard Class 8, minor 3; ERG 132.

Confirmed by: Sulfide Test.

AMMONIUM TETRA-CHLOROPLATINATE(II) (w)

Incompatibilities: Hypochlorites; strong bases.

Used in

Hobbies:
Photography

AMMONIUM THIOCYANATE
NH_4SCN
1762-95-4

Appearance: White or very slightly yellow deliquescent crystals that are soluble in water. Most likely found in brown glass jars. Also sold in 100-lb paper bags, plastic kegs, and drums.

Odor: None.

Hazards: Toxic by ingestion. When heated, can release cyanide gas.

Incompatibilities: Hypochlorites; strong bases.

Used in

Hobbies:
Photography; rockets, liquid fuels

Regulations: Not a DOT-controlled material. RQ 5000 lb.

AMMONIUM THIOSULFATE
$(NH_4)_2S_2O_3$
7783-18-8

Appearance: White or colorless crystals that are soluble in water. Decomposed by heat.

Odor: None.

Hazards: Low toxicity.

Incompatibilities: Hypochlorites; strong bases.

Used in

Hobbies:
Photography, fixing agent; Liquid rocket fuel, Variation 8

Regulations: Not regulated.

AMMONIUM VANADATE
NH_4VO_3
7803-55-8
UN 2859

Appearance: White or slightly yellow crystalline powder that is very slightly soluble in water.

Odor: None.

Hazards: Poisonous! Irritant. PEL 0.5 mg/m^3.

Incompatibilities: Hypochlorites; strong bases.

Used in

Fireworks
Coating

Regulations: PG II; Hazard Class 6.1; ERG 154.

AMY's
See Rush.

AMYL ALCOHOL
$CH_3CH_2CH(CH_3)CH_2OH$
71-41-0
UN 1105

Appearance: Colorless liquid. This is the active ingredient in fusel oil. Amyl alcohol is slightly soluble in water. Sold in 1- and 5- gallon cans.

Odor: Characteristic odor, with a threshold below 10 ppm.

Hazards: Toxic by ingestion. Combustible, flash point 115°F. NFPA Health 3, Fire 3. Vapor pressure 2 mm Hg @ 30°C.

Incompatibilities: Strong oxidizers. Concentrated sulfuric acid, concentrated nitric acid, alkali metals.

Used in

Recreational Drugs:
Rush

Regulations: Hazard Class 3; ERG 129.

AMYL NITRITE
$C_5H_{11}NO_2$
463-04-7

Appearance: A clear or straw-colored liquid that is very slightly soluble in water and that usually comes in a very small bottle (*also see* Rush) or a cloth-covered sealed bulb. Amyl nitrite is used for heart patients and for diagnostic purposes because it dilates the blood vessels and makes the heart beat faster. It is also used to treat people who have inhaled hydrogen cyanide; it induces the formation of methemoglobin, which binds cyanide into non-toxic cyanomethemoglobin. When amyl nitrite became "use by prescription only," butyl nitrite took its place recreationally. *See* Recreational drugs, rush variations 1 and 2.

Odor: Distinct and peculiar; most people do not like it.

Hazards: Amyl nitrite, along with other alkyl nitrites, is a potent vasodilator that functions in situ via the nitric oxide pathway. Physical effects include decrease in blood pressure, headache, flushing of the face, increase in pulse rate, dizziness and relaxation of involuntary muscles, especially the blood vessels walls and the anal sphincter. There are no withdrawal symptoms. Overdose symptoms include nausea, emesis, hypotension, hypoventilation, dyspnea, and syncope.

Regulations: This is a Schedule 6 drug.

ANGELICA ROOT

Used in

Recreational Drugs:
Absinthe

ANHYDROUS ACIDS
Chemical weapon
Facility 15
Also see Corrosive smokes.

ANILINE
$C_6H_5NH_2$
62-53-3
UN 1547

Appearance: Oily liquid, colorless when fresh, darkens upon standing or exposed to air. Will be found in brown glass 1-quart bottles or jugs from a laboratory supply house; flammable; volatile when steamed (boils between 183 and 185°C).

Odor: Characteristic.

Hazards: Very poisonous. PEL 2 ppm.

Incompatibility: Nitric acid, especially red fuming, may ignite aniline spontaneously. Hydrogen peroxide.

Used in

Chemical Weapons:
Facility 14

Designer Drugs:
Group 2, Variation 2

Regulations: Hazard Class 6.1; ERG 153.

Confirmed by: Aniline Test.

ANISE

Description: Anise seed is the dry ripe seed of the fruit of *Pimpinella anisum,* which is present in the United States as both a cultivated and a wild plant.

Used in

Hobbies:
Brewing root beer

Recreational Drugs:
Absinthe

ANISOLE
$C_8H_5OCH_3$
100-66-3
UN 2222

Appearance: Colorless liquid, floats on water (methyl phenyl ether).

Odor: Aromatic.

Hazard: Combustible. NFPA Health 1, Fire 2.

Incompatibilities: Gum arabic, oxidizers.

Used in

Explosives:
PNT

Regulations: Hazard Class 3; PG III; ERG 127.

ANTHRACENE
$C_{14}H_{10}$
120-12-7
UN 3077

Appearance: Off-white-to-pale green crystals insoluble in water.

Odor: Not found.

Hazards: Tumor promoter; PEL 0.2 mg/m³. May act as a sensitizer; irritates eyes, nose, throat, and/or lungs.

Incompatibilities: Strong oxidizing agents, hypochlorites, chromic acid, fluorine.

Used in

Fireworks:
White smoke

Regulations: Hazard Class 9; PG III. Harmful to the environment.

ANTICHLOR®
See Sodium thiosulfate.

ANTIMONY
Sb
7440-36-0
UN 1549

Appearance: Silver-white, lustrous, hard, brittle metal or scale-like crystalline structure. Not tarnished by air.

Odor: None.

Hazards: All of its salts are poisonous.

Incompatibilities: Inert.

Toxins:
Toxic chemicals

Regulations: Hazardous Waste Solid n.o.s.; Hazard Class 6.1; ERG 171.

Confirmed by: Antimony Test.

ANTIMONY PENTASULFIDE
Sb_2S_5
1315-04-4
UN 1549

Appearance: Orange-yellow powder that is insoluble in water.

Odor: Not found.

Hazards: Ignites when heated. Toxic. Irritant.

Incompatibilities: Ignites or explodes in the presence of strong oxidizers.

Used in

Ammunition:
Load 8

Regulations: Hazardous Waste Solid n.o.s.; Hazard Class 4.1; ERG 171.

Confirmed by: Antimony Test.

ANTIMONY SULFATE (w)
$Sb_2(SO_4)_3$
UN 1549

Appearance: Crystalline powder or lumps, deliquescent in moist air.

Hazards: Poisonous!

Used in

Ammunition:
Load 6
Rim-fire primer mix

Toxins:
Toxic chemicals

Regulations: Hazardous Waste Solid n.o.s.;
Hazard Class 6.1; ERG 171.

Confirmed by: Antimony Test.

ANTIMONY SULFIDE
SbS_3
UN 1325

Appearance: Transparent dark ruby-red mass that is insoluble in water.

Odor: None.

Hazards: Toxic.

Incompatibilities: Explodes in contact with strong oxidizers. Concentrated chloric acid, chlorates, thallic oxide.

Used in

Ammunition:
Percussion caps

Fireworks:
Glitter

Toxins:
Toxic chemicals

Regulations: Hazardous Waste Solid n.o.s.;
Hazard Class 6.1; ERG 171.

Confirmed by: Antimony Test.

ANTIMONY SULFITE (w)
UN 1549

Used in

Ammunition:
Load 6
Rim-fire primer mix

Toxins:
Toxic chemicals

Regulations: Hazardous Waste Solid n.o.s.;
Hazard Class 6.1; ERG 171.

Confirmed by: Antimony Test.

ANTIMONY TRISULFATE
See Antimony sulfate.

ANTIMONY TRISULFIDE
See Antimony sulfide.

APCP
See Ammonium perchlorate.

ARSENIC
As
7440-38-2
UN 1558

Appearance: Brittle metallic sponge, insoluble in water.

Odor: Slight metallic garlic.

Hazards: Very poisonous! By ingestion or inhalation. Sublimes in heat.

Incompatibilities: Strong oxidizers, acids, halogens.

Used in

Chemical Weapons:
Facility 11

Toxins:
Toxic chemicals

Regulations: Hazard Class 6.1; PG II; ERG 152; one of the EPA's TCLP chemicals.

Confirmed by: Arsenic Test.

ARSENIC BISULFIDE (w)
As_2S_2
UN 1557
Also see Arsenic sulfide.

Appearance: Red-brown crystals, reacts with water.

Hazards: Poisonous! PEL 0.2 mg/m^3.

Incompatibilities: Water, strong oxidizers.

Used in

Chemical Weapons:
Facility 11

Toxins:
Toxic chemicals

Regulations: One of the EPA's TCLP chemicals.

Confirmed by: Arsenic Test.

ARSENIC PENTASULFIDE (w)
As_2S_5
UN 1557
Also see Arsenic sulfide.

Appearance: Brownish yellow, glassy, amorphous, highly refractive mass that reacts with water, emitting toxic gas.

Hazards: Poisonous! PEL 0.2 mg/m³.

Used in

Chemical Weapons:
Facility 11

Toxins:
Toxic chemicals

Regulations: Hazard Class 6.1; one of the EPA's TCLP chemicals.

Confirmed by: Arsenic Test.

ARSENIC SALTS
There are many arsenic salts; in some the arsenic is the cation, in some it is the anion. Arsenic salts can be gases (e.g., arsine), liquids (e.g., arsenic trichloride), but most are solid.

Odor: Slight, metallic garlic.

Hazards: Very poisonous! PEL 0.2 mg/m³.

Used in

Chemical Weapons:
Facility 11
Facility14

Toxins:
Toxic chemicals
arsine (blood agent), arsenicals (vesicants), arsenic trichloride

Regulations: Hazard Class 6.1; one of the EPA's TCLP chemicals.

Confirmed by: Arsenic Test.

ARSENIC SULFIDE
As_2S_2 (red)
As_2S_3 (yellow)
12612-21-4
UN 1557
A mixture of arsenic trisulfide, arsenic bisulfide, and arsenic pentasulfide.

Hazards: See the hazards for arsenic trisulfide, arsenic bisulfide, and arsenic pentasulfide.

Used in

Fireworks:
White smoke

Toxins:
Toxic chemicals

Regulations: One of the EPA's TCLP chemicals. ICC Poison B, Poison label. Hazard Class 6.1.

Confirmed by: Arsenic Test.

ARSENIC TRICHLORIDE
Chemical weapon
$AsCl_3$
7784-34-1
UN 1560

Appearance: Fuming liquid arsenic. This yellowish oily liquid salt, which fumes in air, is unusual. It will be found in sealed brown glass bottles. Arsenic trichloride is made in Facility 11.

Odor: Unpleasant garlic.

Hazards: Very poisonous. Inhalation hazard. Vapor pressure 10 mmHg @ 23°C. Skin contact can be lethal. Water reactive.

Incompatibilities: High heat, contact with acids produces fumes of HCl and toxic arsenic fumes. Reacts with water or steam.

Used in

Chemical Weapons:
Facility 4
Facility 5

Explosives:
AS-20

Toxins:
Toxic chemicals

Regulations: One of the EPA's TCLP chemicals; ICC Poison B, Poison label. MCA Warning label; Hazard Class 6.1; ERG 157. Australian Group Schedule 2B.

Confirmation Test: Arsenic Test. It would be best NOT to test this material.

ARSENIC TRISULFIDE
As_2S_3
UN 1557
Also see Arsenic sulfide.

Appearance: Yellow-red crystals that are soluble in water.

Hazards: Poisonous!

Used in

Chemical Weapons:
Facility 11
Facility 14

Toxins:
Toxic chemicals

Regulations: One of the EPA's TCLP chemicals; ICC Poison B, Poison label. Hazard Class 6.1.

Confirmed by: Arsenic Test.

ARTEMISIA PONTICA (Roman Wormwood)

Used in

Recreational Drugs:
Absinthe

ARYL COPPER (w)
C_6H_3Cu
Aryl is a compound whose molecule has a ring structure characteristic of benzene.

Used in

Amphetamines:
Synthesis 18

To make allyl benzene using allyl copper in a reaction not discussed in this book.

Confirmed by: Copper Test.

ASCORBIC ACID
$C_6H_8O_6$

Appearance: White crystals or crystalline powder. Stable in air, but deteriorates rapidly in water. Occurs in vegetables and fruits, notably in citrus fruits.

Odor: None.

Hazards: Vitamin C is taken orally.

Used in

Hobbies:
Photography

Illegal Drugs:
Cook 2, Supplementary 4
Cook 2, Storage

Regulations: Not specifically regulated.

ASPHALT
Asphaltum, mineral pitch, Judean pitch, bitumen.

Appearance: A black bituminous substance resulting from petroleum by evaporation of lighter hydrocarbons and partial oxidization of the residue.

Odor: Petroleum.

Hazards: Heated vapors may cause irritation of the nose and throat.

Incompatibilities: Nitric acid, hypochlorites.

Used in

Explosives:
Detonation cord

Fireworks:
Binder; flame deoxidizing

Regulations: Not specifically regulated.

ASPHALTUM
See Asphalt.

ASPIRIN
Although small amounts of aspirin should NOT be considered as an indicator, large amounts could be. Making picric acid out of aspirin has been published in many places and has been seen on the Internet. This would be an expensive way to make these chemicals, but would be a way to avoid detection.

Used in

Chemical Weapons:
Facility 3
Facility 10, Supplementary 1

Explosives:
Picric acid

AURAMINE
492-80-8

Appearance: Yellow, crystalline (sandlike) powder or flaky material. It is used as a dye (Solvent Yellow).

Hazards: Toxic inhalation hazard, absorbed through the skin. Carcinogen, a suspected cardiovascular toxin.

Used in

Fireworks:
Yellow smoke

AZALEAS
See Rhododendrons.

AZIRIDINE
C_2H_5N
151-56-4

Appearance: Colorless mobile liquid that is miscible with water.

Odor: Intense ammoniacal.

Hazards: Flammable. Inhalation hazard; burns eyes and mucous membranes and causes edema of the lungs. This is a severe blistering agent and can cause third-degree burns on the skin. Vapor pressure 160 mmHg @ 20°C.

Incompatibilities: Oxidizers.

Used in

Amphetamines:
Produced as an undesirable by-product when chloroephedrine is subjected to heat or basic solutions. This is the reason that bicarbonate of soda is a better choice to make methamphetamine base than sodium hydroxide (Synthesis 3).

Hobbies:
Rocket fuel, Variation 3

Regulations: Hazard Class 3; Hazardous Waste n.o.s.

B

BAKING SODA
See Bicarbonate of soda.

BANANAS

Used in

Recreational Drugs:
Bananadine

BANGSITE
See Acetylene.

BAQUACIL®
See Biguanide.

BARBITUATE ACID (w)
$C_4H_4O_3N_2 \cdot 2H_2O$

Appearance: White rhombic efflorescent crystals (malonylurea). Not very soluble in cold water, but very soluble in hot water. Made by the condensation of ethylsodiomalonate with urea in ethanol.

Odor: None.

Used in

Designer Drugs:
Group 4

BARITOP PLUS®
See Barium sulfate.

BARIUM CARBONATE
$BaCO_3$
UN 1564

Appearance: Occurs in nature as the mineral witherite. White, heavy powder that is almost insoluble in water.

Odor: None.

Hazards: Poisonous!

Incompatibilities: Effervesces in acids, with evolution of CO_2.

Used in

Fireworks:
Flash delay, strobe formulations

Hobbies:
Ceramics

Regulations: Poison Label n.o.s. Hazard Class 6.1. Barium salts are on the EPA's TCLP chemicals list, due to concern about drilling muds. ERG 154.

Confirmed by: Barium Test.

BARIUM CHLORATE
$Ba(ClO_3)_2·H_2O$
10294-38-4 (anhydrous)
10294-38-9 (monohydrate)
UN 1564 (n.o.s.)
UN 1485

Appearance: Colorless crystals or white powder that is soluble in water.

Odor: Odorless.

Hazards: Poisonous! Oxidizer. NFPA Health 1, Reactivity 1.

Incompatibilities: Organic materials.

Used in

Fireworks

Regulations: PG II; Hazard Class 5.1. Barium salts are on the EPA's TCLP chemicals list, due to concern about drilling muds. ERGs 154 and 140.

Confirmed by: Barium Test.

BARIUM CHLORIDE
$BaCl_2·2H_2O$
UN 1564 (n.o.s.)

Appearance: Colorless crystals or white granules or powder that dissolves in water.

Odor: Odorless.

Hazards: Poisonous!

Incompatibilities: Bromine trifluoride, 1-furan percarboxylic acid (self-igniting).

Used in

Fireworks:
Color

Regulations: Poison Label n.o.s. Hazard Class 6.1. Barium salts are on the EPA's TCLP chemicals list, due to concern about drilling muds. ERG 154.

Confirmed by: Barium Test.

BARIUM CHROMATE
$BaCrO_4$
10294-40-3
UN 1564

Appearance: Lemon chrome or yellow heavy crystals or powder that is not soluble in water.

Odor: None.

Hazards: Oxidizer, PEL 0.5 mg/m^3, irritating to the skin, eyes, lungs, and mucous membranes on contact.

Incompatibilities: Reducing agents.

Used in

Explosives:
Blasting caps

Fireworks:
Fuse-burn-rate adjuster

Hobbies:
Ceramics

Regulations: Poison Label n.o.s. Hazard Class 6.1. Barium salts are on the EPA's TCLP chemicals list, due to concern about drilling muds. ERG 154.

Confirmed by: Barium Test.

BARIUM DIOXIDE
See Barium peroxide.

BARIUM HYDROXIDE
$Ba(OH)_2$
17194-00-2 (anhydrous)
22326-55-2 (monohydrate)
12230-71-6 (octahydrate)
UN 1564

Appearance: Colorless transparent crystals or white masses that absorb CO_2 rapidly from the air. Not soluble in water, becomes less so on standing.

Odor: None.

Hazards: Poisonous! Becomes very alkaline in water. NFPA Health 3.

Incompatibilities: Strong acids.

Used in

Fireworks:
Color

Regulations: Poison label n.o.s. Hazard Class 6.1. Barium salts are on the EPA's TCLP chemicals list, due to concern about drilling muds. ERG 154.

Confirmed by: Barium Test.

BARIUM NITRATE
$Ba(NO_3)_2$
UN 1446

Appearance: Colorless crystals or white crystalline powder that is soluble in water.

Odor: None.

Hazards: Strong oxidizer. Poisonous! NFPA Health 3. Inhalation hazard; irritating to the skin. PEL 0.5 mg/m^3.

Incompatibilities: Combustibles, organic materials. When mixed with finely divided metals or fuels, becomes ignitable and sensitive to friction or impact. A mixture of barium dioxide and zinc with barium nitrate will be explosive.

Used in

Fireworks:
Color, glitter, strobe formulations

Regulations: Hazard Classes 5.1 and 6.1; PG II. Barium salts are on the EPA's TCLP chemical list, due to concern about drilling muds. ERG 141.

Confirmed by: Barium Test.

BARIUM OXALATE
$BaC_2O_4 \cdot H_2O$
516-02-9
UN 1564

Appearance: White crystalline powder, not very soluble in water.

Odor: None.

Hazards: Poisonous! Mild skin irritant.

Incompatibilities: Reactive with strong acids.

Used in

Fireworks:
Flash delay

Regulations: Poison label n.o.s. Hazard Class 6.1. Barium salts are on the EPA's TCLP chemical list, due to concern about drilling muds. ERG 154.

Confirmed by: Barium Test.

BARIUM PEROXIDE
BaO_2
1304-29-6
UN 1449

Appearance: Grayish-tan or white powder. Also called *barium dioxide*. Will be in brown glass jar from a chemical supply house. Lid will probably be sealed.

Odor: None.

Hazards: Oxidizer, toxic by ingestion inhalation. May cause skin burns. PEL 0.5 mg/m³.

Incompatibilities: Reducing agents, organic compounds, finely divided metals, lead dioxide; with solutions of hydroxylamine, causes fuming. There is a risk of fire or explosion on contact with combustible materials. Decomposes on contact with heat, water, or acids to produce oxygen and hydrogen peroxide.

Used in

Ammunition:
Load 6
Rim-fire primer mix

Fireworks

Regulations: Hazard Classes 5.1 and 6.1; PG II. RTECS CR0175000. Barium salts are on the EPA's TCLP chemical list, due to concern about drilling muds. ERG 141.

Confirmed by: Barium Test.

BARIUM SULFATE
$BaSO_4$
UN 1564

Appearance: White, fine, heavy, odorless powder insoluble in water.

Odor: None.

Hazards: Although barium compounds are poisonous, barium sulfate is not toxic by ingestion because of its insolubility. It is used medicinally in the product Baritop Plus®. It is used frequently as a radiocontrast agent for x-ray imaging and other diagnostic procedures. PEL 10 mg/m³.

Used in

Amphetamines:
Synthesis 4

Fireworks:
Color, strobe formulations

Regulations: Poison label n.o.s. Hazard Class 6.1. Barium salts are on the EPA's TCLP chemical list, due to concern about drilling muds. ERG 154.

Confirmed by: Barium Test.

BATH SALTS
See Magnesium sulfate.

BEESWAX
About 80% myricin (ceryl myristate); also contains ester of cerotic acid and a few percent hydrocarbons.

Appearance: White (bleached) or yellow, waxy, soft-to-brittle mass.

Odor: Honey-like.

Incompatibilities: Strong oxidizers.

Used in

Hobbies:
Photography

Regulations: No specific regulations.

BELLADONNA
Poisonous plant

Description: Also called *deadly nightshade*, *banewort*. Belladonna is one of the most toxic plants found in the western hemisphere. It is a perennial plant that has oval dark-green leaves, bell-shaped violet blossoms, and a black cherry-like fruit. Known since ancient times, the tincture of belladonna leaves used to be one of the most popular poisons among professionals. It is used as a decorative plant.

Where found: It is found occasionally in waste places unused land within the confines of human habitation in the eastern United States. May be found anywhere as an ornamental plant.

Deadly parts: The entire plant is poisonous.

Symptoms: Effects occur in several hours to several days. The poison comprises atropine, scopolamine, hyoscyamine, and hyoscine. It works by paralyzing the parasympathetic nervous system, blocking action at the nerve endings. Dilated pupils; blurred vision; increased heart rate; hot, dry, red skin; dry mouth; disorientation; hallucinations; impaired vision; loud heartbeats audible at several feet; aggressive behavior; rapid pulse; fever; convulsions; coma; and death.

Antidotes: Gastric lavage with 4% tannic acid solution and vomiting. Pilocarpine or physostigmine for dry mouth.

Field test: Addition of potassium permanganate produces a slow red color.

Used in

Hobbies:
Decorative plants

Recreational Drugs:
Hallucinatory tea

Toxins:
Atropa belladonna

BENTONITE
$AlSiO_3$
Also called *wikinite*. A colloidal native hydrated aluminum silicate (clay) found in the midwest of Canada and the United States.

Appearance: The color in the massive condition varies from yellowish white to almost black; the powder is cream colored to pale brown. It has the property of forming highly viscous suspensions or gels with not less than 10 times its weight of water.

Odor: None.

Hazards: This material is inert.

Incompatibilities: This material is inert.

Used in

Pathogens:
Used to weaponize anthrax

Regulations: No specific regulations.

Confirmed by: Aluminum Test.

BENZALDEHYDE
C_6H_5CHO
100-52-7
UN 1990

Appearance: Colorless-to-yellowish viscosity 1 liquid. Will usually be in containers from a laboratory supply house; however, it is sold in bulk in the food industry, where it is used as a flavoring agent. Probably sold in brown 1-quart bottles. Although this is a food product, it will most likely be purchased from a laboratory supply house.

Odor: The most incredible odor of almonds. This will make you hungry.

Hazards: Moderately toxic by ingestion or inhalation. LD_{50}(rat) 1300 mg/kg.

Incompatibilities: Incompatibility Group 4-A. Nitric acid, perchloric acid, and chromium trioxide.

Used in

Amphetamines:
Synthesis 1, Supplementary 5
Synthesis 15

Chemical Weapons:
Facility 13
Incapacitating agent, buzz

Designer Drugs:
Group 1, Variation 3

Illegal Drugs:
Cook 1, Supplementary 1

Regulations: CDTA (List I) 4 kg; DOT Class 9; PG III; ERG 171.

Confirmed by: Intense almond odor.

BENZENE
C_6H_6
71-43-2
UN 1114

Appearance: Colorless viscosity 1 liquid. It will be found in 1-pint, 1-quart, or 1-gallon brown glass bottles from a laboratory supply house.

Benzene is heavily regulated under OSHA and is not an over-the- counter product. This is not a good chemical to key on. *Also see* Nonpolar liquids that float.

Odor: Unique mild solvent odor.

Hazards: Flammable liquid, inhalation intoxication hazard; flash point 12°F; LEL 1.2%; UEL 7.8%; LD_{50} 3.8 g/kg; LC_{50} 10,000 ppm @ 7 hours; PEL 1 ppm. Skin notation: chronic exposure to benzene is associated with leukemia and aplastic anemia; benzene is considered to be a carcinogen.

Incompatibilities: Incompatibility Group 4-A. Chromium trioxide.

Used in

Amphetamines:
Synthesis 10
Synthesis 13

Synthesis 14
Synthesis 15
Synthesis 17
Synthesis 18
Synthesis 19
Synthesis 20, Supplementary 1
Synthesis 22

Benzene is a reagent with chloroacetone (*see* Facility 15, Chloroacetone) in the presence of aluminum chloride to produce P-2-P (Synthesis 13). Benzene is also used as an enhancer to produce P-2-P using phenyl acetic acid and acetic acid (Synthesis 14). Benzene is used to extract P-2-P (Synthesis 10). Benzene and acetone are precursors to make P-2-P (Synthesis 17). Benzene is used as a precursor compound for allyl benzene and P-2-P (Syntheses 18, 19, and 22). Benzene is a good nonpolar solvent that works well for methamphetamine hydrochloride extraction. Because it is so heavily regulated, it probably would not be encountered, except that it appears to be the solvent of choice of Uncle Fester (in *secrets of Metharn-phetamine manufacture*). Benzene is used to make phenylacetic acid in Synthesis 20, Supplementary 1.

Chemical Weapons:
Facility 14

Explosives:
Ammonium picrate, DDD, HNB, Hexol, nitrogen sulfide; LNTA, MNA, nitro-PCB, NPF, PCB

Illegal Drugs:
Cook 1, Variation 1
Cook 1, Supplementary 1
Cook 2, Supplementaries 1 and 3
Cook 4, Supplementary 1
Cook 5, Variation 1
Cook 6, Variations 1 and 2
Cook 7
Cook 8, Variation 2
Cook 9

Tear Agents:
Mace

Regulations: RQ 454 kg; Waste Codes D001 (0.5 mg/L), U019, D018, and F005; Hazard Class 3; PG II; ERG 130.

Confirmed by: Red Iodine Crystal Test; Benzene Test.

BENZENE HEXACHLORIDE
C_6Cl_6
58-89-9, 60-87-31, others
UN 2761

Appearance: White crystalline solid. A commercial mixture of the isomers of 1,2,3,4,5,6-hexachlorocyclohexane. It is also used as a pesticide.

Odor: Not found.

Hazards: The gamma portion of the mixture is lindane, which is highly toxic. It is most toxic by ingestion, but moderately toxic by inhalation. Absorbed through the skin. The gamma and alpha isomers are central nervous system stimulants; the beta and delta isomers are depressants of the central nervous system.

Incompatibilities: Strong oxidizers.

Used in

Ammunition

Fireworks:
Chlorine donor

Regulation: Disposal of lindane is highly restricted. The waste must be incinerated. Lindane is on the EPA's TCLP chemicals list.

Confirmed by: Beilstein Test.

BENZILIC ACID
$C_{14}H_{12}O_3$
76-93-7

Appearance: White crystals or white-to-tan powder that is soluble in hot water but only slightly soluble in cold water. Can be made; see Chemical weapons, Synthesis 13. Can be purchased in bulk on the Web.

Odor: None.

Incompatibilities: Can react with strong oxidizers.

Used in

Chemical Weapons:
Facility 13, buzz

Designer Drugs
Variations 1, 2, and 3

Regulations: Australian Group Schedule 2B.

Confirmed by: Becomes reddish violet in sulfuric acid.

BENZINE
See Ligroine; also see Nonpolar liquids that float.

BENZODIAZEPRINES
Date rape
Off-of-the-shelf prescription drug.

BENZOIC ACID
$C_7H_6O_2$
65-85-0

Appearance: Colorless or white lustrous needles or leaflets that begin to sublime at about 100°C. Benzoic acid is soluble in hot water.

Odor: Benzaldehyde or benzoin or almonds.

Hazards: Dust might cause mild respiratory irritation. The salts of benzoic acid are used as food preservatives.

Incompatibilities: Strong oxidizers.

Used in

Explosives:
Benzoyl peroxide. Supplementary 1

Regulations: RQ 5000 lb.

BENZOIC ALDEHYDE
See Benzaldehyde.

BENZOIC ANHYDRIDE
$C_6H_5CO\cdot O\cdot CO\cdot C_6H_5$
93-97-0

Appearance: White crystal powder that is insoluble in water. It is slowly decomposed by cold water or cold alkali.

Odor: Not found.

Hazards: Irritant; flash point 240°F.

Incompatibilities: Can react with oxidizing materials. Bonzoic anhydride is moisture sensitive.

Used in

Illegal Drugs:
Cook 1

Regulations: There are no special requirements for transportation.

BENZOIC TRICHLORIDE
$C_6H_5CCl_3$
98-07-7
UN 2226

Appearance: Colorless-to-yellowish liquid that fumes in the air and hydrolyzes slowly in water. This is a high-volume chemical with production exceeding 1 million pounds annually in the United States.

Odor: Penetrating.

Hazards: Toxic by inhalation, the fumes are highly irritating; passes through the skin. Considered to be among the top 10% of the most hazardous chemicals both for humans and the environment. Carcinogen. Moisture in air will cause benzoic trichloride to off-gas HCl gas.

Incompatibilities: Water, perchlorates, peroxides, permanganates, chlorates, nitrates, chlorine, bromine, and other oxidizing agents.

Used in

Explosives:
Benzoyl peroxide, Supplementary 2

Regulations: Store away from foodstuffs. Hazard Class 8, minor 6.1; PG II; ERG 156.

Confirmed by: Beilstein Test. We recommend not testing.

BENZOL
See Benzene.

BENZOTRIAZOLE
C_6H_4NHNN
95-14-7

Appearance (1,2,3-benzotriazole): Needle-like crystals somewhat soluble in water.

Odor: Not found.

Hazards: Moderately toxic; harmful if swallowed; skin and eye irritant.

Incompatibilities: Strong oxidizers, heavy metals.

Used in

Hobbies:
Photography

Regulations: Nonhazardous for air, sea, and road freight.

BENZOTRICHLORIDE
See Benzoic trichloride.

BENZOYL CHLORIDE
C_6H_5COCl
98-88-4
UN 1736

Appearance: Colorless, fuming, pungent liquid, which breaks down to hydrochloric acid in water. *See* explosives, (benzoyl peroxide, Supplementaries 1 and 2).

Odor: Leather-like, pungent.

Hazards: Dangerous when heated to decomposition; it emits highly toxic fumes of phosgene; will react with water or steam to produce heat and toxic and corrosive fumes. Combustible, flash point 72°C.

Incompatibilities: Water. Reacts vigorously with oxidizing materials, dimethyl sulfoxide, and alkalis.

Used in

Amphetamines:
Synthesis 21, Variation 2

Explosives:
Benzoyl peroxide

Illegal Drugs:
Cook 1, Variation 1
Cook 1, Supplementary 1

Regulations: PG II; Hazard Class 8, minor 6.1; ERG 137; RTECS DM66000000.

Confirmed by: Chloride Test.

BENZOYLFORMIC ACID
C_6H_5COCOO
611-73-4

Appearance: Slightly yellow crystalline powder.

Odor: None.

Hazards: Causes lung, eye and skin irritation. By ingestion causes nausea, vomiting, and diarrhea. No established PEL.

Incompatibilities: Strong oxidizers; strong reducers.

Used in

Amphetamines:
Synthesis 20, Supplementary 2

Regulations: None applicable.

BENZOYL PEROXIDE
$(C_6H_5CO)_2O_2$
94-36-0
UN 2085, UN 2087, UN 2088, UN 2089, UN 2090

Appearance: White crystals, usually seen as a white granular material. *See* Explosives, benzoyl peroxide.

Odor: Mild chlorinated solvent odor.

Hazards: Explodes just above room temperatures @ 114°F. Organic peroxide.

Incompatibilities: Reducing agents; organic materials.

Used in

Hobbies:
Catalyst for plastics

Regulations: Hazard Class 5.2; not acceptable in passenger transportation; ERG 146.

Confirmed by: Peroxide Test.

BENZYL CHLORIDE
$C_6H_5CH_2Cl$
100-44-7
UN 1738

Appearance: Colorless, oily liquid. Most likely found in brown 1-pint to 4-liter bottles purchased from a laboratory supply house not an over-the-counter item.

Odor: Lachrymator; very irritating.

Hazards: Corrosive; irritating to eyes and by inhalation, and absorbed through the skin. This is a suspected carcinogen.

Incompatibilities: Incompatibility Groups 4-A and 3-B. Chromium trioxide.

Used in

Amphetamines:
Synthesis 16
Synthesis 20, Supplementary 8
Synthesis 21, Variations 1 and 2

Regulations: CDTA (List II) 1 kg; RQ 45.4 kg; Waste Codes P028 and D002; Hazard Class 6.1, minor 8; PG II; ERG 156.

Confirmed by: Beilstein Test.

BENZYL CYANIDE
$C_6H_5CH_2CN$
140-29-4
A product of benzyl chloride and sodium cyanide.

Appearance: Colorless or yellowish oily liquid. Since it is not a commercially available product, it will be found in a brown glass bottle from a laboratory supply house.

Odor: Floral odor or almond aromatic.

Hazards: Toxic by ingestion, skin contact, and inhalation. Vapor pressure 0.1 mmHg @ 20°C. May produce hydrogen cyanide gas on burning.

Incompatibilities: Chromium trioxide. May produce HCN in a fire.

Used in

Amphetamines:
Synthesis 16
Synthesis 20

Regulations: CDTA (List I); RQ 1 kg; Hazard Class 6.1; PG III.

Confirmed by: Hydrogen cyanide indicator tube.

BENZYLFORMIC ACID
See Phenylacetic acid.

BENZYL MAGNESIUM CHLORIDE
A Grignard reagent.

This is a classic example of an obvious intermediate listed as an ingredient. This compound is obviously a Grignard reagent, and as such, could not be purchased. Made by adding benzyl chloride to magnesium metal in ether, and probably would be used in the reaction vessel in which it was made. Considered a carcinogen. Not a standard off-the-shelf item; it must be made at the site. A Grignard reagent is used in organic chemistry to add together various parts of organic molecules. Whenever you encounter ethyl ether and magnesium turnings together, you have probably encountered an additional alreaction. Look for chlorinated compounds, the R group of which is to be added to other compounds.

Odor: Ether.

Hazards: This ongoing reaction is very flammable, and if not carefully controlled will heat up considerably, which is dangerous, due to the presence of the ether.

Incompatibilities: Grignard reagents are very reactive.

Used in

Amphetamines:
Synthesis 20
Synthesis 21

Explosives:
TND

Regulations: Should be handled as ether. *See* Ethyl ether.

BICARBONATE OF SODA
$NaHCO_3$
144-55-8

Appearance: White powder. Sold in any store in products such as Arm & Hammer® baking soda. A slightly higher grade can be purchased in drugstores as Rexal Formula III® baking soda. Bicarbonate of soda is a common commodity used as a fire extinguisher powder, and many other products, including Alka-Seltzer™.

Odor: None.

Hazards: None specifically. Bicarbonate of soda has many household uses, including in cooking. NFPA Health 1.

Incompatibilities: Will effervesce in the presence of water and in acids.

Used in

Amphetamines:
Synthesis 3

It is hard to imagine many compounds that have more uses in the manufacture of methamphetamine than does bicarbonate of soda. This compound and the closely related sodium carbonate are used as reagents, and neutralizers, in pH adjustment, to clean up substandard street drugs, as cuts for product, in extraction to remove pill additives, to make methamphetamine carbonate (a smokable form of the drug), and as an enhancement for people using the drug. The most important use of bicarbonate of soda is to make methamphetamine base from any synthesis that involves the production of phenyl-2-chloropropanone. Bicarbonate of soda prevents the formation of aziridine, which is an unwanted compound. The inclusion of bicarbonate of soda in methamphetamine enhances the action of methamphetamine on the body. By taking it simultaneously, it allows the kidneys to reabsorb the methamphetamine, thus increasing the time it is in the body. Bicarbonate of soda can be used to make a product the equivalent of crack cocaine, in that it can be smoked. This is easier to produce than ice, which is the most effective product to smoke, as methamphetamine hydrochloride becomes charred upon heating.

Explosives:
ADN; ADNBF; AS-20; BDPF; CDNTA; DANP,; DINA; DNPU; isoitol nitrate; MPG; MX; nitrocellulose; nitroform; nitroglycerin 2; nitrated milk powder; nitromannitol, Variation 2; NTA; NTND; PEN; PVN; quebrachitol nitrate; TAEN; TEX; TNB; TNEN; TNM; the TNPU; TNT

Baking soda has been specified in production of ADBN, BDPF, DANP, TAEN, and TEX.

Fireworks:
Flash delay, Smoke dye formulations

Illegal Drugs:
Cook 1, Variations 3 and 4
Cook 2, Variations 2, 3, and 6
Cook 2, Supplementary 4

Hobbies:
Photography

Regulations: None applicable.

Confirmed by: Sodium Yellow Flame Test; Carbon dioxide indicator tube.

BIC® LIGHTERS

Used in

Incendiary Devices:
Situation 0.1

BIFORMYL
See Glyoxal.

BIGUANIDE
$C_2H_7N_5$
56-03-1

Is commercially available in plastic bags as BaquaSpa Sanitizer™, SoftSwim Bactericide™, Polyclear®, Baquacil® Chlorine Free Swimming Pool Sanitizer, and Metformin®, which is widely used in the treatment of diabetes.

Used in

Explosives:
BDC

Regulations: Not regulated.

BILORIN
See Formic acid.

BIS(2-CHLOROETHYL ETHER)
Chemical weapon
$ClCH_2CH_2OCH_2CH_2Cl$
542-88-1

Appearance: Colorless, stable liquid. Boiling point 178°C; flash point 131°F; vapor pressure 0.7 mmHg @ 20°C.

Odor: Vapors are an irritant to the mucous membranes.

Hazards: Carcinogen! Mild narcotic, affects the liver and kidneys. Exposures of 1000 ppm for 30 to 60 minutes may cause death in a few days.

Incompatibilities: Can react vigorously with oxidizing materials.

BISMUTH (w)
Bi

Appearance: Crystalline brittle metal with a reddish tinge.

Odor: Mild nitric acid.

Hazards: Flammable in powder form.

Incompatibilities: Strong acids, strong reducing agents, powdered metals, and organic materials that are easily combustible or oxidizable.

Used in

Ammunition:
Load 3
Bullets

Regulations: Not specifically regulated.

Confirmed by: Bismuth Test.

BISMUTH NITRATE
Bi(NO$_3$)$_3$·5H$_2$O
10361-44-1 (anhydrous)
10035-06-0 (pentahydrate)
UN 1477

Appearance: Lustrous, clear, colorless, hygroscopic crystals that are slowly decomposed by water. Comes in bottles, tins, drums, multiwall paper sacks.

Odor: None.

Hazards: Oxidizer, fire risk near organic materials. Bismuth is not easily absorbed into biological systems.

Incompatibilities: Organic materials.

Used in

Ammunition:
Load 3

Regulations: Oxidizer label (rail and air). Hazard Class 5.1; PG II; Nitrates, inorganic, n.o.s.; RQ 300 lb; ERG 140.

Confirmed by: Bismuth Test.

BISMUTH TRIOXIDE
Bi$_2$O$_3$
1304-76-3

Appearance: Heavy yellow powder that is insoluble in water.

Odor: None.

Hazards: Bismuth toxicity is associated with chronic rather than acute exposure. NFPA Health 1.

Incompatibilities: Stable compound.

Used in

Ammunition:
Load 3

Hobbies:
Ceramics

Regulations: Nonhazardous for air, sea, and road freight.

Confirmed by: Bismuth Test.

BLACK POWDER (w)
Can be purchased from gun shops. It comes in three grades: FFF Cannons: FF .36–50 caliber, F .36 caliber or smaller. Grade F is best for making explosives, especially pipe bombs. Can be made; see Explosive, black powder.

Used in

Explosives:
McGyver

Regulations: Hazard Class 1.1D.

BLASTING CAPS

Used in

Explosives:
Initiators for secondary explosives. Note that pre-1930 copper blasting caps are dangerously unstable.

BLEACH
See Sodium hypochlorite.

BLUE DEATH
See Red phosphorus.

BLUING (Blue Boy, Rosebud)

Used in

Amphetamines:
Synthesis 0.5

BMK
See 1-Phenyl-2-propanone.

BORAX
Na$_2$B$_4$O$_7$·10H$_2$0
1330-96-4

Appearance: Sodium borate sold commercially as Borax® a laundry product, in cardboard boxes that look like laundry soap. Borax is a very white granular material.

Odor: None.

Hazards: LD_{50} estimated at 0.1 to 0.5 g/kg. Toxic to all cells, and the excretion rate is very slow. Nonhazardous by directive 67/548/EEC.

Incompatibilities: This product is very stable. Contact with potassium can be explosive. Acid anhydrides.

Used in

Hobbies:
Photography

Regulations: Not regulated.

Confirmed by: Boron Test.

BORIC ACID
H_3BO_4
10043-35-3

Appearance: Colorless scales or white powder soluble in boiling water.

Odor: None.

Hazards: Aqueous solutions of boric acid have been used as eyewash. Boric acid has been spread around kitchens as a roach killer. Children are more susceptible to poisoning by this compound, especially if there are repeated exposures due to the fact that boric acid is eliminated from the body slowly. Inhalation of boric acid can cause irritation, in extreme cases death, due to low blood pressure and renal injury. A fatal dose is between 5 and 30 g for a 180-lb adult. NFPA Health 1. Nonhazardous by directive 67/548/EEC.

Incompatibilities: Potassium, acetic anhydride, alkalis, carbonates, and hydroxides.

Used in

Hobbies:
Photography

Regulations: None applies; nonhazardous by air, sea, and road freight.

Confirmed by: Boron Test.

BOROHYDRIDE
See Sodium borohydride.

BORON (Anhydrous)
B
7440-42-8

Appearance: Monoclinic crystals, yellow or brown amorphous powder.

Odor: None.

Hazards: Very similar to finely powdered metals, can be ignited. Toxic.

Incompatibilities: Ammonia, bromine, cupric oxide, and lead dioxide (explosive).

Used in

Amphetamines:
Synthesis 18

Recreational Drugs:
Rush, Variation 2

Regulations: Hazard Class 4.1.

Confirmed by: Boron Test.

BORON TRIBROMIDE
Industrial chemical weapon
BBr_3
10294-33-4
UN 2692

Appearance: Colorless fuming liquid. Vapor pressure 40 mmHg @ 14°C.

Odor: Irritating.

Hazards: Corrosive. Reacts with water or steam to produce toxic and corrosive fumes.

BORON TRICHLORIDE
Industrial chemical weapon
BCl_3
10294-34-5
UN 1741

Appearance: Colorless fuming liquid reacts violently with water, even the moisture in the air.

Odor: Irritating.

Hazards: Corrosive. Toxic by inhalation, vapor pressure 1 atm @ 13°C. Causes burns.

Incompatibilities: Reacts violently with water or steam to produce toxic and corrosive fumes. Aniline, phosphine, dinitrogen tetroxide.

Regulations: Hazard Class 8; ERG 125.

BORON TRIFLUORIDE
Industrial chemical weapon
BF_3
7632-07-2
UN 1008

Appearance: Colorless gas. This compound will probably be found in a very special laboratory bottle.

Odor: Irritating.

Hazards: Corrosive. May be fatal if inhaled; lung edema. Causes burns. PEL 1 ppm ceiling.

Incompatibilities: Reacts with water or steam to produce toxic and corrosive fumes, including hydrofluoric acid.

Used in

Illegal Drugs:
Cook 4, Variations 1 and 2

Regulations: Hazard Class 2.3, minor 8; ERG 125.

BORON TRIFLUORIDE ETHERATE

Appearance: This is another example of an intermediate being listed as an ingredient. Probably will be made at the site. Clear liquid. This is also called *boron trifluoride-ether complex*. This is a relatively stable compound formed by the combination of boron trifluoride (a gas) with diethyl ether. This complex is a liquid. Boron trifluoride is derived from borax and hydrofluoric acid, or boric acid and ammonium bifluoride and fuming sulfuric acid, to form a gas that can be bubbled through ether.

Odor: Ether.

Hazards: Corrosive and flammable. Produces highly toxic fumes of fluorides when heated to decomposition. Will react with water to produce toxic, corrosive, and flammable vapors. It reacts vigorously with oxidizers!

Incompatibilities: Water.

Used in

Illegal Drugs:
Cook 4, Variations 1 and 2

Regulations: Flammable label, Corrosive label, Poison label, n.o.s. Hazardous Waste.

BROMINE
Chemical weapon
Br_2
7726-95-6
UN 1744

Appearance: Dark red fuming liquid. Vapor pressure 175 mmHg @ 21°C. Will constantly sublime (red fumes) if out of the container or if the container is open; the fumes will ignite organic materials as they move about. Bromine can be made; *see* Explosives, Process 7.

Odor: Strongly irritating.

Hazards: Toxic, oxidizer, corrosive. Extreme inhalation hazard.

Incompatibilities: Spontaneous reactions with reducing agents, organics, ammonia butadiene, butane, hydrogen, sodium carbide, turpentine.

Used in

Chemical Weapons:
Facility 8, Variation 4

Explosives:
TNA

Regulations: Hazard Class 8; PG I; not acceptable in passenger transportations; ERG 154.

BROMINE TRIFLUORIDE
Industrial chemical weapon
BrF_3
7787-71-5
UN 1746

Appearance: Colorless fuming liquid that decomposes in water.

Odor: Strongly irritating.

Hazards: Strong oxidizer. Toxic by inhalation, ingestion or skin absorption. Corrosive. May cause serious burns.

Incompatibilities: Very reactive!!!! Combustible materials, metals, acids, and water.

Regulations: Hazard Class 5.1; ERG 144.

BROMOBENZENE
C_6H_5Br
108-86-1
UN 2514

Appearance: Clear, colorless, mobile liquid.

Odor: Not found.

Hazards: Moderate fire hazard. Combustible; flash point 149°F. Slight inhalation hazard. Vapor pressure 10 mmHg @ 40°C. Skin contact may cause a rash; bromobenzene may be absorbed through the skin. NFPA Health 2, Fire 2.

Incompatibilities: Strong oxidizers.

Used in

Amphetamines:
Synthesis 16

This is not commonly seen in clandestine methamphetamine laboratories.

Explosives:
HNIW

Illegal Drugs:
Cook 3, Variations 1 and 2

Regulation: Hazard Class 3.3; PG III; ERG 129.

Confirmed by: Beilstein Test.

BROMOBENZYL CYANIDE
Chemical weapon
BrC_6H_4CN
5798-79-8
Tear agent.

Regulations: Requires export permits from the DTCC.

2-BROMOETHANOL
$BrCH_2CH_2OH$
540-51-2
UN 1993

Appearance: Hygroscopic liquid that is soluble in water. Will probably come from a laboratory supply house.

Odor: Irritating.

Hazards: Irritant to the eyes and mucous membranes. May be a carcinogen, may cause reproductive damage. Flammable, flash point not found. Vapor pressure 12 mmHg @ 50°C.

Incompatibilities: Forms an azeotrope with water. Strong oxidizers.

Used in

Explosives:
TNEN

Regulations: Hazard Class 3; PG III; ERG 128.

Confirmed by: Alcohol Test.

BROMOMETHANE
See Methyl bromide.

1-BROMO-1-NITROETHANE (w)
$C_2H_4BrNO_3$
563-97-3

Used in

Explosives:
TNP

BUTANE
C_4H_{10}
106-97-8
UN 1075

A gas commonly used in cigarette lighters. The liquid fuel is under pressure. Can be purchased as a compressed gas cylinder from a hardware store or surplus/camping store.

Odor: None.

Hazards: Mildly toxic by inhalation, causes drowsiness. Serious fire hazard.

Incompatibilities: Nickel carbonyl; oxidizers.

Used in

Illegal Drugs:
Cook 4, Variation 1

Explosives:
Methyl ethyl ketone peroxide oxyacetylene

Incendiary Devices:
Situation 1

Regulations: Labeled Liquid Petroleum Gas; Hazard Class 2.1; ERG 115.

Monitored by: Combustible gas indicator.

1,4-BUTANEDIOL
$HOCH_2CH_2CH_2CH_2OH$
110-63-4

Appearance: Colorless viscous liquid that is soluble in water. This is a common industrial chemical.

Odor: Nearly odorless.

Hazards: Combustible, flash point 121°C. May be harmful if swallowed.

Incompatibilities: Reacts with strong oxidizers and mineral acids.

Used in

Date Rape:
GHB

Regulations: Nonhazardous for air, sea, and road freight.

2-BUTANONE
See Methyl ethyl ketone.

BUTYL ALCOHOL
$CH_3 \cdot CH_2 \cdot CH_2 \cdot CH_2OH$
71-36-3
UN 1120

Appearance: Colorless highly refractive liquid that burns with a strongly luminous flame that leaves a greasy spot. Soluble in water, but will float, forming a flat meniscus. Flash point 36–38°C.

Odor: Fusel oil, but weaker. Its vapors irritate and cause coughing.

Hazards: Combustible, flash point 114°F. Vapor pressure 5.5 mmHg @ 20°C.

Incompatibilities: Strong oxidizers.

Used in

Amphetamines:
Synthesis 21, Variation 2

Explosives:
ADN, KDN

Regulations: Hazard Class 3; ERG 129.

Confirmed by: Alcohol Test.

n-BUTYLAMINE
$CH_3(CH_2)_3NH_2$
109-73-9
UN 1125

Appearance: Colorless, watery liquid. This is not an over-the-counter product and will probably be found in a brown glass bottle.

Odor: Amine, ammonia, dead animal.

Hazards: Flammable, flash point 10°F; LEL 1.7%; UEL 9.8%. Toxic by ingestion, skin contact, and inhalation. Corrosive; LD_{50} 0.36 g/kg; PEL 5 ppm ceiling; skin notation.

Incompatibilities: Incompatibility Group 4-A. Chromium trioxide, nitric acid, and perchloric acid.

Used in

Amphetamines:
Synthesis 15

Regulations: RQ 454 kg; Waste Codes D001 and D002; Hazard Class 3.2; PG II; ERG 132.

Confirmed by: Determination by iodine.

tert-BUTYLAMINE
$(CH_3)_3CNH_2$
75-64-9

Appearance: Colorless, watery liquid. This is not an over-the-counter product and will probably be found in a brown glass bottle.

Odor: Amine, ammonia, dead animal.

Hazards: Flammable; flash point 10°F; LEL 1.7%; UEL 9.8%. Toxic by ingestion, skin contact, and inhalation. Corrosive; LD_{50} 0.36 g/kg; PEL 5 ppm ceiling; skin notation.

Incompatibilities: Incompatibility Group 4-A. Chromium trioxide, nitric acid, and perchloric acid.

Used in

Amphetamines:
Synthesis 15

Explosives:
NDTTDATBA, UDTNB

Regulations: RQ 454 kg; Waste Codes D001 and D002; Hazard Class 3.8; PG II; ERG 132.

Confirmed By: Yellow smoky reaction with iodine.

n-BUTYLAMMONIUM IODIDE
$C_{16}H_{36}IN$
311-28-4

Appearance: Tetra-*n*-butylammonium iodide is a white or tan powder that is not soluble in water.

Hazards: Harmful if swallowed.

Incompatibilities: Strong oxidizers.

Used in

Amphetamines:
Synthesis 18, Supplementary 3

Regulations: Can be shipped by air.

n-BUTYL CHLORIDE
$CH_3CH_2CH_2CH_2Cl$
109-69-3
UN 1127

Appearance: Colorless liquid that is insoluble in water. *Also see* 1-Chlorobutane.

Odor: Irritating.

Hazards: Flammable, flash point 20°F. Vapor pressure 80 mmHg @ 20°C.

Incompatibilities: Strong oxidizers and strong bases.

Used in

Illegal Drugs:
Cook 4, Variation 3

Regulations: Hazard Class 3; PG II; ERG 130

Confirmed by: Beilstein Test.

BUTYL NITRITE

Used in

Explosives:
Sodium azide

Recreational Drugs:
Similar to and used as amyl nitrite when amyl nitrite became illegal. *See* Amyl nitrite *and* Rush.

γ-BUTYROLACTONE (w)
$C_4H_6O_2$
96-48-0

Appearance: A hygroscopic, colorless, oily, liquid, aromatic compound, with medium solubility in water. This is a common solvent and reagent in chemistry, and is used as a floor stripper, a stain remover, and paint stripper. The Internet suggests Specialty

Wood Products® Antique Chair paint stripper as a source.

Odor: Weak characteristic odor.

Hazards: Even before being reacted, this can be a substitute for GHB.

Incompatibilities: Strong oxidizers.

Used in

Date Rape:
GHB, Variation 1

Regulations: CDTA listed, but not regulated. No specific DOT or EPA regulations.

C

CAB-O-SIL
SiO$_2$
1129-52-5 (specific)
7631-86-9 (general)

Appearance: Amorphous crystalline chunks. Fumed silica. Available in 5-gallon pails or paper sacks from plastic supply facilities. It is used as a filler for molding plastics.

Odor: None.

Hazards: Ingestion, discomfort, skin drying, irritation to lungs at very high dust levels.

Incompatibilities: This is inert.

Used in

Chemical Weapons:
Facility 3

Regulations: None specified. Labeled Fused Silica, Crystalline-free. Not regulated under SARA.

CADMIUM CHLORIDE (Anhydrous)
CdCl$_2$
UN 2570

Appearance: Colorless-to-white crystals.

Odor: None.

Hazards: Extremely toxic!!

Incompatibilities: None.

Used in

Illegal Drugs:
Cook 4, Supplementary 1

Regulations: Hazard Class 6.1 n.o.s.; one of the EPA's TCLP chemicals; ERG 154.

Confirmed by: Cadmium Test.

CADMIUM IODIDE
CdI$_2$
UN 2570

Appearance: Fine white powder that is slightly soluble in water.

Odor: None.

Hazards: Poisonous!

Incompatibilities: Potassium metal.

Used in

Hobbies:
Photography

Regulations: Hazard Class 6.1 n.o.s.; one of the EPA's TCLP chemicals; ERG 154.

Confirmed by: Cadmium Test.

CADMIUM METAL
Cd
156-62-7
UN 2570

Appearance: Silver-white malleable metal.

Odor: None.

Hazards: Dust is extremely toxic.

Incompatibilities: Hydrazoic acid.

Used in

Chemical Weapons:
Facility 12

Hobbies:
Photography

Regulations: Hazard Class 6.1 n.o.s.; one of the EPA's TCLP chemicals; ERG 154.

Confirmed by: Cadmium Test.

CAFFEINE
$C_8H_{10}N_4O_2 \cdot H_2O$
58-08-2s

Appearance: White fleecy masses that are slightly soluble in water. Caffeine is an alkaloid that is a stimulant in human beings. Is also sold as white tablets which are alertness enhancers. Caffeine can be extracted from coffee beans using benzene, chloroform, trichloroethylene, and dichloromethane. Extraction by water will include many other alkaloids.

Odor: Not found.

Hazards: Large doses (>1 g) cause palpitation, excitement, insomnia, dizziness, headache, and vomiting. Half-life of caffeine in the body can be from 4 to 11 hours. A dose of over 400 mg is considered intoxication: This can be 3 to 4 cups of coffee, or by the addition of caffeine tablets.

Incompatibilities: Oxidizers.

Used in

Amphetamines:
List of materials used to cut methamphetamine

Regulations: None specified. Unlike other psychoactive substances, it is legal and pretty much unregulated.

CALCIUM CARBIDE
CaC_2
75-20-7
UN 1402

Appearance: Calcium carbide (bangsite) is a water-reactive tan rocklike material producing acetylene gas. The primary use of calcium carbide is to generate acetylene gas. *See* Acetylene.

Odor: Slight metallic garlic.

Hazards: Water reactive, producing a very flammable (explosive) gas.

Incompatibilities: Water, iodine, chlorine.

Regulations: Hazard Class 4.3; cannot be shipped by passenger transportations or by air; ERG 138.

Confirmed by: Ignition of Off-Gas; Calcium Test.

CALCIUM CARBONATE
$CaCO_3$
1317-65-3

Appearance: White powder.

Odor: None.

Hazards: Inert. Severe eye irritant. Used as chalk.

Incompatibilities: Acids effervesce.

Used in

Ammunition:
Load 3

Explosives:
Potassium chlorate primer, friction primer

Illegal Drugs:
Cook 5, Supplementary 3

Fireworks:
Color

Regulations: None specific.

Confirmed by: Calcium Test.

CALCIUM CHLORIDE
$CaCl_2$
10043-52-4

Appearance: Cubic, colorless, deliquescent crystals. Used by road departments to keep dirt roads moist.

Odor: None.

Hazards: Used to absorb moisture. Generally inert.

Incompatibilities: Percarboxylic acid.

Used in

Amphetamines:
Synthesis 17
Synthesis 18, Variation 6
Synthesis 18, Supplementary 1
Synthesis 21, Variation 2

Chemical Weapons:
Facility 1
Facility 8
Facility 10, Variation 4

Explosives:
Process 5, Variation 2

Fireworks:
Color

Illegal Drugs:
Cook 1, Variation 1
Cook 4, Supplementary 1

Tear Agents:
Chloroacetone, dichloroacetone, mace

Regulations: None specific.

Confirmed by: Calcium Test.

CALCIUM HYDROXIDE
$Ca(OH)_2$
1305-62-0

Appearance: Soft white granules or powder that is slightly soluble in water. Readily absorbs CO_2 from the air, forming calcium carbonate.

Odor: None.

Hazards: Strongly alkaline, often around pH 12.5 to 13. Ingestion can cause esophageal perforation; inhalation can cause chemical bronchitis; may cause burns to the skin, and blindness. NFPA Health 3.

Incompatibilities: Phosphorus, nitroethane, nitromethane, and other nitroparaffins, and acids.

Used in

Fireworks:
Color

Regulations: Not a listed RCRA hazardous waste; however, does exhibit one or more characteristics of a hazardous waste. Transportation is not regulated.

Confirmed by: Calcium Test.

CALCIUM HYPOCHLORITE
$Ca(ClO)_2$
7778-54-3
UN 1748

Appearance: White granular material. Available from hardware stores as pool chlorine.

Odor: Chlorine.

Hazards: Strong oxidizer. Produces large amounts of chlorine gas when in contact with acid or ammonia.

Incompatibilities: Amines, anthracene, nitroparaffins, organic material, organic sulfides, phenol, and turpentine.

Used in

Amphetamines:
Synthesis 16

Chemical Weapons:
Facility 4, Variation 3
Facility 8, Variation 2
Facility 10, Variations 1 and 4

Explosives:
Process 5, Variation 2

Incendiary Devices:
Situation 1

Regulations: Hazard Class 5.1; PG II; ERG 140

Confirmed by: Oxidizer Test.

CALCIUM NITRATE
$Ca(NO_3)_2 \cdot 4H_2O$
UN 1454

Appearance: White deliquescent mass that is soluble in water.

Odor: None.

Hazards: Strong oxidizer; may ignite in contact with organic materials. May explode if shocked or heated.

Incompatibilities: Heat; organic materials.

Used in

Illegal Drugs:
Cook 2, Supplementary 3

Regulations: Hazard Class 5.1; PG III; ERG 140.

Confirmed by: Oxidizer Test.

CALCIUM OXALATE
$CaC_2O_4 \cdot H_2O$
5794-28-5

Appearance: White crystalline powder that is insoluble in water. When ignited burns to $CaCO_3$ without charring.

Odor: None.

Hazards: Toxic by ingestion. Irritating to eyes, skin, and lungs. Kidney stones are made up of calcium oxalate.

Incompatibilities: Strong oxidizing agents.

Used in

Fireworks:
Color

Regulations: None specific.

Confirmed by: Calcium Test.

CALCIUM PHOSPHATE (Tribasic)
See Tricalcium phosphate.

CALCIUM SULFATE
$CaSO_4$
7778-18-9

Appearance: White powder that is very slightly soluble in water.

Odor: None.

Hazards: Used to absorb moisture. Generally inert. pH near 9.

Incompatibilities: Aluminum.

Used in

Explosives:
Nitroform

Fireworks:
Color

Regulations: Nonhazardous for air, sea, and road freight.

Confirmed by: Calcium Test.

CALGON
$(NaPO_3)_6$
10124-56-8

Appearance: Sodium hexametaphosphate. White flaky crystals or granules that are soluble in water, resulting in an alkaline solution. Used as a pesticide.

Odor: None found.

Hazards: No information is available.

Incompatibilities: No information is available.

Used in

Hobbies:
Photography

Regulations: Although not registered with the EPA, is considered to be toxic to aquatic life.

CALIFORNIUM-252
Radioactive material
Cf

Appearance: A synthetic element that is a silvery-white or gray metal.

Emits: Alpha rays. But, more important is a strong neutron emitter. 1 microgram can emit 170 million neutrons per minute.

Half-life: 2.6 years.

Applications: This is not easily produced, and will be found in very small amounts. It is used as a portable neutron emitter, found as foils, which are used in research. Can also be used as a portable neutron source to identify gold or silver ores, and can be used in moisture gauges to locate water- and oil-bearing layers in oil wells. Does have medical uses for certain types of cancer treatment.

Used in

Dirty Bombs

Regulations: Radioactive elements are highly regulated and have specific requirements for both transportation and disposal.

CALOMEL
Mercury(I) chloride (Hg_2Cl_2); *see* Mercurous chloride.
10112-91-1
UN 3077

Appearance: White-to-gray rhombic crystals or powder.

Odor: Not found.

Hazards: Toxic, but considered the "safe" mercury chloride. *Also see* Mercuric chloride. When exposed to heat or sunlight will give off mercury vapors. *See* Mercury: Hazards.

Incompatibilities: Bromides, iodides, alkali chlorides, sulfates, sulfites, sulfides, carbonates, hydroxides, ammonia, cyanides, silver, lead and copper salts, lime water, iodine, and hydrogen peroxide. Excessive heat should be avoided.

Used in

Fireworks:
Chlorine donor

Regulations: Severely toxic to river, stream and marine life; should not be stored or shipped with food. Hazard Class 9; PG III. Other regulated solids, n.o.s. One of the EPA's TCLP chemicals. ERG 171.

Confirmed by: Mercury Test.

CANADIAN BALSAM

Appearance: Yellowish-to-greenish, viscous, transparent, slightly fluorescent liquid. Canada turpentine. Liquid oleoresin from *Abies balsamea*.

Odor: Agreeable, aromatic.

Incompatibilities: Oxidizers.

Used in

Hobbies:
Photography

CAPROYL HYDRIDE
See Hexane; *also see* Nonpolar liquids that float.

CARBODIHYDRAZIDE
$CO(NHNH_2)_2$
497-18-7

Appearance: Colorless crystals, very soluble in water.

Odor: None found.

Hazards: May explode if heated! Toxicology not fully investigated; skin, eye, and respiratory irritant.

Incompatibilities: Strong acids, strong oxidizers.

Used in

Hobbies:
Photography

Regulations: None specific.

CARBOLIC ACID
See Phenol.

CARBON
Also see Charcoal *and* Lampblack. Carbon comes in many forms. Carbon crystal is a diamond; graphite is the crystalline allotropic form of carbon. Buckyballs (Buckministerfullerene) are still another form of carbon. Carbon lampblack and charcoal are amorphous.

Used in

Explosives:
Nitrogen trichloride

Fireworks:
Fuel additive

Illegal Drugs:
Cook 8, Variation 1

Regulations: ERG 133.

CARBON DIOXIDE
CO_2
124-38-9
UN 1013

Appearance: Colorless gas. *Also see* Dry ice. Found in compressed gas cylinders. Can be generated by dry ice and any carbonate. The atmosphere contains 330 ppm CO_2.

Odor: None.

Hazards: Asphyxiate. Because the body uses CO_2 to regulate oxygen in the blood, high levels of CO_2 can cause panic.

Incompatibilities: Titanium, NaK. Generally, CO_2 is used to make other compounds inert.

Used in

Amphetamines:
Synthesis 14
Synthesis 20, Supplementary 8

Explosives:
AN, ammonium nitride, ammonium perchlorate

Regulations: Cylinders should be empty before disposal. Hazard Class 2.2.

Confirmed by: GasTec Polytec IV indicator tube stripe 7 turns brown.

CARBON DISULFIDE
Industrial chemical weapon
CS_2
75-15-0
UN 1131

Appearance: Colorless liquid that evaporates very quickly (vapor pressure 400 mmHg @ 28°C). This is an inorganic liquid that sinks in water. Flash point –22°F.

Odor: Rotten pumpkins.

Hazards: Extremely flammable! Flash point –30°C. Burns to sulfur dioxide. Inhalation hazards. The anesthetic action is much more powerful than that of chloroform. Affects the nervous system, can cause death considerable time after exposure. NFPA Health 3, Fire 4.

Incompatibilities: Reacts vigorously with oxidizing materials: aluminum, azide solutions, zinc dust, liquid chlorine, hypochlorites.

Used in

Explosives:
Ammonium picramate, nitrocellulose, sodium picramate, TND

Hobbies:
Magic

Incendiary Devices:
Situation 8

Regulations: Hazard Class 3; not permitted on passenger or commercial cargo planes; ERG 131.

CARBON MONOXIDE GAS
CO
160-08-0
UN 1016

Appearance: Colorless gas, most likely to come in a lab bottle, although it is available in large cylinders. Can be produced by incomplete combustion.

Odor: None.

Hazards: Carbon monoxide is a chemical asphyxiant.

Incompatibilities: Bromine trifluoride, cesium monoxide, iodine heptafluoride (but what is not incompatible with these?)

Used in

Chemical Weapons:
Facility 7, Variation 1
Facility 19, Variation 2

Regulations: Hazard Class 2.3; ERG 119.

CARBON TETRACHLORIDE
CCl_4
56-23-5
UN 1846

Appearance: Colorless, heavy, non-flammable liquid.

Odor: Ethereal.

Hazards: Inhalation hazard. Carbon tetrachloride has a narcotic effect (not as strong as chloroform). After exposure to high amounts, death is possible, due to damage to the liver, kidneys, and lungs.

Incompatibilities: Zirconium sponge, burning wax (explodes), magnesium.

Used in

Chemical Weapons:
Facility 8, Variation 3

Explosives:
HNF

Toxins:
Ricin

Regulations: Toxic, n.o.s.; Hazard Class 6.1; one of the EPA's TCLP chemicals; ERG 151.

CARBURETOR CLEANER
Mixture of acetone, methyl ethyl ketone, and ethyl ether.

CARNUBA WAX
8015-86-9
A natural product derived from palm leaves.

Odor: Not found.

Hazards: FDA's GRAS (Generally Recognized as Safe).

Incompatibilities: Oxidizers.

Used in

Amphetamines:
Used as an additive to the final product.

Regulations: None specified.

CASEIN
Casein is the most predominant phosphoprotein found in milk and cheese. Casein consists of a fairly high number of proline peptides, which do not interact. As a result, it has relatively little secondary or tertiary structure; because of this, it can not denature. There is some indication that nitration stabilizes casein. *See* Explosives casein nitrate.

CASTOR BEAN
Poisonous plant

Description: The castor bean is a shrub-like herb with large, long, stemmed leaves. The plant has large palm-shaped leaves with seven to nine portions, cluster-like blossoms, and prickly fruits, each carrying three seeds.

Where found: Since it is considered to be a decorative plant, seeds are available on the Web. The castor bean is an ornamental annual in gardens and grows wild across the entire southern United States.

Deadly parts: Seeds, bark, and leaves are somewhat toxic, but the greatest concentration of toxicity occurs in the seeds and their pods.

Symptoms: Ricin causes diarrhea so severe that victims can die of shock as a result of massive fluid and electrolyte loss. We did not include castor beans in our section on poison plants, because even though ricin is one of the most toxic compounds known, it would require one whole castor bean to poison a child, and the taste is so bad that no adult would be inclined to eat it. However, children have died from eating the beans. If swallowed without chewing, a castor bean would pass through the digestive system without effect. However, if the ricin is extracted out and concentrated, the situation is very different, and the symptoms are very different. *See* the expanded discussion on ricin in Section 5 under "Toxins".

Antidotes: Treated symptomatically.

Used in

Hobbies:
Gardening

Toxins:
Ricin

Regulations: None specific.

CATECHOL
Pyrocatechol
$C_6H_4(OH)_2$
120-80-9
UN 2923

Appearance: Colorless crystals that may discolor to brown, especially when moist. Soluble in water.

Odor: Aromatic.

Hazards: Combustible, flash point 261°F. Strong irritant. Toxic. Called *pyrocatechol* because the addition of inorganic acids evolves heat.

Incompatibilities: Acid chlorides, acid anhydrides, bases, oxidizing agents, and nitric acid.

Used in

Amphetamines:
Synthesis 18, Supplementary 3

Hobbies:
Photography

Regulations: PG II; Hazard Class 8.0; DOT Transport Category 2; ERG 154.

CATHA EDULIS

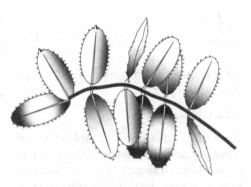

Description: Slow-growing evergreen shrub which can reach a height of 25 feet. It is native to East Africa and Arabia, but is now cultivated in many countries. It can be purchased off the Web as either seeds or cuttings. It is a good houseplant, preferring direct sun.

Where found: Most likely as a houseplant. Although it can grow to be a very large shrub, it grows so slowly that it could remain in the house for years. It is most commonly associated with immigrants from East Africa or Yeman. It is a drug used much like coffee in those cultures. It is very addictive. Although it is illegal to bring *khat* (or *chat*; pronounced "cot") into the country, it is not illegal to own, being a Schedule IV controlled substance. Khat used for recreational purposes looks like this:

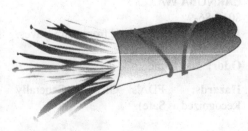

Symptoms: Khat is usually chewed or made into a tea. It is a stimulant. In East Africa it allows people to do two or three jobs.

Active ingredients: Cathanone (much like methcathanone), cathine.

Used in

Recreational Drugs

Regulations: Schedule IV controlled substances.

CATHINE
See d-Norpseudoephedrine.

CAUSTIC SODA
See Sodium hydroxide.

CELLOPHANE
Cellophane is a thin, transparent sheet made of processed cellulose. *Also see* Cellulose.

Used in

Explosives:
Nitrocellulose

CELLOSOLVE®
See Ethyl cellosolve.

CELLULOSE

Appearance: Fibrous material in the cell walls of plants. Usually seen as a white, fluffy, fibrous mass.

Odor: None.

Hazards: FDA GRAS (Generally Recognized as Safe).

Incompatibilities: Concentrated nitric acid.

Used in

Ammunition:
Load 2

Regulations: None specific.

CERESIN
See Ozokerite.

CESIUM-137
Radioactive material
Cs

Appearance: A very reactive alkali metal often used as cesium chloride, a yellow powdery salt.

Emits: Gamma and beta rays.

Half-life: 30 years.

Applications: Large quantities of this material are available. It is available for sale on the Web. Cesium-137 is used for food irradiation, many industrial instruments, and in hospitals for diagnosis and treatment, sterilization of medical equipment, and medical radiotherapy. In Third World countries, where electricity is not easily available for hospitals, x-ray machines often use cesium as a source. Cesium is the most likely source to be used in a dirty bomb.

Used in

Dirty Bombs

Regulations: Radioactive elements are highly regulated and have specific requirements for both transportation and disposal. Exemption Quantity (EQ) 10,000 Bq.

CHALCHLOR
See Mercuric chloride.

CHARCOAL (Activated)
C + impurities
64365-11-3
UN 1361

Appearance: Black amorphous solid, usually in a fine powder form. Charcoal is often used as a filtering agent and can be found in many places, including swimming pool supply houses, hardware stores, laboratory supply houses, and even at grocery stores.

Odor: None.

Hazards: Charcoal is a nuisance dust. The largest danger from activated charcoal is that it can burst into flame spontaneously if wet.

Large amounts of activated charcoal can actually draw the oxygen from an enclosed space. Can be burned, but not easily. Wet charcoal can burn spontaneously. *Also see* Carbon.

Incompatibilities: Liquid air, liquid oxygen, oxidizers.

Used in

Ammunition:
Load 1
Load 9
Load 10

Chemical Weapons:
Facility 4, Variation 5
Facility 9, Variation 1

Explosives:
Black powder, F-TNB, UDTNB

Incendiary Divices:
Situation 6

Fireworks:
Sparks, Glitter, Rockets, black powder

Hobbies:
Rockets, Variation 7

Illegal Drugs:
Peyote

Regulations: ERG 133.

CHARCOAL LIGHTER
Mixed hydrocarbons with formulas near $C_{10}H_{22}$
See Nonpolar liquids that float.

CHAT
See Catha edulis.

CHICKPEA MEAL

Used in

Illegal Drugs:
Cook 2, Supplementary 3

CHLORAMINE
$C_7H_7O_2NSClNa \cdot 3H_2O$
127-65-1

Appearance: Faintly yellow crystalline powder that is soluble in water. Sold by H&S Chemical as a water purifier, Chloramine T®. Chloramine is made in Explosives: Astrolite, Supplementaries 1 and 2, where it is a major step in making hydrazine.

Odor: Slight chlorine.

Hazards: Oxidizer. Is used both externally and internally for medicine. Is a chlorine substitute for water purification. May be harmful if swallowed, inhaled, or absorbed through the skin or eyes.

Incompatibilities: Mixing with other materials hastens the release of chlorine.

Used in

Explosives:
Astrolite, Supplementary 3

Regulations: Hazard Class 5.1.

CHLORINE GAS
Cl_2
7782-50-5
UN 1017

Appearance: A yellow gas, that will probably be found in a laboratory bottle. However, very large chlorine cylinders are used for water purification. Chlorine gas can be made; *see* Explosives, Process 5.

Odor: Chlorine.

Hazards: Chlorine gas is toxic and a strong oxidizer.

Incompatibilities: Ammonia, acetylene, butadiene, benzene, and other petroleum fractions, hydrogen, sodium carbide, powdered metals.

Used in

Chemical Weapons:
Facility 4, Variation 5
Facility 4, N1-2
Facility 4, N2-2

Facility 7, Variation 1
Facility 8, Variation 3
Facility 10, Variation 3
Facility 11

Explosives:
Astrolite, Supplementary 3; nitrogen sulfide; DNFA-P

Incendiary Devices:
Situation 3

Date Rape:
Chloral hydrate

Tear Agents:
Chloroacetone

Regulations: Cylinders should not be disposed of unless they are empty. Hazard Class 2.3; ERG 124.

Monitored by: Chlorine indicator tubes.

CHLOROACETIC ACID
$CH_2ClCOOH$
79-11-8
UN 1750

Appearance: Colorless crystals that are soluble in water.

Odor: Strong plastic, cat urine.

Hazards: Moderate irritant. May cause severe damage to skin and lungs. Death may occur if more than 3% of the skin is emerged in this chemical. NFPA Health 3.

Incompatibilities: Strong oxidizers.

Used in

Amphetamines:
Synthesis 20, Supplementary 1

Explosives:
Binary explosives, Variation 5

Regulations: ERG 153. Generally, there is little regulation of this material.

Confirmed by: Beilstein Test.

CHLOROACETONE
Tear agent
CH_3COCH_2Cl
78-95-5
UN 1695

Appearance: Colorless viscosity 1 liquid. This compound is not available as an over-the-counter product. It will be found in laboratory supply house brown glass bottles with orange lids. *See* "Tear Agents" is Section 5.

Odor: Lachrymator, very irritating.

Hazards: Toxic by ingestion and inhalation. Exposure may cause sleeplessness and in some cases loss of hair. LD_{50} 0.1 g/kg; TLV 1 ppm ceiling; skin notation.

Incompatibilities: Incompatibility Group 4-A. Chromium trioxide.

Used in

Amphetamines:
Synthesis 13

Tear Agents:
Chloroacetic acid

Regulations: Poison Inhalation Hazard Zone B; Hazard Class 6.1; PG III; ERG 131; forbidden on passenger planes; cannot be sent by U.S. mail.

Confirmed by: Carbonyl Test.

CHLOROACETOPHERONE
Tear agent
$C_6H_5COCH_2Cl$
532-27-4
UN 1697s

Appearance: Mace. White crystals. May be found in pepper spray or in pressurized tear gas.

Odor: Floral, strongly irritating, lachrymator.

Hazard: Strong irritant.

Used in

Tear Agents:
Pepper spray

Regulations: Poison Inhalation Hazard
Zone B; Hazard Class 6.1; PG III; ERG 131;
forbidden on passenger planes; cannot be
sent by U.S. mail.

2-CHLOROACETYL CHLORIDE
$ClCH_2COCl$
79-04-9
UN 1752

Appearance: Water-white liquid that
decomposes in water.

Odor: Pungent.

Hazards: Irritating to the eyes; corrosive
to the skin. Tolerance 0.05 ppm in the air.

Used in

Tear Agents:
Mace

Regulations: Forbidden in passenger
transportation.

CHLOROAURIC(III) ACID
See Gold chloride.

2-CHLOROBENZALDEHYDE
C_7H_5ClO
89-98-5

Appearance: Colorless-to-light yellow
liquid that floats on water.

Odor: Not found.

Hazards: Skin contact may cause
blistering or inflammation. Avoid ingestion
(there is no data established at this time).
NFPA Health 2. Combustible flash point
190°F.

Incompatibilities: Oxidizers.

Used in

Tear Agents:
o-Chlorobenzylidene malononitrile

Regulations: None specific.

CHLOROBENZENE
C_6H_5Cl
108-90-7
UN 1134

Appearance: Clear, colorless viscosity
1 liquid. Flash point 85°F. Vapor pressure
10 mmHg @ 20°C. Will be in a clear
glass bottle or jug from a chemical
supply house. Used as a solvent in leather
dye.

Odor: Shoe or leather dyes.

Hazards: Inhalation. Narcotic at levels
between 1200 and 3700 ppm. Flammable.

Incompatibilities: Dimethyl sulfoxide,
silver perchlorate.

Used in

Chemical Weapons:
Facility 14

Explosives:
Picryl chloride

Regulations: Hazard Class 3.1; one of the
EPAs TCLP chemicals; ERG 130.

Confirmed by: Beilstein Test.

o-CHLOROBENZYLIDENE

Tear Agents:
Pepper spray

Regulations: Requires export permits from
the DTCC.

1-CHLOROBUTANE
$CH_3(CH_2)_2CH_2Cl$
203-696-6
UN 1127

Appearance: Colorless liquid that is not
soluble and sinks in water.

Odor: Not found.

Hazards: Flammable, flash point 20°F;
LEL 2%; UEL 10%. Limited animal
experiments indicate low toxicity.

Incompatibilities: Oxidizers.

Used in

Illegal Drugs:
Cook 4, Supplementary 1

Regulations: Hazard Class 3; PG II; ERG 130.

Confirmed by: Beilstein Test.

2-CHLORO-*N*-*N*-DIISOPROPYL-ETHYLAMINE (w)

This is one of those chemicals for which no information could be found, either because it is a very uncommonly used synonym, is an intermediate, or because it went out of commerce when placed on the chemical weapons watch list.

Used in

Chemical Weapons:
V-series nerve agents

1-CHLOROETHANE
See Ethyl chloride.

2-CHLORO ETHANOL
Ethylene chlorohydrin, chloro-ethyl alcohol.
Also see Ethylene oxide.
ClC_2H_4OH
107-07-3
UN 1135

Appearance: Colorless liquid that is completely miscible with water. Most likely found in glass bottles or jugs.

Odor: Faint ethereal.

Hazards: Extremely flammable, flash point –50°C. Boiling point 12.5°C. NFPA Fire 4, toxic by ingestion, inhalation, and skin absorption. May be fatal. Strong irritant. NFPA Health 2. PEL 1 ppm in air. On contact with skin, frostbite will occur. Exposure may cause liver and kidney damage.

Incompatibilities: Oxidizers.

Used in

Chemical Weapons:
Facility 3, Supplementary 1
Facility 4

Regulations: Hazard Classes 3 and 6.1; PG I; ERG 131. Requires special insulated cylinder with special fittings. Australian Group chemical, not listed. Requires export permits from the DTCC.

CHLOROFORM
$CHCl_3$
67-66-3
UN 1888

Appearance: Colorless viscosity 1 liquid. Will be found in brown glass bottles from a laboratory supply house. May be part of a mixture in an orange-red reaction vessel. Chloroform can be made by mixing acetone and calcium hypochlorite.

Odor: Chlorinated solvent, nail polish remover.

Hazards: Intoxication by inhalation, anesthetic. This solvent will almost immediately eat through latex gloves. LD_{50} 0.8 g/kg; TC_{low} 5000 mg/m^3 @ 7 minutes; PEL 50 ppm ceiling; suspected carcinogen; teratogen.

Incompatibilities: Incompatibility Group 4-A. Acetone, aluminum, lithium, magnesium, perchloric acid, sodium, potassium, and a mixture of methanol and sodium hydroxide.

Used in

Amphetamines:
Synthesis 3
Synthesis 19

Chemical Weapons:
Facility 4, N1-1 and 2
Facility 4, N2 1 and 2
Facility 4, N3
Facility 11

Designer Drugs:
Group 1, Variation 1

Explosives:
ADN, A-NPNT, azidoethyl, DMMD, DNFA-P, HNH-3, HNIW, TNA, TNN

Hobbies:
Photography, insect preservation

Illegal Drugs:
Cook 1, Variations 1 and 2
Cook 1, Supplementary 1
Cook 2, Variations 1, 2, 3, 4, and 7
Cook 2, Supplementary 1
Cook 7
Cook 9

Tear Agents:
Chloroacetone (*see* Facility 15); bromobenzyl cyanide

Regulations: RQ 4.54 kg; Waste Codes D023(6 mg/L) and U044; Hazard Class 6.1; PG III; labels 6.1; ERG 151; forbidden on passenger planes; cannot be sent by U.S. mail.

Confirmed by: Beilstein Test.

CHLOROFORM ISOBUTANOL
This is not a chemical, but a mixture of chemicals listed as a chemical in many of the formulas for LSD. *See* Chloroform *and* Isobutanol.

2-CHLOROHEPTANE
$C_7H_{15}Cl$
629-06-1
No UN number, guide 129

Appearance: Clear liquid that floats on water.

Odor: Not found.

Hazards: Highly flammable.

Incompatibilities: Strong oxidizers.

Used in

Illegal Drugs:
Cook 4, Supplementary 2

Regulations: Hazard Class 3; ERG 129.

2-CHLOROHYDRIN (w)
$CH_2ClCHOHCH_2OH$
108-45-2

Appearance: Colorless heavy liquid that is unstable, hygroscopic, and soluble in water.

Odor: Not found.

Hazards: Toxic by ingestion and inhalation.

Incompatibilities: Oxidizers.

Used in

Chemical Weapons:
Facility 4, Variation 1

Explosives:
Used as an antifeezing agent for dynamite.

Regulations: Hazard Class 6.1, n.o.s.

CHLOROMETHANE
See Methyl chloride.

CHLOROPICRIN
Chemical weapon
CCl_3NO_2
760-06-2
UN 1580

Appearance: Slightly oily colorless liquid. Vapor pressure 40 mmHg @ 34°C. Chloropicrin can be made; *see* Chemical weapons, Facility 7.

Odor: Strongly irritating; Lachrymator.

Hazards: Inhalation, extremely toxic! Affects ALL body surfaces. 4 ppm is sufficient to render a person totally incapacitated. Poison B. May decompose violently if heated.

Incompatibilities: Reacts violently with sodium methoxide, propargyl bromide, and aniline. Strong oxidizers. Light metals.

Used in

Chemical Weapons:
Facility 7, Variation 3
Chloropicrin is a choking agent.

Regulations: Hazard Class 6; ERG 154.

1-CHLOROPROPANE-2,3-DIOL
See Chlorohydrine.

1-CHLORO-2-PROPANOL
See Propylene chlorohydrin.

CHLOROPSEUDOEPHEDRINE AND CHLOROEPHEDRINE
$C_6H_5CHCl(CNHCH_3)CH_3$

These are intermediates; made in the first part of Synthesis 3. As seen in clandestine labs, intermediate in Syntheses 3 and 21. Once made, the quality is improved if it is stored in ether overnight in a sealed bottle in a freezer. The presence of these compounds in the final product causes a sick feeling in the end consumer, described as having been poisoned. These are associated with pain in the kidneys and liver. The use of sodium carbonate to make the methamphetamine base can prevent these products from getting to the final product. *See* Synthesis 3.

o-CHLOROSTYRENE
Tear agent
$ClC_6H_4HC=CH_2$
The Nasty Yellow Powder
2039-87-4

Appearance: Yellow liquid that is not soluble in water. Most often seen adsorbed as a yellow powder. Used by the military as a tear agent. There have been several instances of people removing *o*-chlorostyrene from military bases as a practical joke. This has resulted in many hazmat events near military bases.

Odor: Lachrymator. (This is a tear agent.)

Hazards: Insufficient data are available on the effects of this substance on human health. Irritation to eyes, skin, and respiratory system. PEL 75 ppm. Combustible, flash point 58°C, may burn to phosgene.

Polymerization: Can polymerize explosively.

Incompatibilities: Polymerizes under the influence of acids and bases. Can form peroxides with air.

CHLOROSULFONIC ACID
$ClSO_3H$
7790-94-5
UN 1754

Appearance: Clear-to-cloudy colorless-to-pale yellow liquid that reacts violently with water. This is very difficult to containerize, as it is corrosive to most materials.

Odor: Sharp.

Hazards: Chlorosulfonic acid can cause severe acid burns and is very irritating to the eyes, lungs, and mucous membranes. It can cause acute toxic effects in either the liquid or the vapor state. Inhalation may cause loss of consciousness with serious damage to the lung tissue. Very corrosive liquid. This is very difficult to containerize, as it is corrosive to most materials.

Incompatibilities: Water. Most metals, strong bases, carbonates, combustible materials, and strong oxidizers.

Used in

Chemical Weapons:
Facility 6, Variation 2
Dimethyl sulfate, industrial (vesicant)

Regulations: Hazard Class 8; ERG 134.

CHLOROWAX (w)
Chlorowax is used to name many compounds, including PCBs, which are not used in fireworks.

Used in

Fireworks:
Chlorine donor, strobe formulations

Regulations: Chlorinated compounds are generally required to be incinerated under EPA regulations.

Confirmed by: Copper Wire Test.

CHROMIUM^{6+}

Appearance: Strongly oxidizing on the oxidizer test strip; yellow orange or red granular salts or liquids. Chromium trioxide ranges from red to pink to purple and is a very strong oxidizer. Chromium 6 salts, especially chromium trioxide, are used to make chromic acid.

Odor: As a general rule, there is no odor.

Hazards: Oxidizers. Some are very strong oxidizers, some very weak. Carcinogens that can leave a chrome ulcer on the skin if in contact with the skin for a long time.

Incompatibilities: Organic materials.

Used in

Amphetamines:
Synthesis 0.7, Variations 1 and 2

Used to make methcathinone (*see* Synthesis 0.7). People taking this drug have a characteristic unpleasant body odor. The waste is a green sludge that contains chromium. Chromium 6 can be found as chromic acid or potassium dichromate.

Hobbies:
Tanning leather

Regulations: Hazard Class 5.1; one of the EPA's TCLP chemicals.

Confirmed by: Chromium Test.

CITRAL
$C_9H_{15}CHO$

Appearance: A terpene found in the oil of lemon, orange, and lemongrass. It is a pale yellow liquid with a strong lemon odor. It is used in perfumes, as a flavoring agent, and as an intermediate for other organic compounds. May be found as lemongrass oil, from which it can be distilled. Lemongrass oil is 75% citral. It can be synthesized from geraniol, nerol, and linalool by oxidation with chromic acid.

Odor: Lemon.

Hazards: Not hazardous; used as a flavoring!

Incompatibilities: Strong oxidizers.

Used in

Illegal Drugs:
Cook 4, Variations 1 and 2

Regulations: No specific regulations.

CITRIC ACID
$C_3H_4(OH)(COOH)_3 \cdot H_2O$
77-92-9

Appearance: Colorless, odorless crystals.
Odor: None.

Hazards: Very minimum.

Used in

Explosives:
HMTD, TNTPB

Hobbies:
Photography

Illegal Drugs:
Cook 6, Variation 2

Regulations: No specific regulations. FDA's GRAS (Generally Regarded as Safe).

CLAY
$Al_2O_3SiO_2 \cdot xH_2O$

Appearance: A hydrated aluminum silicate. Fine, irregularly shaped crystals ranging from 150 μm to less than 1 μm, reddish brown to pale buff, depending on the iron oxide content. Insoluble in water.

Odor: None.

Hazards: Nontoxic.

Incompatibilities: Regarded as inert. Porous inorganic materials may cause hydrazine to ignite.

Used in

Explosives:
Tetraminecopper(II) chlorate

Fireworks

Regulations: None specific.

CLOVES

Used in

Amphetamines:
Synthesis 18, Supplementary 4

COAL (Soft)

Used in

Amphetamines:
Synthesis 0.5

COBALT-60
Radioactive material
Co

Appearance: Metal.

Emits: Gamma and beta rays.

Half-life: 5.2 years.

Applications: Cobalt-60 is readily available (with restrictions, of course). It is used for sterilization of medical equipment, medical radiotherapy, industrial radiography, radiation source for leveling devices and thickness gauges, food irradiation, and as a source for laboratory use.

Used in:

Dirty Bombs

Regulations: Radioactive elements are highly regulated and have specific requirements for both transportation and disposal.

COCA

Used in

Illegal Drugs:
Cook 1

COKE GRAINS
The carbonaceous residue of the destructive distillation (carbonization) of bituminous coal, petroleum, and coal-tar pitch. The principal type is that produced by heating bituminous coal in chemical recovery or beehive coke ovens.

Used in

Fireworks:
Sparks

Regulations: No specific regulations.

COLEMAN® White Gas
See Nonpolar liquids that float.

COLLODION
A solution of pyroxylin (nitrocellulose) in ether and alcohol.

Appearance: A pale yellow syrupy liquid that is not soluble in water.

Odor: Ether.

Hazards: Very flammable.

Incompatibilities: Oxidizers.

Used in

Hobbies:
Photography

Regulations: Hazard Class 3; PG IV.

COMPOUND 1080
$NaC_2H_3FO_2$
62-74-8

Appearance: White hygroscopic powder, that is slightly soluble in water.

Odor: None.

Hazards: Extremely toxic by ingestion and inhalation. There is NO antidote. This product causes a slow and painful death.

Incompatibilities: None specified.

Used in

Toxins:
Chemical poisons

Regulations: Although banned in 1972, it is still used by federal agents and by persons who inject it into collars worn by livestock, which is legal in the United States. Hazard Class 6.1.

CONGO RED
Sodium diphenyl-bis-α-naphthylamine sulfonates
$C_{32}H_{22}O_6N_8S_2Na_2$

Appearance: Brownish-red powder, soluble in water. Used as a dye, indicator, and in biological stains.

Odor: None.

Incompatibilities: Strong oxidizers.

Used in

Illegal Drugs:
Cook 2, Variation 2
Cook 2, Supplementary 2

Regulations: No specific regulations.

COPPER
Cu
7440-50-8

Appearance: Metal with a distinct reddish color.

Odor: None.

Hazards: As a metal, none.

Incompatibilities: Copper burns in chlorine gas.

Used in

Explosives:
HNBP, TNTPB

Fireworks:
Dark and flash effects

Regulations: No specific regulations. Copper should be recycled.

Confirmed by: Copper Test.

COPPER CARBONATE
$Cu_2(OH)_2CO_3$
1184-64-1

Appearance: Green powder that is insoluble in water.

Odor: None.

Hazards: Toxic by ingestion. Although this compound will be positive on the Oxidizer Test, it does not have sufficient oxidation potential to rate the status of oxidizer for either the EPA or DOT. Harmful if swallowed.

Incompatibilities: Effervesces in acid. Acetylene, hydrazine, nitromethane.

Used in

Fireworks:
Color

Regulations: Nonhazardous for air, sea, and road freight.

Confirmed by: Copper Test.

COPPER(II) CHLORIDE
$CuCl_2 \cdot 2H_2O$
10125-13-0
UN 2802

Appearance: Brownish-yellow or deep blue crystals or powder. Hygroscopic, soluble in water.

Odor: None.

Hazards: Toxic by ingestion and inhalation. Although as is the case of most copper salts, this compound will be positive on the Oxidizer Test, it does not have sufficient oxidation potential to rate the status of oxidizer for either the EPA or DOT.

Incompatibilities: Potassium, sodium acetylene, hydrazine, nitromethane.

Used in

Hobbies:
Photography

Fireworks:
Color

Regulations: Hazard Class 8; PG III; ERG 154.

Confirmed by: Copper Test.

COPPER CYANIDE
$CU(CN)_2$
544-92-3
UN 1587

Appearance: Off-white to yellowish-green powder.

Odor: About 1 people in 10 can smell this as cyanide (almond/chlorine) odor. Most of these people will recognize this compound as cyanide. The other 9 cannot smell this compound.

Hazards: Toxic, probably more toxic because of the copper than of the cyanide. Although as most copper salts, this compound will be positive on the Oxidizer Test, it does not have sufficient oxidation potential to rate the status of oxidizer for either the EPA or DOT.

Incompatibilities: Reacts violently with oxidizing agents, nitrates, magnesium. acetylene, hydrazine, nitromethane. Reacts with concentrated nitric acid to give off hydrogen cyanide gas.

Used in

Hobbies:
Gold mining

Other:
Circuit board recovery
Counterfitting

Regulations: Hazard Class 6.1; PG II; ERG 151.

Confirmed by: Hydrogen cyanide indicator tube.

COPPER(II) NITRATE
$Cu(NO_3)_2 \cdot 3H_2O$
3251-23-8 (anhydrous)
19004-19-4 (semipentahydrate)
UN 1477

Appearance: Blue deliquescent crystals.

Odor: None.

Hazards: Oxidizer, NFPA Health 2.

Incompatibilities: Acetylene, hydrazine, nitromethane.

Used in

Chemical Weapons:
Facility 8, Variation 2
Cyanogen

Explosives:
Copper fulminate

Regulations: Nitrates n.o.s.; Hazard Class 5.1; PG II; RQ 100 kg; ERG 140.

Confirmed by: Copper Test.

COPPER OXIDE
Cu_2O
1317-39-1

Appearance: Fine black powder.

Odor: None.

Hazards: Harmful if swallowed. May act as a skin, eye, or respiratory irritant.

Incompatibilities: Acetylene, hydrazine, nitromethane.

Used in

Fireworks:
Color

Regulations: Nonhazardous for air, sea, and road freight.

Confirmed by: Copper Test.

COPPER(II) OXYCHLORIDE
$CuCl_2 \cdot 2CuO \cdot 4H_2O$
1332-40-7

Appearance: Emerald green to greenish-black powder. Used as a pesticide and as a general disease spray.

Odor: None.

Hazards: Ingestion can result in nausea; an eye irritant; may cause skin allergy in sensitive individuals.

Incompatibilities: Acetylene, hydrazine, nitromethane.

Used in

Fireworks:
Color

Regulations: Nonhazardous for air, sea, and road freight.

Confirmed by: Copper Test.

COPPER SULFATE
$CuSO_4 \cdot 5H_2O$
7758-99-8
UN 3077

Appearance: Blue crystals or blue crystalline granules or powder, soluble in water. Use to be used as a fungicide in ponds.

Odor: None.

Hazards: Toxic by ingestion, strong irritant. May ulcerate membranes.

Incompatibilities: Acetylene, hydrazine, nitromethane.

Used in

Chemical Weapons:
Facility 13

Explosives:
CNTA, copper acetylide, copper fulminate, TADA

Hobbies:
Photography

Regulations: Hazard Class 9; PG III. Environmentally hazardous materials, solid, n.o.s. ERG 171.

Confirmed by: Copper Test; Sulfate Test.

COPPER WIRE
See Copper.

Used in

Recreational Drugs:
Amyl nitrite

Confirmed by: Copper Test.

CORIANDER

Used in

Recreational Drugs:
Absinthe

CORNSTARCH

Appearance: Yellow-to-white fine but grainy powder.

Odor: None.

Hazards: This is food.

Incompatibilities: Oxidizers.

Used in

Explosives:
Nitrostarch

Fireworks:
Black match fuse

Regulations: None specified.

Confirmed by: Flour Test.

COTTON

Appearance: Fluffy, white, stringy material.

Hazards: None.

Incompatibilities: Concentrated nitric acid.

Used in

Explosives:
Gun cotton

Fireworks:
Fuse

m-CRESOL
$CH_3C_6H_4OH$
108-39-4
UN 2076

Appearance: Colorless-to-yellowish liquid that is moderately soluble in water.

Odor: Like phenol but tarry.

Hazards: Toxic and irritant, corrosive to the skin and mucous membranes, absorbed through the skin. May cause serious burns. PEL 5 ppm. Vapor pressure 1 mmHg @ 50°C.

Incompatibilities: Strong oxidizing agents, strong bases, strong acids, aluminum, aluminum alloys, chlorosulfonic acid, and nitric acid.

Used in

Explosives:
Methylpicric acid.

Regulations: Hazard Class 6.1; cresols liquid; PG II; ERG 153. One of the EPA's TCLP chemicals.

CROTONALDEHYDE
Industrial chemical weapon
$CH_3CH:CHCHO$
123-73-9
UN 1143

Appearance: Mobile liquid.

Odor: Pungent.

Hazards: Flammable, toxic, irritating to the eyes and skin. PEL 2 ppm.

Incompatibilities: Butadiene 1–3.

Regulations: Hazard Classes 3 and 6.1. ERG 131P.

CRYOLITE
Na_3AlF
Greenland spar. A natural fluoride of sodium and aluminum.

Appearance: Colorless to white, sometimes red, brown or black, luster vitreous to greasy hardness.

Odor: None.

Hazards: This is a rock.

Incompatibilities: Inert.

Used in

Fireworks:
Color

Regulations: None specific.

CUPRIC NITRATE TRIHYDRATE
See Copper(II) nitrate.

CYANIDE SALTS (n.o.s.)
UN 1588

Used in

Chemical Weapons:
Vomiting agents; blood; agents; tabun (Facility 2)

Hobbies:
Gold mining; entomology

Other:
Circuit board gold recovery

Regulations: Hazard Class 6.1; PG I; ERG 157. Australian Group chemical not scheduled. Requires export permits from the DTCC.

Confirmed by: Cyanide Test.

CYANOGEN BROMIDE
Chemical weapon
CNBr
506-68-3
UN 1889

Appearance: Crystals. Slowly decomposed by cold water. Available in small quantities on the Web. Can be made; *see* Chemical weapons, Facility 8.

Odor: Penetrating odor.

Hazards: Highly toxic, strong irritant to the skin and eyes. Corrodes metal.

Incompatibilities: Strong acids.

Used in

Designer Drugs:
Group 4, Euphoria, Variation 1

Explosive:
CDNTA, MNTA, NTA

Regulations: Hazard Class 6; ERG 157

Monitored by: Hydrogen cyanide indicator tubes.

CYANOGEN CHLORIDE
Chemical weapon
CNCl
506-77-4
UN 1589

Appearance: Colorless gas or liquid, soluble in water. Can be made; *see* Chemical Weapons, Facility 8.

Odor: Penetrating odor.

Hazards: Highly toxic, strong irritant to the skin and eyes. Corrodes metal.

Incompatibilities: Strong acids.

Regulations: Hazard Class 6.1; ERG 125.

Monitored by: Hydrogen cyanide indicator tubes.

CYANOGEN IODIDE (w)
CNI

Appearance: Colorless needles soluble in water. Can be made; *see* Chemical weapons, Facility 8.

Odor: Pungent odor.

Hazards: Highly toxic, strong irritant to eyes and skin.

Incompatibilities: Strong acids.

Used in

Chemical Weapons:
Facility 2
Facility 2, Supplementary 1

Hobbies:
Taxidermy, preservative

Regulations: Hazard Class 6.1; ERG 157

Monitored by: Hydrogen cyanide indicator tubes.

CYANOGUANIDINE (w)
$NH_2C(NH)(NHCN)$
461-58-5

Appearance: Pure white crystals partially soluble in hot water. Will be found in 100-lb multiwall paper bags. Used as a fertilizer.

Used in

Fireworks:

Regulations: No specific regulations.

CYANOTRIMETHYLSILANE
C_4H_9NSi
7677-24-9

Appearance: A liquid with a melting point of 11°C and a boiling point of 114°C, pretty much like water, but with a flash point of 1°C. Reacts in water. The only source we could find was on the Web from China. Can be made using hydrogen cyanide, trimethylchlorosilane, and hexamethyldisilazane in a solvent-free reaction.

Hazards: Very flammable, toxic, water reactive, and a material that is very hazardous to the environment.

Incompatibilities: Water.

Used in

Explosives:
DATBA

Regulations: Hazard Class 3; Flammable label n.o.s.

CYCLOHEXANE
C_6H_{12}
110-82-7
UN 1145

Appearance: Colorless viscosity 1 liquid. *See* Nonpolar liquids that float.

Cyclohexane is most likely to be found in brown glass 4-liter chemical supply house jugs. It is not an over-the-counter commercial product. This compound could be used as an extraction solvent or a carrier solvent. A person using this compound is probably also associated with a legitimate laboratory or is a college student new to the home laboratory scene, as the cost of this purified hydrocarbon is over $70.00 a gallon. It will work just about the same as Coleman® white gas in most cases.

Odor: Mild thinner.

Hazards: Flammable; flash point –4°F; LEL 1.3%; UEL 8%. Intoxication by inhalation; absorbed through the skin; most likely contact problem is drying of the skin, which can cause cracking and infection. LD_{50} 29.8 g/kg; PEL 25 ppm; teratogen.

Incompatibilities: Incompatibility Group 4-A. Chromium trioxide.

Used in

Amphetamines:
Cyclohexane is a solvent that could be used to extract methamphetamine base.

Regulations: RQ 454 kg; Waste Codes D001 and U056; Hazard Class 3; PG II; ERG 128.

Confirmed by: Determination by iodine.

CYCLOHEXANOL
$C_6H_{11}OH$
108-93-0

Appearance: Colorless, oily liquid, not very soluble in water.

Odor: Camphor-like odor.

Hazards: Harmful by inhalation. Vapor pressure 0.98 mmHg @ 25°C.

Incompatibilities: Oxidizers, especially nitric acid and hydrogen peroxide.

Used in

Chemical Weapons:
Facility 1, Cyclosarin, Variations 1 and 2

Regulatons: Nonhazardous for air, sea, and road freight.

Confirmed by: Alcohol Test.

CYCLOHEXANONE
$C_6H_{10}O$
108-94-1
UN 1915

Appearance: Pale yellow viscosity 2 liquid that is slightly soluble in water.

Odor: Acetone and peppermint-like odor.

Hazards: Dermatitis, irritant. Chronic effects on liver and kidney. Flash point 111°C. Class II combustible liquid.

Incompatibilities: Strong oxidizing agents.

Used in

Illegal Drugs:
Cook 3, Variations 1, 2, and 4

Regulations: Waste Codes D001 and U057; Hazard Class 3; PG III; ERG 127.

Confirmed by: Carbonyl Test.

CYCLOHEXENE
C_6H_{10}
110-83-8
UN 2256

Appearance: Colorless liquid.

Odor: Not found.

Hazards: Flammable, flash point 11°F.

Incompatibilities: Oxidizers.

Used in

Explosives:
TNEN

Regulations: Hazard Class 3; ERG 130.

Confirmed by: Determination by iodine.

D

DEATH CAP MUSHROOM

Amanita phalloides. Mushrooms are beyond the scope of this book; however, since this is a fairly easily obtained extremely toxic commodity, we have included it here for information only. The identification of mushrooms is very difficult and requires an expert. The toxin in the death cap mushroom is amatoxic, which must be ingested. Amatoxin destroys the liver and kidneys over several days while the victim remains conscious in excruciating pain. The spores of this mushroom are white, which is an identifying mark and can be seen by holding the mushroom cap over a piece of black paper and thumping it.

Description: This mushroom is found growing below oak trees or other hardwoods. The death cap has a large cap from 2 to 6 inches across. The cap flattens with age. The cap surface can be olive green or brown. The danger of this mushroom is that it is very similar to many widely eaten species.

Taste: Survivors maintain that the mushroom tastes very good.

Effects: The symptoms are delayed, so lethal damage to the liver and kidneys can occur prior to a person knowing that he or she has been poisoned.

Treatment: Frequently, the only cure to poisoning is a liver transplant. Poisoning can be teated by intravenous injection of silibinin dihydrogen disuccinate disodium.

DECACHLORANE
$C_{16}Cl_{12}$
13560-89-9

Appearance: Comes from OxyChem as Dechlorane Plus®, which is a white free-flowing powder. The product is a flame retardant for plastic polymers.

Hazards: Environmental concerns.

Used in

Fireworks:
Chlorine donor

Regulations: No specific regulations.

Confirmed by: Beilstein Test.

DEXTRIN
Gum starch, or British gum.

Appearance: A group of colloidal products formed by the hydrolysis of starches. Soluble in boiling water, less soluble in cold water, may form a gum.

Hazards: Used to make pills and prepare bandages.

Useds in

Explosives:
Lead azide

Fireworks:
Binder, sparks, fuel additive

Regulations: No specific regulations.

Confirmed by: Reddish color with iodine.

DEXTROSE

Appearance: A white powder or crystal. *Also see* Sugar.

Odor: None.

Hazards: A form of sugar.

Incompatibilities: Forms explosives with strong oxidizers.

Used in

Illegal Drugs:
Cook 5, Supplementary 2

Regulations: No specific regulations.

Confirmed by: Sugar Test.

1,2-DIAMINOETHANE
See Ethylenediamine.

2,4-DIAMINOPHENOL·HCl
See Amidol.

DIATOMACEOUS EARTH

Appearance: White to light gray to pale buff powder. Insoluble in water, acid, or dilute alkalies. Siliceous frustules and fragments of various species of diatoms.

Odor: None.

Hazards: Inert.

Incompatibilities: None.

Used in

Explosives:
Dynamite
TNTPB

Fireworks:
Fuse

Hobbies:
Berewing beer

Regulations: No specific regulations.

DIBENZ[*b*, *f*] -1,4-OXAZEPHINE (w)
257-07-8

Regulations: Requires export permits from the DTCC.

DIBROMOBUTYNE-2 (w)
$C_4H_3Br_2$

Hazards: Reactive.

Incompatibilities: Oxidizers.

Used in

Explosives:
HNH-3

DIBUTYLPHTHALATE
$C_6H_4(COOC_4H_9)_2$
84-74-2

Appearance: Colorless, stable, oily liquid that is very slightly soluble in water.

Odor: Slight characteristic ester odor.

Hazards: Toxic, irritant to upper respiratory system.

Incompatibilities: Strong oxidizers.

Used in

Ammuniton:
Load 3

Regulation: There are some air pollution regulations that address this compound.

3,5-DICHLOROANILINE
$C_6H_3NH_2Cl_2$
UN 1590

Appearance: Crystals, slightly soluble in water.

Odor: None found.

Hazards: Toxic.

Incompatibilities: Acids, acid chlorides, acid anhydrides, and oxidizing agents.

Used in

Explosives:
HNTCAP

Hobbies: Photography, developing

Regulations: Hazard Class 6.1; PG II; ERG 153.

3,5-DICHLOROANISOLE
$Cl_2C_6H_3OCH_3$
33719-74-3

Appearance: With a melting point of about 40°C, will probably be a solid, but on a hot day could be very soft to a liquid. Will come from a chemical supply house.

Used in

Explosives:
TATB

Regulations: Currently being tested for the EPA's TCLP chemicals list.

p-DICHLORO BENZENE
See Para-dichlorobenzene.

4,5-DICHLORO-1,3-DIOXOLAN-2-ONE (w)
127213-77-8

Used in

Explosives:
DDD

1,2-DICHLOROETHANE
$ClCH_2CH_2Cl$
107-06-2
UN 1184

Appearance: Colorless oily liquid that sinks in water.

Odor: Chloroform.

Hazards: Toxic by ingestion, inhalation, or skin absorption. Strong irritant to eyes and skin. Flammable, flash point 15°C. Anesthetic, like chloroform.

Incompatibilities: Oxidizers, strong alkalies, strong caustics, magnesium, sodium, potassium, active amines, ammonia, iron, zinc, nitric acid, and aluminum.

Used in

Chemical Weapons:
Facility 2

Explosives:
BDPF, NDTT, HNBP, TNEN, UDTNB

Illegal Drugs:
Cook 2, Variations 5 and 8

Regulations: Hazard Class 3.0, minor 6.1; ERG 131.

Confirmed by: Beilstein Test.

DICHLOROETHYL ETHER
See Bis(2-chloroethyl ether).

α-DICHLOROHYDRIN
See 1,3-Dichloro-2-propanol.

DICHLOROMETHANE
See Methylene chloride.

DICHLOROMETHYL PHOSPHINE (w)
CH_3Cl_2P

Used in

Chemical Weapons:
Facility 3
Facility 3, Sub X

1,3-DICHLORO-2-PROPANOL (w)
$CH_2ClCHOHCH_2Cl$
96-23-1
UN 2750

Appearance: Colorless, slightly viscous, unstable liquid. Not sold as a product, will probably have to be made. Most of the literature is concerned with the effects on the environment due to the formation of this product in processed food.

Odor: Faint chloroform-like odor.

Hazards: Combustible, toxic, and an irritant to the respiratory system.

Incompatibilities: Oxidizers, strong alkalis, strong caustics, magnesium, sodium, potassium, active amines, ammonia, iron, zinc, nitric acid, and aluminum.

Used in

Explosives:
BDPF

Regulations: Hazard Class 6; ERG 153.

Confirmed by: Alcohol Test.

DICYANDIAMIDE
See Cyanoguanidine.

DIESEL FUEL
C_{13} to C_{15} hydrocarbon with additives
UN 1993

Appearance: Viscosity 2 dyed liquid. Dye color tells the source of the diesel fuel.

Odor: Oily.

Hazards: Can be flammable; flash points vary from 100 to 130°F.

Incompatibilities: Strong oxidizers. Mixed with ammonium nitrate becomes tertiary explosive.

Used in

Explosives:
Blasting agents, ANFO

Regulations: Hazard Class 3; ERG 128.

DIETHANOL AMINE
$C_2H_2(OH)_2NH_2$
111-42-2

Appearance: Strongly alkaline liquid that is miscible with water.

Odor: Slightly ammonia-like.

Hazards: PEL 3 ppm, skin. Flash point >1200°F. pH of solution, 10.9. Toxic by ingestion. Very hazardous in case of skin contact. NFPA Health 1.

Incompatibilities: Concentrated acids (reacts strongly), chlorohydrocarbons, oxidizers; attacks copper.

Used in

Explosives:
DNAN

Regulations: Considered a hazardous waste. Not dangerous goods according to international shipping regulations.

Confirmed by: Alcohol Test.

DIETHANOLETHYLAMINE (w)

Appearance: Water soluble.

Used in

Chemical Weapons:
Facility 4, HN1-1 and 2
Facility 4, HN2-2

Regulations: Cannot be shipped to Iraq without permit.

N,N'-DIETHANOLETHYLENEDIAMINE (w)

Used in

Explosives:
DINA, EDT

Confirmed by: Alcohol Test.

DIETHYL ADIPATE
See Dioctyl adipate.

N,N,-DIETHYL ANILINE
$(C_2H_5)_2NC_6H_5$
91-66-7
UN 2432

Appearance: Colorless to yellow liquid that is soluble in water.

Odor: Not found.

Hazards: Highly toxic; may be fatal if inhaled, swallowed, or absorbed through the skin. May cause cumulative effects. May cause central nervous system damage; irritant. Vapor pressure 10 mmHg @ 92°C. Flash point 97°C.

Incompatibilities: Strong acids and strong oxidizers.

Used in

Chemical Weapons:
Facility 3, Variation 3
Facility 3, V-Sub X

Regulations: Hazard Class 6.1; PG III. Toxic to aquatic organisms. ERG 153.

DIETHYL AMINE
$(C_2H_5)_2NH$
UN 1154

Appearance: Colorless liquid miscible in water.

Odor: Like ammonia.

Hazard: Alkaline reactions. Flammable, flash point –9°F.

Incompatibilities: Strong acids.

Used in

Chemical Weapons:
Facility 2, Variation 1

Explosives:
Binary explosives, Variation 3

Illegal Drugs:
Cook 2, Variations 6, 7 and 8

Regulations: Hazard Class 3; ERG 132; not acceptable in passenger, transportation.

Confirmed by: Ammonia indicator tube.

DIETHYL AMINOETHANOL
$(C_2H_5)_2NCH_2CH_2OH$
100-37-8
UN 2686

Appearance: Colorless hygroscopic liquid that is very soluble in water. Absorbs moisture from the air. Used to clean equipment.

Odor: Nauseating, weak ammonia.

Hazards: Toxic by ingestion and skin absorption. Combustible. Flash point 140°F. Vapor pressure 21 mmHg @ 20°C.

Incompatibilities: Strong oxidizers and acids.

Used in

Chemical Weapons:
V-agent series nerve agents.

Regulations: Hazard Class 3; PG III; ERG 132. Australian Group chemical, not listed. Requires export permits from the DTCC.

DIETHYL-N,N-DIMETHYLOS-PHORAMIDATE (w)
2404-03-7
Used as a gasoline additive.

Regulations: Australian Group Schedule 2B.

DIETHYLDIPHENYLUREA (w)
$C_2H_5(C_6H_5)NCON(C_6H_5)C_2H_5$
85-98-3

Appearance: White crystalline solid that is insoluble in water.

Odor: Peppery.

Hazards: Explosive when shocked or heated. Toxic.

Incompatibilities: Strong oxidizers.

Used in

Ammunition:
Load 4, Variation 1

Regulations: Hazardous Waste n.o.s. Hazard Class 1.

DIETHYLENE GLYCOL
$CH_2OHCH_2OCH_2CH_2OH$
111-46-6

Appearance: Colorless hygroscopic liquid with a sharply sweetish taste that is miscible in water. Diethylene glycol is used in antifreezing solutions for refrigerators, sprinkler systems, and water seals for gas tanks. It is also used for finishing and lubricating wool, worsted, cotton rayon, and silk.

Odor: Mild.

Hazards: Toxic by ingestion.

Incompatibilities: Oxidizers.

Used in

Amphetamines:
Synthesis 5

Used to make a very hot bath for a reaction vessel. May contain potassium hydroxide to increase heat even more in Synthesis 5.

Regulations: FDA Hazardous for household use at greater than 10%. Not RCRA or EPA listed. Banned from landfills.

DIETHYL ETHER
See Ethyl ether.

DIETHYLETHYL PHOSPHATE (w)
78-38-6
A gasoline additive.

Used in

Chemical Weapons:
Prealkylated compounds that could be used to make nerve agents.

Regulations: Australian Group Schedule 2B.

DIETHYLETHYL PHOSPHITE (w)

Used in

Chemical Weapons:
Prealkylated compounds that could be used to make nerve agents.

DIETHYL ETHYL PHOSPHONATE (w)

Used in

Chemical Weapons:
Prealkylated compounds that could be used to make nerve agents.

DIETHYLMETHYL PHOSPHONATE

Used in

Chemical Weapons:
Facility 1

DIETHYLMETHYL PHOSPHONATE (w)
683-08-9

Regulations: Australian Group Schedule 2B.

DIETHYLMETHYL PHOSPHONITE (w)
15715-41-0

Used in

Chemical Weapons:
Facility 3, VX

Regulations: Australian Group Schedule 2B.

DIETHYL OXALATE
$C_2H_5 \cdot COO \cdot CH_2 \cdot CO \cdot COOC_2H_3$
95-92-1
UN 2525

Appearance: Colorless, oily, unstable liquid. Anhydrous oxalic acid. Insoluble and gradually decomposed by water. Would be found in brown bottles with sealed caps from a laboratory supply house.

Odor: Irritating.

Hazards: Toxic by ingestion, strong irritant to skin and mucous membranes. Flash point 169°F.

Incompatibilities: Oxidizers, reducers, acids, bases.

Used in

Illegal Drugs:
Cook 5, Variation 1

Designer Drugs:
Group 3, Variation 1

Regulations: Hazard Class 6.1; PG III; shipping label Ethyl Oxalate. ERG 156.

DIETHYL PHOSPHITE (w)
762-04-9
A paint solvent and a lubricant additive.

Used in

Chemical Weapons:
Facility 1

Prealkylated compounds that could be used to make nerve agents.

Regulations: Australian Group Schedule 3B.

o,o-DIETHYL PHOSPHORODITHIOATE (w)
See o,o-Diethyl phosphorothioate.

o,o-DIETHYL PHOSPHOROTHIOATE (w)
2465-65-8

Regulations: Australian Group chemical, not listed.

DIHYDROFURAN-2(3H)-ONE
See γ-Butyrolactone.

DIISOPROPYL AMINE
$[(CH_3)_2CH]_2NH$
108-18-9
UN 1158

Appearance: Colorless liquid, strongly alkaline, completely soluble in water. This is an anti-foam agent for detergents.

Odor: Characteristic strong ammonia. The odor threshold is around 1 ppm.

Hazards: Flammable, flash point 16–30°F. Toxic by ingestion, inhalation, and skin absorption. Vapor pressure 50 mmHg @ 30°C.

Incompatibilities: Strong oxidizers, including perchlorates, nitrates, peroxides, and acids.

Used in

Chemical Weapons:
Amines that can be used to make Agent QL

Explosives:
DITN

Regulations: Hazard Class 3, minor 8; PG II; ERG 132. Australian Group chemical, not listed. Requires export permits from the DTCC.

N,N-DIISOPROPYL-2-AMINOETHANE CHLORIDE (w)
96-79-7

Used in

Chemical Weapons:
Amines that can be used to make Agent QL

Regulations: Australia Group Schedule 2B.

N,N-DIISOPROPYL-2-AMINOETHANOL (w)
96-80-0

Used in

Chemical Weapons:
Amines that can be used to make Agent QL

Regulations: Australian Group Schedule 2B.

Confirmed by: Alcohol Test.

2-DIISOPROPYLAMINOETHANOL THIOL (w)

Used in

Chemical Weapons:
Amines that can be used to make Agent QL

Regulations: Australian Group Schedule 2B.

N,N-DIISOPROPYL-2-AMINOETHYL CHLORIDE HYDROCHLORIDE (w)
4261-68-1

Regulations: Australian Group Schedule 2B.

DIISOPROPYL ETHER
$(CH_3)_2CHOCH(CH_3)_2$
108-20-3
UN 1159

Appearance: Colorless viscosity 1 liquid that floats on water with a sharp curved line. Probably will be in metal containers.

Odor: Characteristic.

Hazards: Very flammable, flash point – 28°C; LEL 1.4; UEL 7.9%. Inhalation will cause drowsiness and sore throat.

Incompatibilities: Oxidizers.

Used in

Designer Drugs:
Group 2, Variation 2

Regulations: Hazard Class 3; ERG 127.

Monitored by: Combustible gas indicator.

DIMETHOXYETHANE
See Dimethyl glycol.

DIMETHOXYMETHANE
$CH_2(OCH_3)_2$
109-87-5
UN 1234

Appearance: Colorless volatile liquid that is soluble in water. Will be found in brown glass bottles from a laboratory supply house. Most commonly called *methylal*. This is an additive to diesel fuel to lower the amount of soot.

Odor: Pungent.

Hazards: Flammable, flash point 0°F. LEL 1.6%; UEL 17.5%. Will easily form explosive mixtures in the air. Irritant, contact with liquid causes defatting of the skin. Narcotic by inhalation; PEL 1000 ppm.

Incompatibilities: May form peroxides upon exposure to air. Strong oxidizers; acids.

Used in

Chemical Weapons:
Facility 3
Facility 8, Variation 2

Illegal Drugs:
Cook 2

Regulations: Hazard Class 3, PG II. ERG 127. Flammable label.

Monitored by: Combustible gas indicator.

DIMETHYL AMINE
$(CH_3)HN$
124-40-3
UN 1032, UN 1160

Appearance: A gas at room temperatures; however, is often used in aqueous solutions of 40% dimethylamine. *Also see* Dimethylamine·HCl. Strongly alkaline and soluble in water. It is used as an accelerator in vulcanizing rubber, tanning, manufacturing of detergent soaps, and is used to attract boll weevils to exterminate them.

Odor: Fishy, ammonia-like.

Hazards: Flammable gas; flash point –6°C. Corrosive.

Incompatibilities: Strong oxidizing agents.

Used in

Chemical Weapns:
Facility 2

Illegal Drugs:
Cook 5, Variation 1
Cook 7

Regulations: Hazard Class 2.1; ERG 118. Australian Group chemical, not listed. Requires export permits from the DTCC.

DIMETHYLAMINE HYDROCHLORIDE
$(CH_3)HN·HCl$
506-59-1

Appearance: Water-soluble white solid if made using a weak hydrochloric acid solution with dimethylamine.

Odor: None.

Hazards: Toxic by ingestion.

Incompatibilities: Oxidizers.

Regulations: Australian Group chemical, not listed. Requires export permits from the DTCC.

2-DIMETHYLAMINOMETHANOL (w)

Used in

Chemical Weapons:
Facility 3, Variation 3

DIMETHYLAMINOPHOSPHORYL DICHLORIDE (w)
667-43-0

Used in

Chemical Weapons:
Various nerve agents.

Regulations: Australia Group Schedule 2B.

DIMETHYL CELLOSOLVE®
See Dimethyl glycol.

DIMETHYL ETHER
CH_3OCH_3
115-10-6
UN 1033

Appearance: Colorless compressed gas or liquid. Boiling point –24.5°F. Soluble in water. Will most likely be found in a laboratory bottle. In reactions the laboratory bottle will probably be in an ice bath, often salted with sodium chloride. It may possibly be in a dry ice bath. In this case the dimethyl ether will be a liquid.

Odor: Ethereal; poor warning qualities at low concentrations.

Hazards: Extremely flammable. Flash point –42°F. Anesthetic by inhalation.

Incompatibilities: Oxygen, chlorine, oxidizers.

Used in

Illegal Drugs:
Cook 4, Variation 3

Regulations: Hazard Class 2.1, not acceptable in passenger transportation; ERG 115.

Monitored by: Combustible gas indicator.

DIMETHYL ETHYL PHOSPHONATE (w)
$C_6H_{18}O_3P$
6163-75-3

Used in

Chemical Weapons:
Prealkylated compounds that could be used to make nerve agents.

Regulations: Australian Group chemical, not listed. Requires export permits from the DTCC.

DIMETHYLFORMAMIDE
$HCON(CH_3)_2$
68-12-2

Appearance: Colorless liquid that is miscible in water.

Odor: Slight ammonia.

Hazards: Possible carcinogen. Harmful by inhalation, ingestion, or skin contact. Irritant. Flash point 136°F, combustible.

Incompatibilities: Strong oxidizing agents, halogenated hydrocarbons, chloroformates, active halogen compounds, strong acids, strong reducing agents, rubber, leather.

Used in

Amphetamines:
Synthesis 18, Supplementary 3

Explosives:
NDTT, TNEN, UNTNB

Illegal Drugs:
Cook 2, Variations 5 and 8

Regulations: Hazard Class 3.

N,N-DIMETHYLFORMAMIDE
HCON(CH$_3$)$_2$
32488-43-0
UN 2265

Appearance: Liquid that has been used to extract pigments from plants.

Odor: Not found.

Hazards: Carcinogen; irritant; flash point 135°F.

Incompatibilities: Oxidizers.

Used in

Designer Drugs:
Group 3, Supplementary 2

Regulations: Hazard Class 3.0; PG III; ERG 129; EPA Flammable.

DIMETHYL GLYCOL
C$_4$H$_{10}$O$_2$
110-71-4
UN 2252

Appearance: Colorless liquid that is miscible in water. Glyme.

Odor: Not found.

Hazards: Flammable, flash point –2°C. Harmful. NFPA Health 2.

Incompatibilities: Hydrofluoric acid.

Used in

Chemical Weapons:
Facility 8, Variation 2

Explosives:
DATB, EDGN, PETN

Toxins:
Enhances ricin and commercial organophosphates.

Regulations: Hazard Class 3; PG II; ERG 127.

Confirmed by: Alcohol Test.

DIMETHYLMETHYL PHOSPHONATE
C$_3$H$_9$O$_3$P
756-79-6

Appearance: Colorless liquid that slowly undergoes hydrolysis in water. Used as a hydraulic fluid, an antifoam, an antistatic, and in the manufacture of flameretardants.

Odor: Not found.

Hazards: Harmful if inhaled or if absorbed through the skin. People who were exposed during an event suffered effects for years.

Incompatibilities: Oxidizers.

Used in

Chemical Weapons:
Prealkylated compounds that could be used to make nerve agents.

Regulations: Australian Group Schedule 2B.

DIMETHYL PHOSPHITE
C$_2$H$_7$O$_3$P
868-85-9

Appearance: Colorless liquid that is moisture sensitive and incompatible with water. Lubricant additive.

Odor: Not found

Hazards: Suspected carcinogen. Flash point 96°C. Harmful by ingestion, inhalation, and if absorbed through the skin. Vapor pressure 64 mmHg @ 95°C (not high).

Incompatibilities: Water; strong oxidizing agents; acid chlorides; strong bases.

Used in

Chemical Weapons:
Prealkylated compounds that can be used for nerve agents.

Regulations: No specific regulations for DOT. Australian Group Schedule 2B.

DIMETHYL PHOSPHONATE
See Dimethyl phosphite.

N,N-DIMETHYLPHOSPHOSAMIDIC DICHLORIDE (w)

Used in

Chemical Weapons:
Prealkylated compounds used to make nerve agents.

Regulations: Australian Group Schedule 2B.

DIMETHYL POLYSULFIDE
624-92-0
A polymer of $C_2H_6S_2$

Appearance: A yellow liquid; consistency will depend in, on degree of polymerization. Used as an additive to make ink rollers for printing.

Used in

Chemical Weapons:
Facility 3

Regulations: Degree of contamination is determined by absolute sulfur content.

DIMETHYL SULFATE
Chemical weapon
$(CH_3)_2SO_4$
77-78-1
UN 1595

Appearance: Colorless oily liquid that sinks in water. Insoluble in water, but decomposes gradually in water. A powerful corrosive to the skin. Can be made; *see* Facility 6.

Odor: Onions.

Hazards: Vapors are poisonous; passes through the skin with a strong corrosive action as it becomes sulfuric acid. Massive exposures will cause liver and kidney damage. Carcinogen. Combustible, flash point 182°F.

Incompatibilities: Water reactive.

Used in

Chemical Weapons:
Facility 6

Designer Drugs:
Group 3, Variation 1

Explosives:
MNTA

Illegal Drugs:
Cook 4, Supplementary 2
Cook 5, Variation 1
Cook 6, Variation 1

Regulations: Hazard Class 8, not acceptable in passenger transportation; ERG 156; ICC Classification Corrosive Liquid, white label; MCA Warning label.

DIMETHYL SULFONE
67-71-0
$(CH_3)_2SO_2$

Appearance: White crystalline flakes that are not very soluble in water. It is a naturally occurring, organic, sulfur-containing compound related to dimethyl sulfoxide. It is found in small amounts throughout nature and has been found in small amounts of human blood and urine. Could possibly be purchased as a health food additive.

Used in

Amphetamines:
Products used as a cut for final product

DIMETHYL SULFOXIDE
$(CH_3)_2SO$
67-68-5

Appearance: Clear colorless liquid that is hydroscopic and miscible in water. Although easily available at a chemical supply house, probably will be found purchased from hardware stores, where it is sold as a folk medicene.

Odor: Strong garlic-like.

Hazards: Although listed as nonhazardous, it is considered a mild poison and a skin irritant; however, it is a folk medicine, and is used as a veterinary medicine, where it is rubbed on the skin to cure a number of deep pains, such as arthritis. The danger is that it is absorbed quickly through the skin and will also transport things mixed with it through the skin. Flash point 85°C, vapor pressure 0.42 mmHg @ 20°C.

Incompatibilities: Strong oxidants, strong reducing agents, acyhalides, phosphorus halides, arylhalides, bromobenzoyl acetanilide, magnesium perchlorate, perchloric acid, and sodium hydroxide. Reacts vi-olently with a number of compounds.

Used in

Amphetamines:
Synthesis 18, Supplementary 2

Explosives:
BDPF, DIANP, hexaditon, HNS

Regulations: Nonhazardous for air, sea, and road, transportation.

Confirmed by: Alcohol Test.

DIMETHYLUREA
$(CH_3NH)_2CO$
598-94-7 and 96-31-1

Appearance: Colorless prisms soluble in water. We found two listings for dimethylurea: 1,1-dimethylurea (CAS 589-94-7), which is used to treat hypoglycemia; and 1,3-dimethylurea (CAS 96-31-1), which is more widely used and is listed as very hazardous in skin and eye contact and toxic by inhalation and ingestion. It is toxic to the kidneys, nervous system, liver, and mucous membranes. Since the formulas did not specify, it is equally likely that you may encounter either. I would recommend that you treat it as the 1,3 unless the label specifies otherwise.

Used in

Explosives:
MNA

Regulations: Not a DOT-controlled material. Listed in OSHA (29 CFR 1910.1200).

3,5-DINITROBENZOIC ACID
$(NO_2)_2C_6H_3COOH$
99-34-3

Appearance: White or off-white solid that is very slightly soluble in water.

Odor: None.

Hazards: Skin and eye irritant. Melting point 206°C.

Incompatibilities: May ignite in contact with strong bases. Incompatible with oxidizing agents and strong bases.

Used in

Illegal Drugs:
Cook 4, Supplementary 2

Regulations: None are listed specifically. Note incompatibilities!

3,5-DINITROBENZOYL CHLORIDE
$(NO_2)_2C_6H_3COCl$
99-33-2
UN 1759

Appearance: Yellow-to-brown crystals that are decomposed by water. Should be refrigerated.

Odor: None.

Hazards: Severe skin and eye irritant. May act as a respiratory irritant; however, the melting point is high, 69°C.

Incompatibilities: Water; strong oxidizers, including nitrates; combustible materials.

Used in

Illegal Drugs:
Cook 4, Supplementary 1

Regulations: Hazard Class 8; Corrosive liquid, n.o.s.; ERG 154.

4,4-DINITROBUTYLACETATE (w)

Used in

Explosives:
ADNB

2-DINITRODIPHENYL AMINE (w)
Will be difficult to obtain; could not find this even in specialty catalogs.

Used in

Hobbies:
Rocket fuel, Variation 6

DINITROGEN TETROXIDE
UN 1067

Appearance: A red gas. Will be in stainless steel containers similar to DOT-110-AW, unless being made at the site. If released, will form a red cloud which is way above the IDLH. It is unlikely that it is being used to make phenyl acetic acid. If this were available in bulk, it would be considered a chemical weapon.

IF YOU ENCOUNTER THIS MATERIAL, THE SITUATION IS VERY DANGEROUS! IF YOU CAN SEE THE RED CLOUD, IT CAN KILL YOU!

Odor: Green grass, sharp.

Hazards: Very strong oxidizer; very dangerous inhalation hazard.

Incompatibilities: All organic material, reducing agents. Reactions may be explosive.

Used in

Amphetamines:
Synthesis 20, Supplementary 5

Hobbies:
Rocket fuel
No hobby formula listed this material, but it is what the professionals use for rocket fuel and might be considered by an enthusiastic amateur.

Regulations: 49 CFR parts 107, 171, 180, most specifically 173.336 and 834(m). ERG 152. Would require special permission to transport through any community.

2,6-DINITROTOLUENE
$(CH_3)C_6H_3(NO_2)_2$
606-20-2
UN 2038

Appearance: Tan crystals.

Odor: None found.

Hazards: Stable but shock sensitive. Toxic by inhalation, quickly absorbed through the skin. Highly toxic, possible carcinogen, reproductive hazard, and neurological hazard. May cause sensitization by inhalation or skin contact. Corrosive. Heating may cause explosion.

Incompatibilities: Oxidizers, reducing agents, and strong bases.

Used in

Illegal Drugs:
Cook 5, Variation 1

Designer Drugs:
Group 3, Variation 1

Regulations: Hazard Class 6.1; RCRA and CERCLA waste by virtue of being toxic; ERG 152.

DIOCTYL ADIPATE
$C_2H_5OCO(CH_2)_4OCOC_2H_5$
103-23-1

Appearance: Colorless-to-amber liquid that is insoluble in water. Used as a plasticizer.

Odor: Slight aromatic.

Hazards: NFPA Health 0, Fire 1, and Reactivity 0.

Incompatibilities: Strong oxidizers.

Used in

Hobbies:
Rocket fuel, Variation 2
Rocket fuel, Variation 3

Regulations: Not specified.

DIOCTYL PHTHALATE (w)
$C_6H_{10}[COOCH_2CH(C_2H_5)C_4H_9]_2$

Appearance: Light-colored liquid that is insoluble in water. Used as a plasticizer.

Odor: Not reported.

Hazards: Flash point 425°C.

Incompatibilities: Oxidizers.

Used in

Explosivies:
Plastic explosives, Variation 2

Hobbies:
Casting resins

Regulations: Not specified.

DIOXANE
$C_4H_5O_2$
123-91-1
UN 1165

Appearance: Colorless liquid. Solvents for oils and wax.

Odor: Faintly pleasant; ethereal.

Hazards: Toxic by inhalation; PEL 25 ppm in the air; absorbed by the skin; possible carcinogen. Flammable, flash point 65°F; dangerous fire hazard, may form explosive peroxides.

Incompatibilities: Oxidizers.

Used in

Explosives:
A-NPNT, DANP, KNF, TATB

Regulations: Hazard Class 3; ERG 127.

1,3-DIOXOLANE
$OCH_2CH_2OCH_2$ (cyclic)
646-06-0
UN 1166

Appearance: Water-like liquid that is soluble in water.

Odor: Very low odor threshold; ethereal.

Hazards: Flammable, flash point 35°F. NFPA Fire 3, moderately toxic. NFPA Health 2. TLV 20 ppm. There are chronic effects on the kidneys, bladder, and brain.

Incompatibilities: Oxidizers.

Used in

Explosives:
TNEN

Regulations: Hazard Class 3; PG II; ERG 127.

DIPARATOLUIDINITRO-ANTHROQUINONE (w)
Used to make green smoke in fireworks. No commercial source found.

Used in

Fireworks:
Green smoke

DIPHENYL AMINE
$CH_{12}H_{11}N$
122-39-4

Appearance: Colorless or white crystals that are slightly soluble in water, sublimes. Sold in 50-lb polyethylene-lined paper bags and fiber drums. Also available from laboratory supply houses in brown glass jars

Odor: Floral.

Hazards: Toxic by ingestion, absorbed through the skin. Target organs kidney, liver, bladder. TLV 10 ppm.

Incompatibilities: Strong oxidizers.

Used in

Ammunition:
Load 3, Variation 1
Load 3, Variation 3

Chemical Weapons:
Facility 14

Regulations: Not specifically controlled.

Confirmed by: Adamsite Test.

DIPHENYLMETHANE-4-DIISOCYANATE
Industrial chemical weapon
$CH_2(C_6H_4NCO)_2$
101-68-8
Methylene-diparaphenylene isocyanate

Appearance: Light yellow fused solid flakes or crystals that are not soluble in water.

Odor: Not detectable at warning levels. Detectable at 35 ppm; PEL 0.002 ppm.

Hazards: Despite a vapor pressure that is very low, the tolerance for this compound is 0.02 ppm, and inhalation is a strong possibility if the compound is heated, which is how it is usually used. Highly toxic by inhalation. The reaction in the lungs is thought to be allergic, which is why the tolerance level is so low. A person who has been sensitized to this product has an even lower threshold, and is very susceptible to this compound and other isocyanates.

Incompatibilities: Strong oxidizers.

Used in

Hobbies:
Art, urethane plastic

Regulations: No specific regulations.

DIPHENYL UREA
$C_6H_5NH·CO·NH·C_6H_5$
102-07-8
603-54-3

Appearance: White crystals that are very slightly soluble in water.

Odor: None.

Hazards: The toxicology of this compound has not been reported.

Incompatibility: Strong oxidizers.

Used in

Explosives:
TNPU

Regulations: None required on air, sea, or road transport.

DIRT

Appearance: Standard dirt, may have red overtones.

Odor: When a concern, chemical.

Used in

Amphetamines:
Synthesis 0.4

Precursor: Dirt-to-dope lab; an offshoot of the Mexican national laboratories. *See* Synthesis 1, Variation 3, where carelessness with large amounts of product produces a large amount of spilled waste. Such labs are not common and may be a one- or two-time occurrence. *See* Synthesis 0.4, Dirt lab.

This method has been used and the Internet keeps the cooks abreast of new ideas. The dirt will be from large-scale and careless laboratories such as those operated by Mexican national cartels. Methamphetamine is very stable once formed. Although at this time we do not have access to how persistent the drug is, we are aware of cases where blood was taken from the seats of cars in accidents occurring 6 months prior to be tested for methamphetamine, and the compound was found.

Regulations: Depend on the type of contamination.

DISODIUM SULFATE
See Sodium sulfate.

DMF
See Dimethylformamide.

DMSO
See Dimethyl sulfoxide.

DOA
See Dioctyl adipate.

DRILL BITS

Used in

Ammunition:
Armor-piercing bullets

DRY ICE
CO_2
124-38-9
UN 1845

Appearance: Four- to six-inch square slabs of white vaporizing material in thin cardboard boxes. Although dry ice is used as an ingredient in some processes, it will generally be seen as a form of temperature control for reaction vessels or reflux columns. In some cases, such extreme temperature control can be a good key to determine what is being done. *Also see* Carbon dioxide.

Odor: None.

Hazards: Skin-contact hazard, can cause immediate frost bite. An asphyxiant gas. At levels near 10,000 ppm CO_2 can cause anxiety, which for unhealthy persons could be dangerous. Levels like these are unlikely except in the most enclosed spaces. PEL 5000 ppm.

Incompatibilities: Generally considered as inert; will react with the alkali metals.

Used in

Amphetamines:
Synthesis 2
Synthesis 14
Synthesis 16
Synthesis 20, Supplementary 7
Synthesis 21, Variation 2

Chemical Weapons:
Facility 5

Explosives:
AN, ammonium perchlorate, McGyver, dry ice bomb

Illegal Drugs:
Cook 4, Supplementary 2

Regulations: Cannot be shipped by air without previous permission from the airline.

Confirmed by: Carbon dioxide indicator tube.

DURAFLAME LOGS®

Used in

Incendiary Devices:
Situation 0.4

E

EDTA
See Ethylenediaminetetraacetic acid.

EMULTEX R®
See Lecithin.

ENGINE-STARTING FLUID
See Ethyl ether.

EPHEDRINE (Pseudoephedrine)
$C_6H_5(CHOH)C_2H_4NHCH_3$
299-42-3

Appearance: A white crystalline powder as seen in clandestine labs. Ephedrine is a bronchial dilator. Like methamphetamine, it is found in two forms, as the base and as the hydrochloride. It can be purchased in products such as Dexatrine® or Primatine® mist in the form of tablets, powder, or in bronchial inhalers. It can be found on cotton or be extracted from cotton with hydrochloric acid. Mexican national laboratories may have pure ephedrine powder instead of pills. The powder may become yellow upon standing. Ephedrine may be found dissolved in alcohol or water, found on filters, and found in makeshift filtration systems such as mop bucket wringers. Although in theory, ephedrine and pseudoephedrine are the same in reactions, it appears that ephedrine is slightly more effective, making it the preferable product to begin with.

Odor: None; may smell musty on standing.

Hazards: A large amount may be toxic by ingestion. LD_{50} 0.001 g/kg.

Used in

Ampetamines:
Synthesis 0.4
Synthesis 0.7
Synthesis 1
Synthesis 2
Synthesis 3
Synthesis 4
Synthesis 5

Regulations: CDTA (List I) 0 kg. This product is not allowed to be sold in large quantities over the counter.

Confirmed by: Ephedrine Test.

EPSON SALTS
See Magnesium sulfate.

ERGOT
Claviceps purpures
129-51-1

Appearance: From *Secale cornutum*, spurred rye. 15–30% Fatty oils; also lactic succinic and other acids; a fungus.

Used in

Illegal Drugs:
Cook 2, Variations 4 and 6
Cook 2, Supplementaries 2, 3, and 4

ERYTHRITOL
$C_4H_{10}O_4$
564-00-1

Appearance: White sweet crystals, soluble in water. Erythritol is made from lichens.

Odor: None found.

Hazards: May act as an irritant. Can be eaten in large amounts; acts as a laxitive.

Incompatibilities: Strong oxidizing agents.

Used in

Explosives:
ETN

Regulations: Nonhazardous for air, sea, and road freight.

ETHANAL
See Acetaldehyde.

ETHANE
C_2H_6
74-84-0

Appearance: Clear, flammable gas. Will come in a laboratory bottle.

Odor: None.

Hazards: Flammable gas, like natural gas, propane, and butane. NFPA Health 1, Fire 4.

Incompatibilities: Oxidizers.

Used in

Explosives:
Methyl ethyl ketone peroxide and oxyacetylene

Regulations: Liquefied petroleum gas, Hazard Class 2.1. Laboratory bottles should be empty before disposal.

Monitored by: Combustible gas indicator.

ETHANOL
See Ethyl alcohol.

ETHANOL AMINE
$H_2N·CH_2·CH_2·OH$
141-43-5
UN 2491

Appearance: Colorless viscous liquid that is miscible with water. Used as a solvent for fats and oils; also used in dry cleaning.

Odor: Faint.

Hazards: Toxic by ingestion and inhalation. Causes severe burns to the skin and eyes.

Incompatibilities: Strong oxidizers.

Used in

Explosives:
Binary explosives, Variation 3
NENA

Regulations: Hazard Class 8; ERG 153.

ETHER
See Ethyl ether.

ETHYL ACETATE
$CH_3COOC_2H_5$
141-78-6
UN 1173

Appearance: Colorless viscosity 1 liquid that floats on water. Slowly decomposed by moisture, becomes acidic. Used for artificial fruit essences, and a solvent for cleaning textiles. Ethyl acetate may be found in rectangular metal hardware store 1-gallon containers, 5-gallon cans, or from a laboratory supply house in glass 4-liter jugs.

Odor: Artificial fruit flavors (reminiscent of bananas or Juicy Fruit gum).

Hazards: Intoxication by inhalation, absorbed through the skin; however, will probably cause drying of the skin. PEL 400 ppm. When dilute, used as a food additive. Flammable, flash point 24°F. Concentrated is irritating to the eyes and skin. LEL 2%; UEL 11.5%; Group D atmosphere. Intoxication by inhalation; absorbed through the skin; most likely contact problem is drying of the skin, which can cause cracking and infection; LD_{50} 11 g/kg.

Incompatibilities: Group 6-B. Chromium trioxide and nitric acid. Chlorosulfonic acid, lithium aluminum hydride, strong oxidizers.

Used in

Amphetamines:
Synthesis 16
Synthesis 18, Supplementary 3

Ammunition:
Load 3

Explosives:
ADN, ADNB, HDIW, KDN, TNTC, BDC, NDTT

Illegal Drugs:
Cook 8, Variation 1

Regulations: RQ 2270 kg; Waste Codes D001 and U112; Hazard Class 3; PG II; ERG 129.

Confirmed by: Carbonyl Test.

ETHYL ALCOHOL
CH_3CH_2OH
64-17-5
UN 1170

Appearance: Colorless viscosity 1 liquid. Since this is used in so many reactions, it is probably not a good chemical to key for a person's, intentions. *Also see* Alcohol (nonspecific). Ethanol is easily available in recreational beverages, including Everclear® and vodka (alcoholic beverages can easily be distilled to make reagent-grade >95% ethanol). Ethanol is also found in household items such as cologne, and from laboratory supply houses in 4-liter brown glass jugs labeled as proprietary solvent. Check whiskey or vodka bottles stored in other parts of the house. Storing a "stash" in recreational beverages is a favorite way of hiding methamphetamine in plain sight.

Odor: Recreational alcoholic beverages.

Hazards: The primary hazard is its flammability, flash point 55° F; LEL 3.3%; UEL 19%; Group D atmosphere. Intoxication by inhalation; absorbed through the skin; LD_{50} 7 g/kg; PEL 1000 ppm; Proposition 65 carcinogen (California). However, some people drink it.

Incompatibilities: Incompatibility Groups 3-A and 4-A. Chromium trioxide and nitric acid.

Used in

Amphetamines:
Synthesis 1, Supplementary 5
Synthesis 3
Synthesis 4
Synthesis 9
Synthesis 15
Synthesis 16

Synthesis 18, Supplementary 5
Synthesis 19

The primary use of ethanol in the clandestine methamphetamine laboratory is as a polar solvent for the purification and extraction of ephedrine. In use it may be a clear, pink, or slightly blue liquid over a powder, or heated to form a paste. Color is from dyes used in the tablets. It can be used as the alcohol to enhance the reaction in step 2 of Synthesis 3. Look for an orange-red liquid in a sealed reaction vessel. It is used to enhance the reaction of perchloric acid in the hydrogenation method (Synthesis 4). It can be used as the alcohol to enhance the reaction in Synthesis 5. Used with sodium metal to enhance the production of methamphetamine from P-2-P and methylamine (Synthesis 9). Used to enhance the reaction between formaldehyde and ammonium chloride to make methylamine (Synthesis 15). Used to react benzyl chloride and sodium cyanide to make benzyl cyanide (Synthesis 16).

Ammunition:
Load 3, Variations 1 and 3
Load 4, Variation 1

Chemical Weapons:
Facility 1, Variation 1
Facility 2, Variations 1 and 2
Facility 3
Facility 3, Variation 3
Facility 4, Variation 3
Facility 13

Date Rape:
Chloral hydrate

Explosives:
BDC; copper fulminate; DDD; diazodinitrophenol; DNAN; DNFAP; EDDN; EDI; ETN; Hexanitrate; Isoitol nitrate; KDN; lead nitroanilate; mercury fulminate; MOM, NTND, PETN; nitromannitol, Variation 2; picric acid; PVN, silver NENA; sodium azide, Variation 2; styphnic acid; TNAD; TNN

Illegal Drugs:
Cook 1, Variation 1
Cook 2, Variations 1 and 3

Cook 2, Supplementary 3
Cook 4, Variations 3 and 5
Cook 5, Variation 1
Cook 4, Supplementary 2
Cook 6, Variations 1, 2 and 3
Cook 9

Recreational Drugs:
Bananadine

Regulations: WasteCodeD001; HazardClass 3; ERG 127. Heavily regulated under ATF.

Confirmed by: Alcohol Test.

ETHYL CELLOSOLVE®
$C_4H_{10}O_2$
110-80-5

Appearance: Clear liquid that is miscible in water. It will probably be found in a commercial solvent container marked Cellosolve®. It is actually 2-ethoxyethanol.

Odor: None.

Hazards: Combustible. The flash point is 44°C.

Incompatibilities: Strong oxidizers.

Used in

Hobbies:
Rockets, Variation 6

Regulations: None specific.

ETHYL CHLORIDE
C_2H_5Cl
75-00-3
UN 1037

Appearance: Gas at ordinary temperatures and pressures; when compressed is a colorless, very volatile liquid, Boiling point 12.5°C, which is soluble in water and burns with a green smoky flame. Usually found as a liquid in sealed glass tubes.

Odor: Characteristic odor, which is ether-like.

Hazards: Flammable, flash point −58°F. Anesthetic. Vapor pressure 100 mmHg @ 20°C. NFPA Health 2, Fire 4.

Incompatibilities: Oxidizers.

Used in

Chemical Weapons:
Ethyldichloroarsine

Explosives:
Plastic explosives, Variation 2

Regulations: Hazard Class 2.1, not acceptable in passenger transportation ERG 115.

ETHYL CHLOROCARBONATE
$ClCOOC_2H_5$
UN 1182

Appearance: Water-like liquid that decomposes in water.

Odor: Strongly irritating to eyes and mucous membranes..

Hazards: Toxic; strong irritant to eyes and skin. Flammable, flash point 61°F.

Incompatibilities: Oxidizers.

Used in

Explosives:
DATB, NENA

Regulations: Hazard Class 3, minor 6.1; forbidden in passenger transportation ERG 155.

ETHYL CHLOROFORMATE
See Ethyl chlorocarbonate.

ETHYL CYANIDE
See Propionitrile.

ETHYL DIETHANOL AMINE
$C_2H_5N(CH_2CH_2OH)_2$
139-87-7

Appearance: Water-like liquid.

Odor: Amine.

Hazards: Combustible; low toxicity.

Incompatibilities: Oxidizers.

Used in

Chemical Weapons:
Facility 4, N1-3

Regulations: Australian Group Schedule 3B.

O-ETHYL 2-DIISOPROPYLAMINOETHYL-METHYLPHOSPHONITE (w)
57856-11-8

Used in

Chemical Weapons:
Facility 3

Regulations: Australian Group Schedule 2B.

ETHYLENE (Ethene)
$H_2C=CH_2$
74-85-1
UN 1962

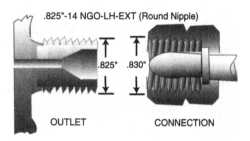

.825"-14 NGO-LH-EXT (Round Nipple)

.825"   .830"

OUTLET   CONNECTION

Appearance: Colorless flammable gas. Will be in a laboratory bottle.

Odor: Suffocating, ethereal.

Hazards: Flammable gas. Forms explosive atmosphere.

Incompatibilities: Chlorine, oxygen.

Used in

Chemical Weapons:
Facility 4, Variation 5

Regulations: Liquefied flammable gas, Hazard Class 3; cylinders must be empty before disposal. ERG 116P.

Monitored by: Combustible gas indicator.

ETHYLENE BROMOHYDRIN
See 2-Bromoethanol.

ETHYLENE CHLORIDE
See 1,2-Dichloroethane.

ETHYLENE CHLOROHYDRIN
$CH_2Cl \cdot CH_2OH$
107-07-3
UN 1135

Appearance: Colorless liquid, miscible with water.

Odor: Ethereal.

Hazards: Combustible, flash point 140°F. Toxic by ingestion and inhalation. Vapor pressure 4.8 mmHg @ 20°C.

Incompatibilities: Oxidizers.

Used in

Chemical Weapons:
Facility 4

Regulations: Hazard Class 6.1; must display poison label on rail or in air.

ETHYLENEDIAMINE
$NH_2CH_2 \cdot CH_2NH_2$
107-15-3
UN 1604

Appearance: Colorless, clear, thick, strongly alkaline liquid that is soluble in water. Will be in brown glass bottles or jugs. Readily absorbs carbon dioxide from the air.

Odor: Ammonia.

Hazards: Corrosive, liquid pH of solution 12. Strong irritant. Toxic by inhalation and skin absorption. Although flammable, poses only a minor fire risk. Flash point 93°F.

Incompatibilities: Explosive with methylene chloride.

Used in

Explosives:
Binary explosives, Variation 3
DNFA-P, EDDN, EDT, TNAD

Regulations: Must display corrosive label in air. Hazard Class 8; ERG 132.

ETHYLENEDIAMINETETRAACETIC ACID
$(HOOCCH_2)_2NCH_2CH_2N(CH_2COOH)_2$
25102-12-9
6381-92-6
7558-79-4
10378-23-1
60-00-4

Appearance: Colorless crystals. Has many laboratory uses; also has many medical uses, such as chealating metals from the body. Because of its medical applications, comes in several grades, each with its own CAS number, which is why there are so many.

Odor: None.

Hazards: Very low toxicity; can be used as a food additive.

Incompatibilities: Strong oxidizers.

Used in

Hobbies:
Photography

Regulations: None specific for this compound.

ETHYLENE DICHLORIDE
See 1,2-Dichloroethane.

ETHYLENE GLYCOL
$HO \cdot CH_2 \cdot CH_2 \cdot OH$
107-21-1

Appearance: Colorless, sweet, syrupy liquid (viscosity 2). Very soluble in water. Reagent grade will be in clear glass bottles or jugs; however, for many purposes, radiator fluid is sufficient. Automotive ethylene glycol will probably be colored with fluoriscene dye (green).

Odor: Sweet, almost like sugar.

Hazards: Toxic by ingestion. TLV 50 C.

Incompatibilities: Oxidizers. Chlorosulfonic acid.

Used in

Amphetamines:
Used as a heating liquid to replace a heating mantle in any synthesis.

Explosives:
DNFA-P

Incendiary Devices:
Situation 4

Regulations: Ethylene glycol is considered to be a contaminant in motor oil waste streams.

Confirmed by: Alcohol Test.

ETHYLENE GLYCOL DIMETHYL ETHER
See Dimethyl glycol.

ETHYLENE OXIDE
CH_2CH_2O epoxide
75-21-8
UN 1952
UN 3070
UN 3298
UN 2983
UN 3299
UN 1040

Appearance: Clear gas. Most likely to be found in laboratory bottles. Hospital containers often look like compressed consumer containers (shaving cream). The twenty-seventh highest-volume chemical produced in the United States. Used to sterilize medical and laboratory equipment. Becomes a liquid at 12°C and below.

Odor: People who work with this gas claim that they can smell it.

Hazards: Carcinogen. Highly flammable, flash point −17.7°C. Irritating to the eyes.

Incompatibilities: Water, air, bases, oxidizing metals, acids, alcohols, alkali metals, and ammonia.

Used in

Chemical Weapons:
Facility 4, Variation 2

Explosives:
Acetone peroxide and oxyacetylene
EDT

Hobbies:
Rocket fuel
Although not typical for amateurs, this has been used for rocket fuel.

Regulations: Not acceptable in passenger transportation, Hazard Class 3. 6% ERG 126, 9% ERG 126, 8.8% ERG 126, 12% ERG 126, propylene oxide with not more than 10% ERG 129P, with nitrogen ERG 119P.

Monitored with: Infrared device.

ETHYL ETHER
$C_2H_{50}C_2H_5$
60-29-7
UN 1155

Appearance: Colorless viscosity 1 liquid. Ether is most readily available as engine-starter fluid, which comes in aerosol cans. Standard laboratory ether comes in distinctive 1-pint or 1-quart metal bottles from a laboratory supply house. It can also be found in brown glass jugs. It would be prudent to refrigerate ether normally; however, in a non-spark-proof refrigerator, this can be very dangerous. If magnesium filings and chlorinated compounds are also present, a person may be making a Grignard reagent. *Also see* Grignard reagent.

Odor: Hospital.

Hazards: Extremely flammable; flash point −49°F; LEL 1.9%; UEL 36%. With a 34% flammable range, ethyl ether is a serious fire hazard. Ether is a central nervous system depressant. On standing for a long period, it may form explosive peroxide crystals. If possible, do a Peroxide Test on the ether. If the test is positive, be careful. The more contaminated the ether, the less likely it is to form peroxide. If fine white crystals are formed where the ether has evaporated, first consider peroxide. Do not open the container! Finding ether, magnesium shavings, and a chlorinated solvent together indicates the possible production of a Grignard reagent. This procedure can be very explosive. TLV 400 ppm.

Incompatibilities: Chromium trioxide and nitric acid.

Used in

Amphetamines:
Synthesis 0.7, Variations 1 and 2
Synthesis 1, Supplementary 5
Synthesis 2, Variation 2
Synthesis 18, Variations 5 and 6
Synthesis 18, Supplementaries 2 and 3
Synthesis 20, Supplementary 8
Synthesis 21, Variations 1 and 2

The primary use of ether is as a solvent for methamphetamine extraction. Because it can be used over after salting, and is difficult to obtain, saving used ether is not uncommon. The use of ether decreased in favor of Freon® because of the odor of ether, which often helped police detect clandestine laboratories. The nonavailability of Freon has created a need to return to ether, with an attendant increase in the number of fires associated with clandestine labs. Ether enhances the making of the Grignard reagent (Synthesis 21). Used to make MDA, in the neo-Nazi method, and in the Chewbacca method for making meth. Synthesis 18, Supplementaries 2 and 3; Synthesis 0.7, Variations 1 and 2.

Ammunition:
Load 3, Variations 1 and 3
Load 4, Variation 1

Chemical Weapons:
Facility 1, Variation 1
Facility 3, Variation 1

Facility 3, V-Sub X
Making VX (see Facility 19); diimine, HVX, V-Sub X.

Explosives:
Acetone peroxide, ADN, DENA, DNAN, EDT, EGDN, HNF, isoitol nitrate, KDN, NNTA, MON, NDTT, nitroform, PEN, silver NENA, silver nitroform, TNP, TNTPB.

Illegal Drugs:
Cook 1, Variation 1
Cook 1, Supplementary 1
Cook 2, Variations 3 and 6
Cook 3, Variations 1 and 2
Cook 4, Variations 1, 2, and 3
Cook 4, Supplementaries 1 and 2
Cook 5, Variation 1
Cook 6, Variation 1
Cook 7
Cook 9

Recreational Drugs:
Bananadine

Regulations: CDTA (List II) 135.8 g; RQ 45.4 kg; Waste Codes D001 and U117; Hazard Class 3; PG I; ERG 127.

Confirmed by: Carbonyl Test.

O-ETHYL METHYLPHOSPHONOTHIOIC ACID (w)

Used in

Chemical Weapons:
Facility 3, Variation 2

ETHYL NITRITE (w)
$C_2H_5NO_2$
109-95-5
UN 1194 (solution)

Appearance: Yellowish volatile liquid that decomposes in water. Boiling point 16.4°C.

Odor: Peculiar.

Hazards: Very flammable liquid, flash point −31°F. Narcotic and toxic at high concentrations.

Incompatibilities: Oxidizers.

Used in

Amphetamines:
Synthesis 18, Variations 3 and 4

Regulations: Hazard Class 3; ERG 113.

ETHYL OXALATE
See Diethyl oxalate.

ETHYLPHOSPHINYL DICHLORIDE (w)
1498-40-4

Used in

Chemical Weapons:
Prealkylated compound that can be used to make nerve agents.

Regulations: Australian Group Schedule 2B.

ETHYLPHOSPHINYL DIFLUORIDE (w)
753-98-0

Used in

Chemical Weapons:
Prealkylated compound that can be used to make nerve agents.

Regulations: Australian Group Schedule 2B.

**ETHYLPHOSPHONOTHIOIC
DICHLORIDE** (w)
Used in the manufacture of pesticides.

Used in

Chemical Weapons:
Prealkylated compound that can be used to make nerve agents.

ETHYLPHOSPHONYL DICHLORIDE (w)
1066-50-8

Used in

Chemical Weapons:
Prealkylated compound that can be used to make nerve agents.

Regulations: Australian Group Schedule 2B.

ETHYL SILICATE RESIN

Appearance: As the product Sylgard®.

Used in

Explosives:
Plastic explosives, Variation 5

2-ETHYLTHIOETHANOL
$C_2H_5SC_2H_4OH$
110-77-0

Appearance: Pale straw-colored liquid that is soluble in water.

Odor: Pungent.

Hazards: Combustible, flash point 172°F. Causes severe eye and skin irritation. NFPA Health 3.

Incompatibilities: Strong oxidizers.

Used in

Chemical Weapons:
Facilities 3, V-Sub X

Regulations: Combustible liquid n.o.s., NA 1993. Spills that will reach waterways are reportable.

Confirmed by: Alcohol Test.

EUGENOL
$C_{10}H_{12}O_2$
202-58-91

Appearance: Colorless or pale yellow liquid that becomes darker and thicker on exposure to air. Insoluble in water. May be stored under nitrogen. Available from food manufacturing establishments. Can be made; *see* Synthesis 18, Supplementary 4.

Odor: Spicy cloves.

Hazards: May be harmful if swallowed; causes eye, skin, and respiratory tract irritation. Vapor pressure 1 mmHg @ 78.4°C. Used to provide a pungent taste to foods.

Incompatibilities: Strong oxidizers.

Used in

Amphetamines:
Synthesis 18, Supplementary 3

Regulations: Not regulated as a hazardous material.

EVERCLEAR®
An alcoholic beverage that is sold over the counter in many but not all states. It is close to 200 proof, which makes it very concentrated ethyl alcohol available commercially. *See* Ethyl alcohol.

F

FERRIC AMMONIUM OXALATE
$(NH_4)Fe(C_2O_4)\cdot 3H_2O$
13268-42-3

Appearance: Green crystals that are soluble in water. They are sensitive to light.

Odor: Not found.

Hazards: Moderate irritant to skin, and moderately toxic. TLV 1 mg/m^3. NFPA Health 2.

Incompatibilities: No information found.

Used in

Hobbies:
Photography; blueprinting

Regulations: Not controlled. By definition included under OSHA Hazard communication standard.

Confirmed by: Oxalate Test Iron Test.

FERRIC CHLORIDE (Anhydrous)
FeCl
7705-08-0
UN 1773

Appearance: Black/brown crystalline solid. Although ferric chloride is not difficult to get, it has several industrial uses, especially in the photographic industry. It will probably be purchased from a laboratory supply house in brown glass jars.

Odor: May be irritating, due to release of hydrogen chloride gas.

Hazards: Corrosive; forms an acid solution in water. Irritating to the skin; hazardous to the eyes; toxic upon ingestion. LD_{50} 1.28 g/kg. Ferric chloride is oxidizing, and oxidizer is positive on the Oxidizer Test.

Incompatibilities: Sodium and potassium metal.

Used in

Amphetamines:
Synthesis 15
Synthesis 20, Supplementary 1
Synthesis 22

Illegal Drugs:
Cook 1, Variation 1
Cook 4, Supplementary 2
Cook 5, Supplementary 4

Regulations: RQ 454 kg; Hazard Class 8; PG III; ERG 157.

Confirmed by: Iron Test; Chloride Test; Oxidizer Test.

FERRIC OXALATE
$Fe_2(C_2O_2)_3$

Appearance: Pale-yellow amorphous scale or powder, soluble in water. Changes to ferrous in sunlight, making it attractive in photography. Decomposes upon heating.

Odor: None.

Hazards: Toxic by ingestion and inhalation.

Used in

Hobbies:
Photography

Regulations: None specific.

Confirmed by: Iron Test.

FERRIC OXIDE
Fe_2O_3
1309-37-1

Appearance: Red iron rust. Sold as paint pigment or through chemical supply houses.

Odor: None.

Hazards: This is rust. May be harmful if swallowed, especially by children.

Incompatibilities: Inert.

Used in

Ammunition:
McGyver gunpowder

Fireworks

Regulations: None specific.

Confirmed by: Iron Test.

FERRIC SULFATE (Anhydrous)
$Fe_2(SO_4)_3$
10028-22-5

Appearance: Grayish-white very hygroscopic powder. May be yellowish in color. Ferric sulfate is slowly soluble in water.

Odor: None.

Hazards: Harmful if swallowed and especially if eaten by children. Causes respiratory irritation upon inhalation of the powder.

Incompatibilities: Can form explosives with strong oxidizers.

Used in

Explosives:
Process 3, Supplementary 1
Process 3, Variation 3

Illegal Drugs:
Cook 2

Regulations: None specific.

Confirmed by: Iron Test.

FERROALUMINIUM
FeAl

Appearance: A metal alloy made specifically for fireworks manufacture.

Used in

Fireworks:
Branching sparks

FERROUS OXIDE
Fe_2O_3
1332-37-2
1309-37-1

Appearance: Fine black powder that is used as a pigment and is insoluble in water. The mineral form is called *wulsite*. This is a non-stoichiometric compound; the ratio of the elements iron and oxygen can vary due to the crystal formations. This may account for why two CAS numbers were found in the research.

Odor: None.

Hazards: This compound is FDA-approved for cosmetics and is used in some tattoo inks.

Incompatibilities: Aluminum metal.

Used in

Fireworks

Incendiary Devices:
Situation 5

Regulations: None specific.

Confirmed by: Iron Test.

FERROUS SULFATE (Heptahydrate)
$FeSO_4·7H_2O$
7782-83-0

Appearance: Pale bluish-green odorless crystals or granules that are water soluble. Will be in brown glass jars or square plastic jars. On standing without protection from the air, will slowly change to ferric sulfate.

Odor: None.

Hazards: Low. Harmful if swallowed. Death has resulted from swallowing close to an ounce of iron sulfate.

Incompatibilities: Strong oxidizing agents.

Used in

Designer Drugs:
Group 3, Variation 1

Hobbies:
Photography

Illegal Drugs:
Cook 5, Variation 1

Regulations: None specific.

Confirmed by: Iron Test.

FERROUS SULFIDE
See Iron sulfide.

FLAMMABLE LIQUIDS AND SOLIDS
See Incendiary devices, Situation 0.1; *also see* Nonpolar liquids that float.

FLAXSEED OIL
See linseed oil.

FLESHETTS

Fleshetts are small, sharp pieces of metal that have been grooved in order to enhance their ability to carry powdered poisons in an explosion. Terrorists prefer nails because they are more available. Although nails have the same penetrating capability, they do not carry poisons as effectively.

Used in

Toxins:
See Ricin.

FLOOR STRIPPER
See γ-Butyrolactone.

FLORENCE FENNEL

Description: Also known as finocchio; an annual known mainly for the stem, which swells to a "bulb" as it grows. It is similar to celery and can be used raw or cooked. It is very popular in Italian cooking.

Used in

Recreational Drugs:
Absinthe

FLOUR

Appearance: Polysaccharide, white, fine powder.

Odor: Only when cooking.

Hazards: This is food. Children can create an explosive atmosphere by throwing flour in the air and igniting it. Some people have foolishly done this as a type of mosquito control.

Incompatibilities: Can form explosives similar to nitrostarch when mixed with strong oxidizers.

Used in

Ammunition:
Load 5
Load 9

Regulations: None specific.

Confirmed by: Flour Test.

FLUORINE
Industrial chemical weapon
778-24-14
UN 1045

Appearance: VERY reactive, clear, oxidizing gas. Will be found in unusual gas cylinders.

Hazards: This is the strongest OXIDIZER known; it will react with EVERYTHING!!!

Incompatibilities: Everything.

IF YOU ENCOUNTER THIS MATERIAL, THE SITUATION IS VERY DANGEROUS! LEAVE IF THERE IS

ANY INDICATION OF RELEASE!!!!
THERE IS NOTHING THAT YOU CAN
DO! THIS CAN KILL YOU!

Regulations: Hazard Class 2.4; TLV 1 ppm;
PG I; ERG 124.

FLUORISIL COLUMN

Used in

Illegal Drugs:
Cook 4, Variations 1 and 2

1-(4-FLUOROPHENYL)PROPAN-2-ONE (w)

Used in

Designer Drugs:
Group 4 chemicals to replace P-2-P.

FORMALDEHYDE
HCHO
5-07-0
UN 1198 and UN 2209

Appearance: Colorless gas; will be found as
formaline in the liquid form. Generally,
formaldehyde is 37–50% formaline in a water
solution which also contains 15% methanol.
A 50% solution will off-gas considerable
amounts of formaldehyde. *Also see*
Paraformaldehyde, Formaline, *and* Trioxane.

Odor: The smell of a biology laboratory.

Hazards: A strong upper respiratory and
skin irritant. At levels as low as 3 ppm in
the air, the gas can totally incapacitate most
people. Some people are sensitive to this
compound and have adverse respiratory
reactions at levels as low as 0.5 ppm. The
gas is flammable and can polymerize
spontaneously. Flammability is not a
concern, as the biological effects will occur
far sooner. LD_{50} 0.8 g/kg; PEL 0.75 ppm.
Suspected carcinogen. Bischloromethyl
ether (BCME), one of the strongest
carcinogens known, is formed when

formaldehyde gas and hydrogen chloride gas
are mixed (possible during Synthesis 16).

Incompatibilities: Acids. With nitrogen
dioxide, there is a slow reaction that forms
an explosive.

Used in

Amphetamines:
Synthesis 1, Supplementary 5
Synthesis 16

Chemical Weapons:
Hydrogen chloride and formaldehyde react
to make BCME.

Explosives:
DMMD; DPT; HMTD, Variation 2

Regulations: RQ 45.4 kg; Waste Codes
D001 and D002; Hazard Classes 8 and 3;
ERG 132.

Confirmed by: Aldehyde Test.

FORMALINE
Also see Formaldehyde.

Used in

Illegal Drugs:
Cook 6, Supplementary 1

FORMIC ACID
HCOOH
64-18-6
UN 1779

Appearance: Colorless viscosity 1 to 2
refractive liquid. Will be purchased in clear
glass bottles or jugs with orange lids from
laboratory supply houses.

Odor: It has a strong smell of ants or bees
for those people who are very sensitive to
formic acid. For most people, it has a strong,
irritating, disagreeable odor.

Hazards: Primary hazard is as a corrosive.
Skin contact and extreme eye contact hazard.
Formic acid is very difficult to wash from
the eyes. A splash may blind a person even
if there is an eyewash station nearby. Lung

and eye irritant in the atmosphere. LD_{50} 1.21 g/kg; PEL 5 ppm.

Incompatibilities: Incompatibility Group 6-B. Nitric acid, chromium trioxide, potassium hydroxide, sodium hydroxide, perchloric acid, and sulfuric acid.

Used in

Amphetamines:
Synthesis 7, Variations 1 and 2
Synthesis 7A

Explosives:
HNIW, MNA, NTO, TADA

Regulations: RQ 2270 kg; Hazard Class 8; PG II; ERG 153.

Confirmed by: Acetic acid indicator tube.

FREON®
$FCCl_2CClF_2$
UN 1078

Appearance: Colorless viscosity 1 liquid that separates and sinks in water. May be a colorless gas. Freon is a name used generically for a variety of fluorochlorohydrocarbons, and there are many products that are sold as Freon. Freons 112 and 113 are liquids that can be used in place of any of the nonpolar liquids that float. Since Freon has become highly regulated, it is obtained primarily from illegal sources. It is not easy to say what types of containers it may be in, however, it was sold in plastic 1-gallon containers, which are very heavy because of the weight of the Freon. Five-gallon Freon gas cylinders have been found in laboratories where the cook did not understand that the gas is useless for extraction. The end product of extraction is a white gel-like sludge made up of water, sodium hydroxide, and covered with Freon. The Freon prevents the sodium hydroxide from becoming wet during the pH Test. Simple heating can break this surface, bringing the pH up to 14.

Odor: Very faint; some consider it to be sweet, many people cannot smell Freon.

Hazards: Freons are simple asphyxiates made more dangerous by their high vapor pressures: 285 mmHg @ 30°C for liquids, higher for gases, a vapor density of 6, and no strong odor. Freons have caused many fatalities in an enclosed space. LC_{50} 80,000 ppm 30 min; PEL 1000 ppm.

Incompatibilities: Perchloric   acid, chromium trioxide, and nitric acid.

Used in

Amphetamines:
Freon is used as a solvent only for methamphetamine base extraction. Inaccessibility of this compound has caused the reemergence of ethyl ether for extraction.

Regulations: Use, quantities, and releases are restricted (ozone depletion). Waste Codes U121 and F002; Hazard Class 2.2; PG (requires special label); ERG 126.

Confirmed by: Beilstein Test; Freon Test.

FULLER's EARTH
See Bentonite.

FUMING SULFURIC ACID
See Oleum.

FURAN
C_4H_4O (cyclic)
110-00-9
UN 2389

Appearance: Colorless liquid, insoluble, and floats on water. Turns brown on standing. Will be in a container from a laboratory supply house.

Hazards: Possible carcinogen. Toxic by ingestion and inhalation. Vapor pressure 493 mmHg @ 20°C, absorbed through the skin. Flammable, flash point less than −35°C with a wide explosive range, LEL 2.3%, UEL 14.3%. Forms peroxides in air.

Incompatibilities: Strong oxidizing agents, specifically oxygen and hydrogen peroxide.

Used in

Illegal Drugs:
Cook 3, Variation 1

Regulation: Hazard Classes 2.1 and 6.1; ERG 128.

FUSEL
See Isoamyl alcohol.

G

GABBO
See Norite.

GALLIC ACID
See 3,4,5-Trihydroxybenzoic acid.

GASOLINE
C$_6$ hydrocarbon mixture + additives
8006-61-9
UN 1203

Appearance: Dyed viscosity 1 solvent (often yellow). Insoluble in water and floats; additives are often soluble in water. *Also see* Nonpolar solvents that float. TLV 300 ppm.

Odor: Characteristic.

Hazards: Very flammable.

Incompatibilities: Strong oxidizers.

Used in

Explosives:
TNT

Incendiary Devices:
Situation 0.2
Situation 10

Regulations: Hazard class 3; ERG 128.

Confirmed by: Determination by iodine.

GAT
See Catha edulis.

GELATIN
Mixed proteins

Appearance: Dry, fine, white, colorless to tan powder or moist gel.

Odor: None.

Hazards: None, used in foods.

Incompatibilities: Strong oxidizers.

Used in

Explosives:
Astrolite, Supplementary 3

Hobbies:
Photography

Regulations: None specific.

Confirmed by: Protein Test.

GERANIOL
See Citral.

GLACIAL ACETIC ACID
See Acetic acid.

GLASS (Ground)

Used in

Explosives:
Potassium chlorate primer
Friction primer

GLOVE BOX
Also see Agar.

Used in

Chemical Weapons:
Pathogens

GLYCERIN
C$_3$H$_5$O$_3$
56-81-5

Appearance: Clear viscosity 6 liquid that is slowly soluble in water upon shaking.

Odor: None.

Hazards: Glycerin is very inert (GRAS).

Incompatibilities: Strong oxidizers, especially concentrated nitric acid.

Used in

Explosives:
Nitroglycerin
Smokeless powder

Hobbies:
Photography

Tear Agents:
Acrolien

Regulations: FDA GRAS (Generally Regarded as Safe).

Confirmed by: Alcohol Test.

GLUE
Glue is used in the manufacture of hydrazine. Unfortunately, the specific type or general species of glue is not specified in the formula provided.

Used in

Explosives:
Astrolite, Supplementary 3

GLYME
See Dimethyl glycol.

GLYOXAL SOLUTION
OHCCHO
107-22-2

Appearance: A solid yellow crystal that is soluble in water, the solution is light yellow. Polymerizes on standing. Probably found in a 40% solution. Vapor has a green color and burns with a violet flame.

Odor: Characteristic.

Hazards: Mixtures of vapor, with the air, may explode. Harmful by inhalation. Solutions tend to be acidic, pH 2 to 3.

Incompatibilities: Polymerization with ammonia, amines, alkalis. Reacts with chlorosulfonic acid, ethyleneimine, sodium hydroxide, nitric acid, and concentrated sulfuric acid.

Used in

Explosives:
HNIW

Hobbies:
Photography, leather tanning

Regulations: None specific.

GOLD CHLORIDE
$HAuCl_4 \cdot 3H_2O$
16961-25-4
Au_2Cl_6
13453-07-1
UN 3260

Appearance: Yellow-to-red crystals that are soluble in water. Decomposed by heat.

Odor: None.

Hazards: Irritant becomes acid in water.

Incompatibilities: Volatile acids.

Used in

Hobbies:
Photography; gold plating; jewelry; ceramics

Regulations: Hazard Class 8; PG II; Corrosive Solid Acid, n.o.s.

Confirmed by: Gold Test.

GRAMOXINE®
See Paraquat.

GRAPHITE
C
7782-42-5

Appearance: A black, shiny, crystalline powder. Graphite is the crystalline allotropic form of carbon. It is often used as a solid lubricant.

Odor: None.

Hazards: No significant hazards with normal handling.

Incompatibilities: Potassium, chlorine trifluoride.

Used in

Ammunition:
Load 3

Fireworks:
Flame deoxidizer

Regulations: None specific.

GREEN ANIS
See Anis.

GRIGNARD REAGENT
R–MgCl

Any compound with the functional group MgCl is a Grignard reagent. This is considered a carcinogen. This is not an off-of-the shelf item; it must be made at the site. Grignard reagent is used in organic chemistry to add various parts of organic molecules together. Whenever a responder encounters the presence of ethyl ether and magnesium turnings together, it is probably an addition reaction requiring a Grignard reagent. Look for chlorinated compounds whose R group is to be added to another compound. Grignard reagents are often listed as reagents, but they should be considered as intermediates.

Odor: Ether.

Hazards: This ongoing reaction is very flammable and if not controlled carefully will heat up considerably, which is dangerous due to the presence of the ether. If the magnesium is too fine, there is a high chance that the reaction could blow up.

Incompatibilities: This is a highly reactive chemical and should not be mixed with other chemicals.

GUAIFENESIN
93-14-01

Used in

Amphetamines:
Guaifenesin is a fast-acting decongestant. It is used in conjunction with ephedrine in certain decongestant preparations. Persons who make meth often claim that it is added by manufacturers of over-the-counter pills to mess up the cook. This compound mimics the actions of ephedrine and methamphetamine in various reactions. It is used in over-the-counter asthma medications. The FDA has cited manufacturers for putting in too much guifenesin, as they feel that it constitutes false advertising for people trying to buy ephedrine. Some people prefer the guaifenesin, and it is sold separately as under various tradenames.

In the manufacture of meth it tends to appear as a major contaminant (called a *milk shake*) at final salting of the product. It necessitates double extraction at the beginning of the ephedrine-to-methamphetamine process.

GUANIDINE (w)
$C=NH(NH_2)_3$

Appearance: Colorless crystals that are soluble in water and decompose when heated. We could find no source for guanidine, but several salts are available; guanidine thiocyanate (593-84-0), guanidine hydrochloride (50-01-1), guanidine carbonate (593-85-1), and guanidine nitrate. Guanidine nitrate was the cheapest. All could be made into the base by adding sodium hydroxide.

Odor: Not found.

Hazards: Combustible. All of the salts were listed as irritants. It is likely that guanidine will be very basic.

Used in

Explosives:
Nitroguanidine

GUANIDINE NITRATE
$H_2NC(NH)NH_2 \cdot HNO_3$
UN 1467

Appearance: White granules that are soluble in water.

Hazards: Shock- and heat-sensitive explosive. Strong oxidizer may ignite organic materials on contact. Toxic.

Incompatibilities: Organic materials. May produce toxic or explosive substances in water.

Used in

Fireworks:
Dark and flash effects, strobe formulations

Regulations: Hazard Class 5.1; ERG 143.

GUM ARABIC
9000-01-5

Appearance: Dried gummy exudation from *Acacia senegal.* Yellowish white to light amber, more or less spheroidal; tears or lumps of various sizes, or angular fragments.

Odor: Almost odorless.

Incompatibilities: Solutions of ferric salts, dialyzed iron, lead subacetate.

Used in

Ammunition:
Load 2, Variation 2
Load 9
Load 10

Explosives:
Potassium chlorate primer

Hobbies:
Photography

Regulations: None specific.

GUM SANDARAC

Appearance: As beige rounded granules; the powdered form is used for white areas to resist ink when printing.

Odor: Most like incense.

Used in

Hobbies:
Photography

Regulations: None specified.

GUN BLUING
See Hydrogen selenide.

GYPSUM
See Calcium sulfate.

H

HAWAIIAN BABY WOOD ROSE

Description: Looks like a wooly morning glory.

Used in

Illegal Drugs:
Cook 2

HEAVENLY BLUE (Morning Glory) Seeds

Used in

Illegal Drugs:
Cook 2

HEET®
See Methanol.

HEMLOCK
Poisonous plant

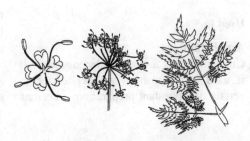

Description: Hemlock is an annual or biennial weed. It has an unpleasant smell, a hollow stem with red spots, and small white blossoms. The lethal dose for a human being is 150 mg.

Where found: Transplanted to this continent. It is common on waysides and in waste places around farm buildings, especially in the eastern United States and on the Pacific coast.

Deadly parts: The entire plant is poisonous, especially the fruit.

Symptoms: The symptoms start in half an hour, but it takes several hours for death. The poison is the alkaloid conlin. Gradual weakening of muscle power. The pulse is rapid and weak. There is quite a bit of pain in the muscles as they deteriorate and die. Sight is often lost, but the mind remains clear until death, which comes from paralysis of the lungs.

Antidotes: Gastric lavage works only if done immediately after ingestion..

Field test: Heat with thick phosphoric acid to evaporation; conlin turns blue green.

Used in

Toxins:
Conium maculatum

HEPTANE
C_7H_{16}
142-84-5
UN 1206

Appearance: Clear viscosity 1 flammable liquid that floats on water. Hepane is most likely to be found in brown 4-liter laboratory supply house glass jugs. *Also see* Nonpolar liquids that float. It is not an over-the-counter commercial product. This compound could be used as an extraction solvent. A person using this compound is probably also associated with a legitimate laboratory or is a college student new to the laboratory scene, as the cost of this purified hydrocarbon is over $70 a gallon and it will work no more effectively than Coleman® white gas in most situations.

Odor: Solvent.

Hazards: The primary hazard is flammability; flash point −7°F; LEL 1.1%; UEL 7.5%; Group D atmosphere. Intoxication by inhalation; absorbed through the skin; most likely skin contact problem is drying of the skin, which can cause cracking and infection. LD_{50}(rat) 28,700 mg/kg; PEL 500 ppm.

Incompatibilities: Incompatibility Group 6-B. Chromium trioxide.

Used in

Amphetamines:
Synthesis 20, Supplementary 6

Explosives:
UDTNB

Regulations: RQ 2270 kg; Waste Code D001; Hazard Class 3; PG II; ERG 128.

Confirmed by: Determination by iodine.

HEXACHLOROBENZENE
C_6H_6

Appearance: White needles that are insoluble in water. Used as a fungicide for seeds and a wood preservative.

Hazards: Moderately toxic by ingestion. Combustible.

Incompatibilities: Strong oxidizers.

Used in

Explosives:
Nitro PCB

Fireworks:
Chlorine donor

Regulations: Chlorinated compounds generally require incineration.

Confirmed by: Beilstein Test.

HEXACHLOROCYCLOHEXANE
See Benzene hexachloride.

HEXACHLOROETHANE
$CCl_3 \cdot CCl_3$
67-72-1
UN 9037

Appearance: White crystals that are insoluble in water.

Odor: Camphor.

Hazards: Carcinogen. Irritating to the skin and eyes on contact. Inhalation will cause irritation to the lungs.

Incompatibilities: Strong oxidizing agents, strong bases.

Used in

Chemical Weapon:
Facility 14

Regulations: Hazardous under OSHA Communication Standard, toxic under TSCA. Toxic Solid n.o.s. Hazard Class 6.1; PG III; RQ 100 lb.

HEXACHLOROPLATINIC ACID(IV)

Used in

Hobbies:
Photography

HEXAMETHYLDISILANE
$C_6H_{18}Si_2$
1450-14-2

Appearance: Clear liquid that reacts with water. This product is used to thinly coat small parts and jewels with silicon carbide.

Hazards: Water reactive, flammable.

Incompatibilities: Water, oxidizers.

Used in

Explosives:
TNEN

HEXAMETHYLENE TETRAMINE
See Hexamine.

HEXAMINE
$C_6H_{12}N_4$
00100 07 0
UN 1328

Appearance: White waxy flammable solid. Often most easily obtained in tablet form as fire starters or as heating element tablets from surplus/camping stores. Also available from laboratory supply houses in brown glass jars.

Odor: Amine fishy death.

Hazards: Flammable. May be harmful by inhalation. May be skin irritant. Has been used medicinally as an anthiseptic.

Incompatibilities: Oxidizers, especially hydrogen peroxide and nitric acid.

Used in

Amphetamines:
Used along with methyl iodide to make methamphetamine in a method not discussed in this book.

Explosives:
Cyclotrimethylene trinitriamine, HMTD, RDX

Fireworks:
Fuel deoxidizer

Illegal Drugs:
Cook 6
Mescaline

Regulations: Hazard Class 4.1; PG II, flammable solid; ERG 133.

HEXAHYDROPHTHALATE
See Dioctyl phthalate.

HEXANAPHTHENE
See Cyclohexane; *also see* Nonpolar liquids that float.

HEXANE
C_6H_{14}
100 54 3
UN 1208

Appearance: Colorless viscosity 1 liquid. *Also see* Nonpolar liquids that float. Hexane is most likely to be found in brown 4-liter laboratory supply house glass jugs. It is not an over-the-counter commercial product. This compound could be used as an extraction solvent. A person using this compound is probably also associated with a legitimate laboratory or is a college student new to the laboratory scene, as the cost of this purified hydrocarbon is over $70 a gallon and it will work no more effectively than Coleman® white gas.

Odor: Mild thinner.

Hazards: The primary hazard is flammability, flash point −7° F; LEL 1.1%; UEL 7.5%; Group D atmosphere. Intoxication by inhalation; absorbed through the skin; most likely skin-contact problem is drying of the skin, which can cause cracking and infection. LD_{50} 28.7 g/kg; PEL 500 ppm.

Incompatibilities: Incompatibility Group 6-B. Chromium trioxide.

Used in

Amphetamines:
Synthesis 0.7, Variation 2

Hexane is an unlikely solvent for methamphetamine base extraction. The waste will usually have a pH of 14, from the meth base residual from the extraction process.

Chemical Weapons:
Diphenylchloroarsine, methyl dichloroarsine, ethyl dichloroarsine

Explosives:
ADNB, DMMD, F-TNB, HNTCAB, NDTT, TATB, TNEN

Illegal Drugs:
Cook 4, Variations 1 and 2
Tetrahydrocannabinol

Tear Agents:
Mace

Regulations: RQ 2270 kg; Waste Code D001; Hazard Class 3; PG II, Label 3; ERG 128.

Confirmed by: Determination by iodine.

HONEY OIL
Making hash; mixture of marijuana and butane. Quickly off-gases the butane gas on standing. Extremely flammable.

HYDRAZINE (Hydrazine and Hydrazine Hydrate)
H–N=N–H
302-01-2
UN 2029

Appearance: Hydrazine is a clear inorganic viscosity 1 liquid. Hydrazine is a corrosive, very hazardous, very flammable liquid that is miscible in water. It is most likely to be found in a brown glass 1-quart bottle. Although it has been shipped in 55-gallon metal drums, this somewhat regulated material is not likely to be found in large containers. Hydrazine is used in agricultural chemicals (maleic hydrazide), as a polymerization catalyst, for plating metals on glass and plastics, as solder fluxes, and as photographic developers. Can be made; *see* Explosives, Astrolite, Supplementaries 1 and 2.

Odor: Slight ammonia.

Symptoms: Inhalation: sore throat, coughing, labored breathing, dizziness. Ingestion: headache, nausea, abdominal pain, diarrhea, dizziness. Skin contact: redness, pain. Absorbed in eyes: temporary blindness, redness, and pain.

Notify safety personnel of large spills or leaks. Issue warning: flammable corrosive poison.

Recommended actions: Remove all sources of heat and ignition. Provide maximum explosion-proof ventilation.

Evacuate all personnel from the area, except cleanup personnel

Dike to contain material. Pump to containers, including the contaminated bottom sediments. Notify environmental authorities to discuss disposal and cleanup of contaminated materials.

Extinguishing media: Carbon dioxide, dry chemical, alcohol foam. Flooding with water may be necessary to prevent reignition.

Hazards: May ignite spontaneously under various situations. Decomposes to oxides of nitrogen. Corrosive, poison, highly toxic by all routes of entry. May explode when heated. Hydrazine is considered a carcinogen; PEL 0.1 ppm; LD_{50}(rat) 0.41 mg/kg; IDLH 80 ppm; boiling point 236°F; specific gravity 1.004; vapor pressure 10 mmHg; vapor density 1.1; water solubility complete. The flash point of 99°F would normally be considered barely flammable, but the LEL 2.9%, UEL 98% range is so broad that hydrazine is almost certain to ignite. Auto ignition temperature 518°F.

Incompatibilities: Can ignite spontaneously on contact with oxidizers or porous materials such as earth, wood, and cloth; iron oxide can increase the potential for ignition.
Does not polymerize.

Used in

Amphetamines:
Synthesis 5
Used as a controlled and regulated precursor to make P-2-P. These are flammable carcinogens and dangerous materials.

Designer Drugs:
Group 3, Supplementary 2

Explosives:
Astrolite G, A-1-5
CDNTA, HNF, NTA, MNTA
Sodium azide

Hobbies:
Photography, developer

Illegal Drugs:
Cook 2, Variations 3 and 6

Regulations: More than 37% Hazard Class 8; ERG 132; less than 36% Hazard Class 6.1; ERG 152.

Confirmed by: Determination by iodine.

HYDRAZINE HYDRATE
UNs 2030
Similar to hydrazine, may often be used in place of hydrazine.

Used in

Amphetamines:
Synthesis 20, Supplementary 2

HYDRIODIC ACID
HI
10034-85-2
UN 1787

Appearance: Reagent-grade hydriodic acid may be clear or yellow liquid. Since hydriodic acid is very highly regulated, it is probably made on the spot. Home-made HI is red to purple in color and may be in any container. Will sometimes be oxidizer positive, due to the free iodine. Although hydriodic acid is difficult to purchase, it may still be available from a laboratory supply house, in which case it will be a clear-to-yellow liquid in a clear glass 4-liter jug. Older HI will become dark red-purple or brown. Can be made; *see* Synthesis 1, Supplementary 1.

Odor: Irritating at 2 ppm in the air.

Hazards: The primary hazard is as a corrosive. Skin-contact and extreme eye-contact hazards. Lung and eye irritant.

Incompatibilities: Incompatibility Group 6-B. Magnesium, potassium, sodium, lithium, sodium hydroxide, and ammonia.

Used in

Amphetamines:
Synthesis 2, Supplementary 1

Synthesis 1
Synthesis 20, Supplementary 4

Hydriodic acid is probably made in the clandestine lab by the addition of iodine and red phosphorus to water under acetylene or hydrogen (look for a balloon on the top of the neck of a reaction vessel). Hydriodic acid has been found in 5-gallon plastic gasoline containers at Mexican national laboratories. The HI has probably been smuggled in from Mexico. HI could be in any container, but it is probably going to be used in a reparatory or addition funnel. Hydriodic acid can also be made by adding iodine to hypophosphorous acid, and distilling hydriodic acid is a reagent in the hydrogenation of ephedrine to make methamphetamine using red phosphorus and hydriodic acid (Synthesis 1). HI can also be made by bubbling hydrogen sulfide gas through a water suspension of iodine.

Explosives:
Isoitol nitrate

Illegal Drugs:
Cook 2
Cook 4, Variation 3
Cook 5, Variation 1

Regulations: CDTA (List I) 57% 1.7 kg; Waste Code D002; Hazard Class 8; PG II and III; ERG 154.

Confirmed by: Iodine Test.

HYDROBROMIC ACID
HBr (aqueous)
10035-10-6
UN 1788

Appearance: Colorless-to-slightly yellow liquid. This is usually a 48% solution.

Odor: Sharp, irritating.

Hazards: HBr is a STRONG acid. Toxic, by inhalation and ingestion, also strong irritant. Causes severe burns upon skin contact. TLV 3 ppm.

Incompatibilities: Strong bases, strong oxidizers, and ammonia.

Used in

Illegal Drugs:
Cook 3, Variation 2

Regulations: Hazard Class 8; PG II; forbidden in air and passenger transportation; corrosive label.

Confirmed by: Bromide Test.

HYDROCHLORIC ACID
HCl (aqueous)
7647-01-0
UN 1789

Appearance: *Also see* Hydrogen chloride gas. HCl is a clear liquid with white fumes above it or an invisible gas that fumes when released to the atmosphere. This product is used in so many formulations, legal and illegal, that it is not a good one to use to key on. Reagent-grade hydrochloric acid may be a clear fuming liquid in a clear 1-gallon container with a blue lid. Muriatic acid (commercial-grade HCl that contains traces of iron) is a clear-to-yellow liquid usually found in 1-gallon plastic containers.

Odor: Irritating; almost takes your breath away at 2 ppm in the air.

Hazards: Primary hazard is corrosive. Skin-contact, extreme eye-contact, and inhalation hazard. Lung and eye irritant. LD_{50} 0.9 g/kg; LC_{50} 3124 ppm @ 30 min; PEL 5 ppm ceiling.

Incompatibilities: Magnesium, potassium, sodium, lithium; strong bases, including sodium hydroxide and ammonia.

Used in

Amphetamines:
Synthesis 1, Supplementary 5
Synthesis 7
Synthesis 13
Synthesis 18, Variation 5

Synthesis 18, Supplementaries 3 and 4
Synthesis 21, Variation 2

Hydrochloric acid is used to "salt" methamphetamine base. This is a low-yield method and will not be found in sophisticated laboratories. Salting is more effective using hydrogen chloride. Hydrochloric acid is also used in a reaction vessel in Synthesis 7 to enhance the reaction of P-2-P and methylamine to make methamphetamine using sodium cyanoborate. Hydrochloric acid is used to enhance the reaction of benzene and chloroacetone in the presence of aluminum chloride to make P-2-P (Synthesis 13). Hydrochloric acid is used in step 3 of Synthesis 15 to make P-2-P from benzaldehyde and nitroethane. Synthesis 18, Supplementaries 3 and 4. MDA.

Chemical Weapons:
Facility 1, Variation 1
Facility 1, Cyclosarin, Variations 1 and 2
Facility 4, N1-2
Facility 1, Soman
Facility 7, Variation 3
Facility 8, Variation 2
Facility 13
Facility 14
Facility 17

Designer Drugs:
Group 1, Variation 1
Group 4, Euphoria, Variation 2

Explosives:
Process 5, Variation 2
Process 6
ADBN, A-NPNT, diazodinitrophenol, DINA, 4,4-DNB, DNAT, hexaditon, HNS, NINHT, nitroform, nitrogen trichloride, McGyver bottle bombs, TNA, TNAD, TND, TNEN, TNTPB, TPG.

Illegal Drugs:
Cook 1, Variation 1
Cook 2, Variations 2, 3, and 6
Cook 3, Variation 2
Cook 4, Suplementaries 1 and 2
Cook 5, Variation 1
Cook 6, Variation 1
Cook 7
Cook 9

Tear Agents:
Mace

Regulations: Transport Category 2; Hazard Class 8.0; PG II; ERG 157. Acids must be treated as a corrosive waste.

HYDROFLUORIC ACID
HF
7664-39-4
UN 1790

Appearance: Colorless viscosity 1 liquid. HF will always be found in plastic bottles. It has been seen in mustard bottles because the long, pointed pour spouts make for an easy way to apply the acid.

Odor: Harsh biting, faintly reminiscent of vinegar.

Hazards: A toxic corrosive liquid with devastating effects on the human body, made even more dangerous because it does not cause pain on contact. Considerable damage can occur before the effects of contact with this compound are noticed by the victim.

Incompatibilities: Potassium, sodium, lithium, sodium hydroxide, and ammonia.

Used in

Amphetamines:
This chemical is *often* found in meth labs. Hydrofluoric acid is not a meth chemical. Look for indications of hobby arts and crafts. It is most likely associated with hobbies (tweekers have a lot of spare time).

Chemical Weapons:
Facility 1
Facility 1, Cyclosarin, Variation 2
Also see Compounds used to fluoridate G-Agents at the end of Facility 1

Often used to make G-agents, sarin/soman, sarin-isopropyl, cyclosarin, etc. If phosphorus compounds are present, look for phosphorus trichloride and phosphorus oxychloride, two very nasty hydrolyzers.

Hobbies:
Decorative and stained glass

Other:
Gangbangers use HF to etch graffiti onto large plate glass windows.

Regulations: Waste Codes D002; Hazard Class 8.0; PG II; ERG 157. PEL 3 ppm.

Confirmed by: Fluoride Test.

HYDROGEN
H_2
1333-74-0
UN 1049

OUTLET   CONNECTION

Appearance: Colorless flammable gas. Hydrogen is generally found in laboratory cylinders. Hydrogen gas can be generated by placing aluminum foil in hydrochloric acid in plastic pop bottles. Lithium aluminum hydride is also used to generate hydrogen inside reaction vessels. Hydrogen gas regulators become warm or even hot when the gas is escaping.

Odor: None.

Hazards: A flammable nontoxic gas with a flammable range of 12% LEL to 75% UEL, making ignition of a gas leak almost certain. Hydrogen burns with an invisible flame. Group B atmosphere; asphyxiant.

Incompatibilities: Platinum, lithium, sodium, oxygen.

Used in

Amphetamines:
Synthesis 1

Synthesis 3
Synthesis 10

Hydrogen gas is used as a chemical blanket when producing hydriodic acid (Synthesis 1). Hydrogen gas is used as a chemical blanket to make chloropseudoethedrine into methamphetamine (Synthesis 3) and as a chemical blanket in the reaction of ethanol and platinum oxide to make methamphetamine from P-2-P and methylamine (Synthesis 10). Look for lecture bottles or industrial cylinders of hydrogen gas with a hose going to the top of a reaction vessel. Mexican national laboratories will use full industrial-size cylinders.

Designer Drugs:
Group 3, Supplementary 2

Illegal Drugs:
Cook 1, Variation 1

Regulations: Waste Code D001; Hazard Class 2.1; ERG 115.

Confirmed by: GasTec Polytec IV stripe 7 turns gray to dark brown at very high levels;

Monitored by: Combustible gas indicator.

HYDROGEN BROMIDE
Industrial chemical weapon
HBr
1003-51-06
UN 1048

Appearance: Reddish gas. This is one of the very few colored gases.

Odor: Biting.

Hazards: Toxic by inhalation.

Incompatibilities: Organic gases and liquids.

Used in

Chemical Weapons:
List of chemicals used as acid smokes

Regulations: Waste Code D001; Hazard Class 2.1.

Monitored by: Draeger or GasTec bromine indicator tubes.

HYDROGEN CHLORIDE (Gas)
HCl
7647-01-0
UN 1050

Appearance: Invisible-to-white fumes when released to the atmosphere. This product is used in so many formulations, legal and illegal, that it is not a good one to use to key on. Cylinders of hydrogen chloride are generally available.

Odor: Irritating; almost takes your breath away at 2 ppm in the air.

Hazards: Primary hazard is corrosive. Inhalation and extreme eye contact, Lung irritant, skin contact. LD_{50}(rat) 0.9 g/kg; LC_{50} 3124 ppm @ 30 min; PEL 5 ppm ceiling.

Incompatibilities: Magnesium, potassium, sodium, lithium, sodium hydroxide, and ammonia.

Used in

Amphetamines:
Synthesis 18, Variations 5 and 6
Synthesis 20, Supplementary 7
Synthesis 21, Variation 2
MDA

Hydrogen chloride gas is used to "salt" methamphetamine base and therefore will be found in every methamphetamine synthesis. Although hydrochloric acid will also work, this is the preferred product to salt methamphetamine HCl. The cylinders are one of the more hazardous of the abandoned methamphetamine wastes, especially if they are leaking. Homemade hydrogen chloride generators are made from Hudson sprayers, cylinders, 5-gallon gasoline containers, and fire extinguishers. These will contain table salt and sulfuric acid. Hydrogen chloride gas cylinders are always part of a Mexican national laboratory.

Chemical Weapons:
Facility 2
Facility 4, Variation 2
Facility 13

Explosives:
DANP, DINA, nitroform, nitrogen trichloride, lead nitroanilate

Illegal Drugs:
Cook 1
Cook 3, Variation 1
Cook 4
Cook 5
Cook 6

Tear agents:
Mace

Regulations: RQ 2270 kg; Waste Code D002; Hazard Class 8; PG II and III (gas 2.3,8); ERG 157.

Confirmed by: GasTec Polytec IV stripe 2 turns red.

HYDROGEN CYANIDE (Liquid or Gas)
HCN
74-90-8
UN 1051

Appearance: Hydrogen cyanide is a colorless liquid or an invisible gas with a bitter almond odor, depending on the temperature. Hydrogen cyanide gas can be generated by adding acid to various alkali cyanide salts, or can be made into a liquid by distilling it over a salted ice bath. Platinum can catalyze ammonia, sodium hydroxide, and methane to form hydrogen cyanide. The addition of concentrated nitric acid to various heavy metal cyanide salts, such as potassium ferrocyanide, will form hydrogen cyanide gas.

Odor: NOT detected by everyone. Bitter almonds at levels near 5 ppm; chlorine-like at levels near 5 ppb.

Hazards: PEL hydrogen cyanide 10 ppm; IDLH 50 ppm. Vapor density 0.9; boiling point of hydrogen cyanide liquid is 26°C. Hydrogen cyanide is toxic by inhalation. At levels greater than 20,000 ppm it can be absorbed into the skin at lethal levels. The gas is also flammable at that level.

Incompatibilities: Strong acids.

Used in

Chemical Weapons:
Facility 8, Variation 2

Explosives:
TADA

Regulations: Cylinder must be empty before disposal. Forbidden on passenger air transportatoin. Hazard Classes 2.3 and 6.1; ERG 117.

Confirmed by: Cyanide Test.

Monitored by: Hydrogen cyanide indicator tubes.

HYDROGEN FLUORIDE (Gas)
HF
7664-39-3
UN 1740

Appearance: Colorless gas or liquid (boiling point 67°F) that is completely soluble in water.

IF YOU ENCOUNTER THIS MATERIAL, THE SITUATION IS VERY DANGEROUS! IF THERE IS ANY RELEASE AT ALL, IT CAN KILL YOU!

Odor: Sharp, irritating; will take your breath away. At lower levels may have a vinegar odor.

Hazards: Toxic, corrosive, oxidizing liquid and gas. Causes eye, skin, and respiratory tract burns. NFPA Health 4. PEL ceiling 3 ppm.

Incompatibilities: Contact with organic materials, or silica silicates, glass, etc. may cause fire.

Used in

Chemical Weapons:
Facilities 1 and 2

Regulations: Hazard Classes 8 and 6.1; forbidden on passenger and air transportation. ERG 154. RQ any amount. Australian Group chemical, not scheduled. Requires Export Permits from the DTCC.

HYDROGEN PEROXIDE
H_2O_2
7722-84-1
UN 2984, UN 2015, UN 2014

Appearance: Colorless viscosity 1 liquid. 30% Hydrogen peroxide is purchased from a laboratory supply house, although some hair bleach preparations contain 30% hydrogen peroxide. 3% hydrogen peroxide can be purchased at any drugstore. Higher concentrations of hydrogen peroxide are VERY STRONG OXIDIZERS. All concentrations appear like water. Hydrogen peroxide can be made using barium peroxide and dilute phosphoric acid, or by electrolysis utilizing ammonium persulfate.

Odor: None.

Hazards: A moderate to VERY strong oxidizer, depending on the concentration.

Incompatibilities: Organic materials, copper, chromium, iron, most metals, aniline, most flammable liquids, nitromethane.

Used in

Amphetamines:
Synthesis 1, Supplementary 4
Synthesis 20, Supplementary 5

30% Hydrogen peroxide is used to remove iodine from tincture of iodine in red phosphorus labs. This is probably a "Bevis and Butthead" operation.

Explosives:
Acetone peroxide

Almost any concentration of hydrogen peroxide above 3 % can be mixed with acetone and either sulfuric acid or hydrochloric acid to make acetone peroxide, a highly explosive compound.

Astrolite, Supplementary 3
HMTD, Variation 2
TADA

Regulations: Hazard Class 5.1; > 60% ERG 143, > 20% ERG 140, > 8% ERG 140.

Confirmed by: Oxidizer Test; Peroxide Test.

HYDROGEN SELENIDE
H_2Se
2319-78-9 (anhydrous)
UN 2202 (anhydrous)

Appearance: Colorless gas that is soluble in water. Most likely to be seen as gun bluing an aqueous solution about 2.5% hydrogen selenide in water. Also found as a 2.5% solution in some dandruff shampoos. The gas would be in laboratory bottles.

Odor: Very pungent, but has poor warning qualities at low concentrations.

Hazards: Highly toxic by inhalation. TLV 0.05 ppm. Strong irritant to the skin, damaging to the lungs and the liver. The gas is a fire and explosion hazard. The aqueous solution is very toxic by ingestion.

Incompatibilities: Reacts violently with oxidizers.

Used in

Hobbies:
Gun bluing compound

Toxins:
Chemical toxins

Regulations: Pertain mostly to the gas. Not acceptable in passenger transportation. Hazard Classes 2.3 and 6.1. Gun bluing solutions would be treated as poisons.

HYDROGEN SULFIDE
H_2S
7783-06-4
UN 1053

Appearance: Colorless flammable gas. Most likely found in a lecture bottle. Can be generated in many ways. *See* Amphetamines, Synthesis 1, Supplementary 1 (iv).

Odor: Hydrogen sulfide has the odor of rotten eggs, which cannot be detected at toxic levels over 140 ppm, as it kills the sense of smell at these levels.

Hazards: A toxic gas that may be fatal if inhaled. Will very quickly knock a person out, so that they lose conscientiousness before realizing the situation. Flammable at high levels; LEL 4.3%, UEL 46%; LC_{50} 444 ppm; PEL 20 ppm ceiling.

Incompatibilities: Strong oxidizing agents, many metals. May react violently with metal oxides, copper, fluorine, sodium, and ethanol.

Used in

Amphetamines:
Synthesis 1, Supplementary 1

Hydrogen sulfide can be used to make hydriodic acid by bubbling the gas through a suspension of iodine in water. The gas for this operation is usually made by adding sulfuric acid to iron sulfide.

Apparently, this gas has been purchased by people who did not understand the difference between hydrogen sulfide gas and hydrogen chloride gas. Although this is not a useful component of the clandestine laboratory, it has been a fatal mistake for cooks.

Chemical Weapons:
Facility 1, Supplementary 5
Facility 4, Variation 2

Explosives:
A-NPNT, HMTD, LNTA

Regulations: Waste Code U135; Hazard Classes 2.3 and 2.1, Zone B; ERG 117; forbidden on passenger planes; cannot be sent by U.S. mail.

Confirmed by: GasTec Polytec IV tube stripe 4 turns from yellow to brown.

HYDROQUINONE
$C_6H_4(OH)_2$
123-31-9
UN 2662

Appearance: White crystals that are soluble in water. Will be seen in glass bottles and jars; may be sold in 100-lb paper sacks.

Odor: Aromatic.

Hazards: Possible carcinogen. Toxic by ingestion and inhalation; irritant. May cause sensitization.

Incompatibilities: Incompatible with strong oxidizers, strong bases, oxygen, and ferric salts.

Used in

Hobbies:
Photography, developer

Regulations: Hazard Class 6.1; PG III; ERG 153.

HYDROXYLAMINE·HCl
NH₂OH·HCl
5470-11-1

$NH_2OH \cdot HCl$

Appearance: Viscosity 2 liquid. Most likely in 1-gallon, or 4-liter brown glass jugs. Colorless hygroscopic crystals that are soluble in water, usually found in solution. Solution will be acidic.

Odor: None.

Hazards: Toxic by ingestion, causes methemoglobin, which affects kidney and liver. Corrosive, strong irritant to tissue. Vapor pressure negligible.

Incompatibilities: Strong oxidizers, strong bases strong acids.

Used in

Amphetamines:
Synthesis 20

Chemical Weapons:
Facility 7, Variations 1 and 2
Phosgene oxime

Hobbies:
Photography, developer

Regulations: None specific; treat as toxic and corrosive.

HYDROXYLAMINE SULFATE
(NH₂OH)₂·H₂SO₄

$(NH_2OH)_2 \cdot H_2SO_4$

Appearance: Colorless crystals soluble in water. The solution has a corrosive action on the skin.

Odor: None.

Hazards: Irritant, toxic.

Incompatibilities: Strong oxidizers, strong bases, strong acids.

Used in

Hobbies:
Photography, developer

Regulations: None specific; treat as toxic and corrosive.

HYDROXYL-TERMINATED POLYBUTADIENE
A product of Alco Co.; sold as HTPB, R-45 M.

Used in

Explosives:
Plastic explosives C3

Hobbies:
Rocket fuel, Variation 2
Rocket fuel, Variation 8

2-(HYDROXYMETHYL)-2-METHYL-PROPANE-1,3-DIOL (w)

Used in

Explosives:
EGDN

3-HYDROXY-1-METHYLPIPERIDINE (w)
(HO)C₅H₈NCH₃
3554-74-3

$(HO)C_5H_8NCH_3$

Used in

Chemical Weapons:
Facility 13
Buzz

Designer Drugs:
Group 1, Variation 1

Regulations: Australian Group precursor chemical , no specific list.

HYPOPHOSPHOROUS ACID
H_3PO_2
6303-21-5
UN 1053

Appearance: Colorless (very heavy) viscosity 2 liquid. Clear glass jugs with white caps.

Odor: None.

Hazards: Corrosive. Use of this material to reduce ephedrine is associated with high levels of phosphine gas.

Incompatibilities: Strong bases.

Used in

Amphetamines:
Synthesis 1, Variation 4

Regulations: Waste Code D001; Hazard Class 2.3, Zone B, Labels 2.3 and 2.1; ERG 117; forbidden on passenger planes; cannot be sent by U.S. mail.

HYSSOP

Description: An bushy evergreen herb, growing 1 to 2 feet high, with a square stem, linear leaves, and 6 to 15 flowers in whorls.

Used in

Recreational Drugs:
Absinthe

I

ICE
This is usually not a reagent; it is an indication of reaction temperature. Reaction temperatures are NOT described in this book; however, since this one is so obvious, we have included it as an important key point.

Used in

Chemical Weapons:
Facility 1
Facility 8, Variations 1, 2, and 3

Illegal Drugs:
Cook 2

Explosives:
Copper nitride, MNA

Keeping explosives cool during their manufacture is very common.

Toxins:
Ricin

INDOLE
C_8H_7N
120-72-9

Appearance: Colorless-to-yellowish scales that are soluble in water.

Odor: Strong fecal odor.

Hazard: Slightly toxic, NFPA Health 2. Harmful in contact with the skin and if swallowed. May cause serious eye damage.

Incompatibilities: Strong oxidizing agents; iron and iron salts.

Used in

Designer Drugs:
Indole-based-drugs.

Illegal Drugs:
Cook 7

Regulations: Hazard Class 6.1; PG III, label Toxic Solid n.o.s.; ERG 154.

Confirmed by: Indole Test.

INFUSORIAL EARTH
See Diatomaceous earth.

IODINE
I_2
7553-56-2

Appearance: Reagent-grade iodine may be dark purple-gray metallic flakes or pills (some are the size of peas). Iodine is usually purchased from laboratory supply houses in brown glass jars. It is also available from horse veterinarians, where it is used as a disinfectant for shoeing horses. Some people will use hydrogen peroxide to remove iodine from tincture of iodide. Due to the use of iodine in the most common meth method, it is a highly regulated material.

Odor: Strong, biting odor, usually noted only if too close for safety.

Hazards: Iodine is a corrosive and a poison by ingestion. Skin contact will result in discoloration of the skin, a benign condition that can be cleaned off by washing with isopropyl alcohol. Low volatility makes lung and eye irritation unlikely under normal conditions. Vapor pressure 1 mmHg @ 38.7°C. LD(acute) 14 g/kg; PEL 0.1 ppm; teratogen. Iodine is the weakest of the halogen oxidizers.

Incompatibilities: Incompatibility Group 2-A. Acetaldehyde, acetylene carbides, aluminum, ammonia, ammonium salts, lithium, sodium, potassium, and sodium hydride.

Used in

Amphetamines:
Synthesis 1
Synthesis 18
Synthesis 18, Variation 6
Synthesis 21, Variation 2

May be found as a contaminant of a red hydriodic acid made with iodine and red phosphorus. Iodine is used primarily to produce hydriodic acid (Synthesis 1), which has become increasingly difficult to obtain due to regulation. Hydriodic acid is produced by placing iodine and red phosphorus in water under a hydrogen blanket in a reaction vessel. Dry iodine with red phosphorus and ephedrine in a sealed vessel (usually, a sports cup) is used in Synthesis 1, Variation 3, or with hypophosphorous acid in Synthesis 1, Variation 4. Iodine enhances the reaction of the Grignard reagent to produce methamphetamine (Synthesis 21). Synthesis 18, Supplementary 3.

Chemical Weapons:
Facility 2

Explosives:
Ammonium triiodide, TND, TNEN

Illegal Drugs:
Cook 3, Variation 2
Cook 4, Supplementaries 1 and 2

Regulations: Iodine is shipped as a poison. Hazard Class 6.1. Iodine is listed as a methamphetamine precursor.

Confirmed by: Iodine Test.

IRIDIUM-192
Radioactive material
Ir

Appearance: A platinum group metal which resembles platinum but with a slight yellowish cast.

Emits: Beta rays.

Half-life: 74 days.

Applications: This is another element that is not available in large quantities: It is used in nondestructive testing in the oil industry, and in the airline industry. It has some medical applications.

Used in:

Dirty Bombs

Regulations: Radioactive elements are highly regulated and have specific requirements for both transportation and disposal.

IRIDIUM(IV) CHLORIDE (w)
$IrCl_4$

Appearance: brownish-black hygroscopic mass that is soluble in water.

Used in

Hobbies:
Photography

IRON (Iron Filings)
Fe
7439-89-6

Appearance: Iron appears as black, sometimes slightly rusty or slightly shiny powder, granules, or filings.

Odor: None.

Hazards: Primary hazard is as a flammable solid, but only if in the form of a very fine powder; otherwise, incompatible storage poses the only real risk.

Incompatibilities: Aluminum and hydrogen peroxide greater than 50%.

Used in

Amphetamines:
Synthesis 15

Explosives:
PNT

Incendiary Devices:
Situation 5

Illegal Drugs:
Cook 4, Supplementary 2

Regulations: No specific regulations. Large amounts of very fine iron can ignite and are difficult, if not impossible, to extinquish.

Confirmed by: Iron Magnet Test.

IRON OXIDE
See Ferrous oxide *and* Ferric oxide.

IRON OXIDE (Black)
1309-37-1

IRON OXIDE (Red)
1332-37-2

IRON SULFIDE
FeS
1317-37-9

Appearance: Black-to-metallic chunks, graduals, or rods that are insoluble in water. However water can release some H_2S gas. Also called iron(II) sulfide and ferrous sulfide.

Odor: Rotten eggs. May be some rotten egg odor associated with the material, especially if it becomes wet. Totally dry, there will be no odor associated with the product.

Hazards: Primary hazard is as a flammable solid, but only under certain conditions; otherwise considered as nonflammable. Incompatible storage poses the only real risk. Not particularly toxic, NFPA Health 1. However, even weak acids produce hydrogen sulfide gas. *See* Hydrogen sulfide.

Incompatibilities: Should not be stored with acids (gives off H_2S gas), or lithium metal (can explode). Both could be used in Synthesis 2.

Used in

Amphetamines:
Synthesis 1, Supplementary 1

Preparation of hydriodic acid from water and iodine. Added to sulfuric acid to generate hydrogen sulfide gas to make hydriodic acid.

Regulations: None specific.

Confirmed by: Iron Magnet Test; Sulfide Test.

IRON WOOL
See Steel wool.

ISOAMYL ALCOHOL
$(CH_3)_2CH\cdot CH_2CH_2\cdot OH$
123-51-3

Appearance: Colorless clear liquid that is soluble in water.

Odor: Characteristic disagreeable odor that can be choking.

Hazards: Toxic by inhalation. NFPA Health 1, Fire 2. Combustible, EPA flammable, flash point 109–114°F.

Incompatibilities: Reactive with reducing agents.

Used in

Recreational Drugs:
Rush, Variation 2

Regulations: Hazard Class 3, label Flammable Liquids n.o.s.; PG II.

Confirmed by: Alcohol Test.

ISOAMYL NITRITE
See Amyl nitrite.

ISOBUTANOL
$(CH_3)_2CHCH_2OH$
78-83-1

Appearance: Clear oily liquid. Partially soluble in water. This is a laboratory chemical and will come from a laboratory supply house. May be sold in drums.

Odor: Sweet.

Hazards: Moderately toxic by ingestion and inhalation. Vapor pressure 9 mmHg @ 20°C, TLV 50 ppm in air. Strong irritant! Flash point 100°C. May cause narcosis.

Incompatibilities: Strong oxidizers, aluminum.

Used in

Illegal Drugs:
Cook 3

Regulations: Hazard Class 3; Waste Code U140.

Confirmed by: Alcohol Test.

ISOBUTYL PROPANE
See 2-Amino,2-methyl propane.

ISOCYANOGEN TETRABROMIDE (w)

Used in

Explosives:
Tetrazide

ISO-HEET®
See Isopropyl alcohol.

ISOPENTANE
C_5H_{12}
78-78-4
UN 1265

Appearance: Viscosity 1 liquid solvent that separates and floats on water. *Also see* Nonpolar liquids that float. This is a standard C5 hydrocarbon.

Odor: Solvent.

Hazards: Flammable. Will defat the skin. NFPA Health 1, Fire 4.

Incompatibilities: Strong oxidizers.

Used in

Explosives:
AS-20

Regulations: Hazard Class 3.1; PG I. There is no established RQ.

Confirmed by: Determination by iodine.

ISOPHORONE DIISOCYANATE
A product of VEBA Chemical Co.

Used in

Hobbies:
Rocket fuel, Variation 2

ISOPROPYL ALCOHOL
$CH_3CHOHCH_3$
167-63-0
UN 1219

Appearance: Colorless viscosity 1 liquid. Most likely to be found as rubbing alcohol, 70% or 90%. Isopropyl alcohol can be purchased in plastic 1-pint bottles from a drugstore. Laboratory supply houses can provide this product in almost pure form in clear glass bottles or plastic jugs. Has far too many legal applications to provide much of a clue by itself.

Odor: Rubbing alcohol!

Hazards: Primary hazard is flammable liquid, flash point 53°F; LEL 2%; UEL 12.7%; LD_{50} 5.8 g/kg; PEL 400 ppm.

Incompatibilities: Incompatibility Group 3-A. Chromium trioxide and nitric acid.

Used in

Amphetamines:
Synthesis 18, Variation 6

Isopropyl alcohol is a polar solvent that can be used for the extraction of ephedrine. *Also see* Alcohol (Nonspecific).

Chemical Weapons:
Facility 1

Explosives:
ADN, CDNTA, KDN, MNTA, NTA, nitroform

Illegal Drugs:
Cook 4, Supplementary 1

Regulations: Waste Code D001; Hazard Class 3; PG II; ERG 129.

Confirmed by: Alcohol Test.

ISOPROPYLAMINE
$(CH_3)_2CH \cdot NH_2$
75-31-0
UN 1221

Appearance: Colorless viscosity 1 flammable liquid that is a strong base and miscible in water.

Odor: Ammonia; odor threshold around 1 ppm.

Hazards: Flammable, flash point −35°F. NFPA Fire 4. Very irritating by inhalation, ingestion, and eye or skin contact. Absorbed through the skin. Vapor pressure 478 mmHg @ 20°C. Causes burns.

Incompatibilities: Strong acids; contact will cause explosive spattering. Contact with strong oxidizers may cause fires and explosions; specifically, perchloryl fluoride, or 1-chloro-1,3-epoxypropane. TLV 5 ppm.

Used in

Chemical Weapns:
Facility 1, Variations 1 and 2
Facility 1, Soman

Used as a scavenger for the binaries of sarin/soman.

Regulations: Hazard Class 3; ERG 132.

ISOPROPYL CHLORIDE
$CH_3 \cdot CHCl \cdot CH_3$
75-29-6
UN 2356

Appearance: Colorless liquid, slightly soluble in water.

Hazards: Flammable, flash point −26°F. Irritant, harmful by inhalation and ingestion, burns to produce phosgene.

Special considerations: Cool to −32°F before opening the container. High vapor pressure.

Incompatibilities: Oxidizers; strong bases.

Used in

Chemical Weapons:
Sarin-isopropyl

Regulations: Hazard Class 3; PG I; ERG 129.

ISOPROPYL ETHER
$(CH_3)_2CH \cdot OCH(CH_3)_2$
108-20-3
UN 1159

Appearance: Colorless liquid that is slightly soluble in water; will float on water.

Odor: Not found.

Hazards: Extremely flammable, flash point −18°F. Irritant, narcotic in high concentrations. TLV 250 ppm.

Incompatibilities: Oxidizers. Air (isopropyl ether) readily forms peroxides on storage. These concentrate over time, creating a very explosive compound.

Used in

Chemical Weapons:
Facility 1
Facility 1, Soman

Regulations: Transport Category 2, Hazard Class 3; PG II; ERG 127.

ISOPROPYL NITRITE (w)
$C_3H_7NO_2$
541-42-4

Appearance: Clear pale yellow liquid. Made with an alkyl nitrite and isopropanol.

Odor: Not found.

Hazards: A chemical that can be used as a popper.

Incompatibilities: Oxidizers.

Used in

Explosives:
Sodium nitrite

J

JEQUIRITY BEANS
Poisonous plant

The jequirity bean is not native to the United States and should not be found unless imported illegally. Tourists will sometimes bring in the beans in handmade jewelry purchased in Brazil. There is no reason ever to find the entire plant.

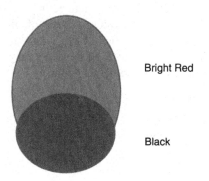

Bright Red

Black

Appearance: The jequirity bean is very hard, bright red and black, about the size of a pea or pinto bean.

Hazards: Very toxic. Considered 75 times more toxic than ricin. Effects may not appear until 8 hours after exposure.

Used in

Toxins:
Abrin

Regulations: It is illegal to bring agricultural products, especially seeds, into the United States without inspection and permission.

JIMSONWEED
Poisonous plant

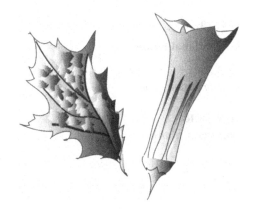

Description: Also called *Jamestown weed, gypsum weed, locoweed, thorn apple, zombie's cucumber*. 30 to 150 cm tall with erect, forking, purple stems. The leaves are large, 7 to 20 cm long, and have irregular teeth similar to those of an oak leaf. The flowers are one of the most distinctive characteristics. They are trumpet-shaped, white to purple, 5–12 cm long. All parts of the plant emit a foul odor, especially when crushed or bruised. The poison is made up of tropane alkaloids, atropine, hyoscyamine, and scopolamine. Tropanes make up as much as 0.7% of the plant.

Where found: Found in most of the continental United States, from New England to Texas, Florida to the far western states. Common in overgrazed pastures, barnyards, and wasteland, preferring rich soils and warm climates.

Symptoms: There is a mnemonic: "blind as a bat, mad as a hatter, red as a beet, hot as hell, dry as a bone, the bowel and bladder lose their tone, and the heart runs alone." In case of overdose, the person will experience hypothermia, coma, respiratory arrest, seizures, hallucinations, and death.

Antidotes: Is treated symptomatically.

Regulations: Consumption is illegal in some states. The FDA in 1968 banned over-the-counter sales of products prepared from jimson-weed. The plant is generally considered too toxic for medical applications.

Field test: Slowly becomes red in 0.1 N potassium permanganate.

Used in

Recreational Drugs

Toxins:
Datura stramonium

JUNIPER
Poisonous plant

Description: The leaves have two forms: juvenile needle-like leaves and adult scale

leaves. Juniper is a shrubby plant, variable in shape, reaching 1–4 m tall. Grows best at 1000 to 3300 m. The cones are berry-like. This is probably the least toxic of the poisonous plants discussed. At low doses it enhances water loss and encourages menstruation to start. The cones are used to make gin.

Where found: Everywhere.

Deadly parts: All parts are poisonous.

Symptoms: At high doses, it causes convulsions. Death from respiratory arrest occurs within 10 hours to several days. On the skin, the oil causes blisters and sometimes necrosis. When swallowed, the irritant causes gastroenteritis with hemorrhages and vomiting of greenish masses with an ether-like odor. Polyurea may occur with bloody urine, followed by oliguria and anuria, convulsive coma, and acute kidney problems.

Antidotes: Milk to protect from gastric irritation, then stomach lavage and vomiting to remove the material.

Used in

Hobbies:
Gardening

Toxins:
Juniperus sabina

K

KELLER REAGENT
A mixture of glacial acetic acid and ferric chloride is used to test for the potency of psilocybe mushrooms. There is a product used in metallurgical engineering called Keller's reagent, which consists of nitric acid, hydrochloric acid, and hydrofluoric acid. That reagent is NOT used in any of the formulas noted in this book

Made in

Illegal Drugs:
Cook 5, Supplementary 4

KEROSENE
Mixed hydrocarbons with formulas near
$C_{13}H_{54}$
8008-20-6
UN 1223

Appearance: Clear viscosity 2 liquid insoluble in water. *Also see* Nonpolar liquids that float.

Odor: Oily.

Hazards: Combustible. Flash point from 100 to 150°F, depending on the mixture. Kerosene is moderately toxic by inhalation.

Incompatibilities: Strong oxidizers, especially the halogens, with which it forms explosives.

Used in

Explosives:
Diazodinitrophenol

Illegal Drugs:
Cook 1

Regulations: Hazard Class 3; ERG 128.

Confirmed by: Determination by iodine.

KETAMINE
Date rape
Over-the-counter prescription used as a date rape drug.

KHAT
See Catha edulis.

KRATON 1107
Plastic elastimer

Used in

Hobbies:
Rocket fuel, Variation 3

L

LACTOSE
$C_{12}H_{22}O_{11} \cdot H_2O$
63-42-3

Appearance: Milk sugar. White crystalline powder slowly soluble in water. *See* Sugar.

Odor: None.

Hazards: This is milk sugar.

Incompatibilities: All sugars are very good fuels, and therefore incompatible with oxidizers. If they do not react immediately, they will form explosive mixtures.

Used in

Amphetamines:
See Chemicals used for final cut.

Fireworks:
Smoke dye formulations

Regulations: None specific.

Confirmed by: Sugar Test.

LAMPBLACK
Also see Carbon *and* Charcoal.

Appearance: Very fine black powder.

Odor: None.

Hazards: Treat as nuisance dust. PEL 10 mg/m³.

Incompatibilities: Relatively inert. Can be part of explosive oxidizer fuel mixtures.

Used in

Fireworks:
Burnrate reducer

Hobbies:
Photography; drawing

Regulations: Paint pigment n.o.s.

Confirmed by: Carbon monoxide indicator tube.

LANOLIN

Wool fat, anhydrous lanum, oesipes, agnin, alapurin, agnolin, lanalin, laniol, lanain, lanesin, lanicholo. Purified wool fat of sheep.

Appearance: Yellowish, tenacious semisolid fat.

Odor: Very slight, if any.

Hazards: Used in lotions.

Incompatibilities: Strong oxidizers.

Used in

Hobbies:
Photography

Regulations: None specific.

LAVENDER OIL

Used in

Hobbies:
Photography

LEAD ACETATE

$Pb(C_2H_3O_2)_2 \cdot 3H_2O$
7439-92-1
UN 1616

Appearance: Reagent grade may be white crystals or flakes; however, may be yellow or brown and can become very lumpy. Lead acetate will be purchased from a laboratory supply house. The container should feel heavy.

Odor: None.

Hazards: Primary hazard is poison by ingestion, or skin absorption. Any form of organic lead is very dangerous, as it can easily be absorbed into the body chemistry. Lead is a suspected carcinogen and a known reproductive hazard. Levels of lead exposure that are tolerable to adults can be destructive to children. Permanent retardation can occur from exposure to lead. TD_{low} 8.52 g/kg, PEL 0.05 mg/m³, and 30 μg/whole liter blood.

Incompatibilities: Reacts with perchloric acid to form explosives.

Used in

Amphetamines:
Synthesis 12

Lead acetate is a reagent in the production of P-2-P from phenyl acetic acid in a dry distillation, Dark blue-gray to black paste in a reaction vessel. The vessel may be damaged, as the reaction is destructive to vessels, which can only be used about three times.

Explosives:
Lead azide

Regulations: RQ 4.54 kg; Waste Codes D008 and U144; Hazard Class 6.1; PG III; ERG 151.

Confirmed by: Lead Test.

LEAD(II) HYDROXIDE

$Pb_2O(OH)_2$

Appearance: White powder; does not dissolve in water, does absorb CO_2 from the air. The only product offered for sale was called lead sub-carbonate, $(PbCO_3)_2Pb(OH)_2$, which is basic lead carbonate, lead carbonate hydroxide. 1319-46-6

Odor: None.

Hazards: Poisonous!

Incompatibilities: Strong oxidizers and strong acids.

Used in

Explosives:
LNTA

Regulations: Nonhazardous for air, sea, and road freight. One of the EPA's TCLP chemicals.

Confirmed by: Lead Test.

LEAD NITRATE

$Pb(NO_3)_2$
1009-74-8
UN 1469

Appearance: White or colorless translucent crystals that are soluble in water.

Odor: None.

Hazards: Poisonous by ingestion and inhalation! Oxidizer.

Incompatibilities: May react explosively with hydrocarbons and organic compounds. May form explosive compounds with ammonium thiocyanate potassium acetate, or lead hypophosphite.

Used in

Explosives:
Lead azide, Variation 2; lead nitroanilate; lead picrate; lead styphnate; lead TNP

Regulations: Hazard Classes 5.1 and 6.1; marine pollutant; ERG 141; One of the EPA's TCLP chemicals.

Confirmed by: Lead Test.

LEAD OXIDE
Pb_3O_4
1314-41-6
UN 1479

Appearance: Bright red-to-orange heavy powder that is insoluble in water. Red lead was used as paint pigment; however, it is not as common as it used to be.

Odor: None.

Hazards: Poisonous if inhaled or swallowed!

Incompatibilities: May react vigorously with reducing agents.

Used in

Chemical Weapons:
Facility 13

Designer Drugs:
Group 1, Variation 3

Explosives:
Nitrogen trichloride
Lead rods from battery

Fireworks:
Goldschmidt thermite

Regulations: Hazard Class 5.1; PG II, ERG 140; oxidizing solid n.o.s. One of the EPA's TCLP chemicals. Persistent in the environment.

Confirmed by: Lead Test.

LEAD THIOCYANATE
$Pb(SCN)_2$
502-87-0
UN 3802

Appearance: White crystalline powder that is insoluble in water (0.05%) and decomposes in hot water. Used in safety matches. Becomes increasingly yellow on exposure to light.

Odor: None.

Hazard: Poison! Irritation to the eyes and skin on contact. Lead salts are very toxic to a fetus.

Incompatibilities: Strong reducing agents and finely powdered metals.

Used in

Ammunition:
Load 8

Regulations: Hazard Class 9; PG III; ERG; one of the EPA's TCLP chemicals.

Confirmed by: Lead Test.

LECITHIN
$C_{43}H_{88}O_9NP$

Appearance: Brownish-yellow waxy hygroscopic mass. It is insoluble in water, but swells in water. Sold as Emultex R by A.E.Staley Mfg. Co. A mixture of the diglycerides of stearic, palmitic, and oleic acids, linked to the choline ester of phosphoric acid. Made from eggs. Lecithin is available from health food stores and drugstores.

Incompatibilities: Strong oxidizers.

Used in

Explosives:
Plastic explosives, Variation 7

Hobbies:
Rocket fuel, Variation 4

Regulations: None specific.

LEMONGRASS OIL
See Citral.

LETTUCE
Opium can be obtained by boiling and concentrating endive or escarole lettuce hearts. This is more of a folk recipe than a production procedure.

LIDOCAINE
$C_6H_3(CH_3)_2NHCOCH_2N(C_2H_5)_2$

Appearance: White or slightly yellow crystalline powder that is insoluble in water.

Odor: Characteristic odor.

Hazards: USP grade used in medicine as a local anesthetic.

Incompatibilities: Strong oxidizers.

Used in

Amphetamines:
See Chemicals used for final cut

Regulations: None specific.

LIGHT BULBS (Colored)
Generally, we avoid listing equipment, but the manufacture of LSD is very dependent on special lighting, which is an obvious indicator. Red, yellow, and black lights are a necessary part of the operation. Purple filters in front of tungsten bulbs can be used in place of black lights.

LIGHTER FLUID
Cigarette lighter fluid usually comes in small tins from any market (e.g., Ronson®, Zippo®). The liquid is a mixture of C_5 hydrocarbons, similar to Ligroin and petroleum ether. *See* Petroleum ether.

LIGROIN
CnHn near C_5
8032-32-4
UN 1271
Ligroin is a product name for petroleum distillates and petroleum ether. *Also see* Nonpolar liquids that float.

Appearance: Colorless liquid with a hydrocarbon odor. Not soluble and floats in water. Boiling point 35–60°C. Floats on water with a sharp curved line. Flash point varies but is typically –48°C. Consists mostly of *n*-pentane and hexane, with small amounts of methylpentane, cyclopentane, dimethylbutane, and other short-chain hydrocarbons. This material is changeable for many other products, such as hexane and pentane.

Hazards: The primary hazard is flammability, flash point –7° F; LEL 1.1%; UEL 7.5%; Group D atmosphere. Intoxication by inhalation; absorbed through the skin; the most likely skin contact problem is drying of the skin, which can cause cracking and infection. LD_{50} 28.7 g/kg; PEL 500 ppm.

Incompatibilities: Incompatibility Group 6-B. Chromium trioxide.

Used in

Amphetamines:
Like all hydrocarbons, this could be used as a nonpolar solvent to extract methamphetamine from the cooking process.

Illegal Drugs:
This product was listed specifically for production of LSD; however, many other products could replace this in the synthesis.

Regulations: RQ 2270 kg; Waste Code D001; Hazard Class 3; PG II; ERG 128.

Confirmed by: Determination by iodine.

LILY OF THE VALLEY
Poisonous plant

May lily; our lady's tears.

Description: The stems grow to 30 cm tall, with two leaves to 25 cm long, and a raceme of 5 to 15 flowers on the stem apex. Flowers are white bell-shaped, sweetly scented. The fruit is a small red berry.

Where found: Western North America, northern Rocky Mountain states, Pacific coast, and the midwest.

Deadly parts: The entire plant is deadly, especially the leaves.

Symptoms: Immediate. There about 20 toxins, glycosides, such as convalatoxin, convalarin, and convalamarin, as well as saponins. Hot flashes, tense irritability, headache, hallucinations, red skin patches, cold clammy skin, dilated pupils, vomiting, stomach pains, excess urination, nausea, excess salivation, and slow heartbeat, sometimes leading to coma and death from heart failure.

Antidotes: Stomach lavage is recommended as well as cardiac depressants such as quinidine to control cardiac rhythm.

Used in

Hobbies:
Gardening

Toxins:
Convallaria majalis

LINALOOL
See Citral.

LINDANE
See Benzene hexachloride.

LINSEED OIL
8001-26-1

Appearance: Sticky yellow viscosity 6 oil.

Odor: Strong odor, like oil-based paints.

Hazards: Can combust spontaneously on bunched-up-rags.

Incompatibilities: Oxygen, oxidizers.

Used in

Fireworks:
Coating, sparks

Incendiary Devices:
Situation 9

LITHARGE
See Lead monoxide.

LITHIUM
Li
7439-93-2
UN 1415

Appearance: Lithium metal is a shiny soft silvery metal when fresh. It will discolor to black as you watch it once exposed to the air.

Odor: None.

Hazards: Primary hazard is water reactive. Lithium is the least reactive of the alkali metals (the reaction is a strong, quick effervescence). Many people in the field report only a hissing sound when lithium is in contact with water as opposed to the open flame seen with sodium and potassium. The resulting solution is very caustic. The vapors escaping from the water reaction are very caustic and will cause coughing. Free lithium may be found as part of a sludge in meth lab wastes.

Incompatibilities: Incompatibility Groups 2-A and 3. Air, chloroform, chromium trioxide and heat, hydrogen, iodine, nitric acid, and water.

Used in

Amphetamines:
Synthesis 2, Variations 1 and 2

Usually found as thin strips of foil removed from over-the-counter lithium batteries. Usually, the ribbon is cut as it is pulled from the battery and dropped into a solvent. It is shiny when first removed, but slowly becomes dark dull gray. May be found in white powder (ephedrine) in plastic baggies used to store the lithium and ephedrine ready for use. During the process the mixture turns blue in an open dish when mixed with anhydrous ammonia. The dish also contains ephedrine, ammonia, and methamphetamine. Lithium hydroxide is the end product of this reaction and is likely to be part of the waste stream of a Nazi lab. Lithium is a catalyst in the hydrogenation of ephedrine to methamphetamine (Synthesis 2). It is also used in the neo-Nazi method, with sodium hydroxide and an ammonium salt in a pressure plastic pop bottle.

Illegal Drugs:
Cook 4, Variation 3

Regulations: Waste codes D001 and D003; Hazard Class 4.3 (lithium metal), Class 9 (lithium batteries); PG I; ERG 138; cannot be sent by U.S. mail; 5 kg of lithium batteries is allowable on passenger planes.

Confirmed by: Flame Test.

LITHIUM ALUMINUM HYDRIDE
$LiAlH_4$
16853-85-3
UN 1410

Appearance: Reagent-grade lithium aluminum hydride may be white or yellow-gray powder. This commodity is stored in brown glass chemical reagent jars. The lid will be very tightly sealed, perhaps waxed or taped. Lithium aluminum hydride is used to add hydrogen to a compound. It is a reducing agent in organic chemistry.

Odor: None.

Hazards: Primary hazard is water reactivity and an extremely flammable solid.

Incompatibilities: Moist or warm air, alcohol, liquid acids, ethyl ether, and water.

Used in

Amphetamines:
Synthesis 14
Synthesis 18, Variation 5

Although it is difficult to obtain lithium aluminum hydride, it is a valuable reagent in the clandestine laboratory. Lithium aluminum hydride is a component used for the generation of hydrogen gas for chemical blankets for the thionyl chloride chloroform hydrogenation method (Synthesis 3) and could be used in any hydrogenation method. It is especially important in reactions where phenyl 2-chloropropane or other chlorinated by-products or intermediates are formed, as it works as a chemical scavenger for these compounds. It can also be used in a tube furnace (Synthesis 14).

Illegal Drugs:
Cook 5
Cook 6, Variation 1
Cook 7

Regulations: Waste Codes D001 and D003; Hazard Class 4.3, label Dangerous When Wet; PG I; ERG 138; forbidden on passenger planes; cannot be sent by U.S. mail.

Confirmed by: Flame Test; Aluminum Test.

LITHIUM BATTERIES
UN 3090

Lithium batteries are a source of lithium metal. *See* Lithium.

LITHIUM CARBONATE
Li_2CO_3
554-13 2

Appearance: White light alkaline powder that is slightly soluble in water.

Odor: None.

Hazards: In small amounts, is used medicinally; ingestion of large amounts can lead to kidney failure.

Incompatibilities: Water acids, effervesces. Incompatible with fluorine.

Used in

Fireworks:
Flash delay

Hobbies:
Ceramics

Regulations: None specified.

Confirmed by: Flame Test.

LITHIUM HYDROXIDE
$LiOH \cdot H_2O$
1310-66-3
UN 2680

Appearance: White caustic soluble powder.

Hazards: Strongly alkaline. Toxic by ingestion and inhalation. Causes burns. NFPA Health 3.

Special: Air sensitive; must be stored under nitrogen or other inert gas.

Incompatibilities: Carbon dioxide; acids.

Used in

Illegal Drugs:
Cook 2, Variations 5 and 8

Regulations: Hazard Class 8; PG II; label Lithium Hydroxide, Solid; ERG 154. Disposal by incineration.

LITHIUM HYPOCHLORITE
LiClO
13840-33-0
UN 1471

Appearance: White granular material. May be purchased from a swimming pool supply house as SHOCK in plastic bags, or as a swimming pool sanitizer from Biogard®, or from a laboratory supply house in brown glass jars.

Odor: Chlorine.

Hazards: Strong oxidizer. Corrosive. Causes irreversible burns to eyes and skin burns. Releases chlorine gas. Fatal if swallowed.

Incompatibilities: Acids (will release chlorine gas), Organic compounds, especially alcohols (will spontaneously burst into flame), ammonia and ammonium salts (will release chlorine gas).

Used in

Illegal Drugs:
Cook 2, Variation 8

Incendiary Devices:
Situation 1

Regulations: Hazard Class 5.1; label Lithium Hypochlorite, Dry; PG II; ERG 140.

Confirmed by: Oxidizer Test.

LITHIUM OXALATE
$Li_2C_2O_4$
553-91-3

Appearance: White crystals that are soluble in water.

Odor: Not found.

Hazards: Toxic by ingestion. NFPA Health 2.

Incompatibilities: Oxidizers.

Used in

Fireworks:
Flash delay

Regulations: None specific.

Confirmed by: Flame Test.

LITHIUM PERCHLORATE
$LiClO_4$
7791-03-9

Appearance: Colorless deliquescent crystals that are soluble in water. Upon standing, may form $LiClO_4 \cdot 3H_2O$.

Odor: Not found.

Hazards: Oxidizer. Explosive and/or fire hazard in contact with organic compounds. Causes irritation to the skin, mucous membranes, and respiratory system on inhalation.

Incompatibilities: Organic materials. Strong reducers. Can form explosives in contact with metals and metal salts and organic materials.

Used in

Fireworks:
Strobe formulations

Regulations: Hazard Class 5.1; ERG 140.

Confirmed by: Perchlorate Test; Flame Test.

LIVER OF SULFUR
See Potassium sulfide.

d-LYSERGIC ACID
The user must prepare any lysergic acid. *See* Cook 2, Supplementary 2. *Also see* Ergot, Hawaiian baby word rose, Morning glory seeds, and Rye grain.

M

MAGNALIUM
MgAl
A metal alloy made specifically for the fireworks industry.

Used in

Fireworks:
Colored stars, dark and flash effects, strobe effect

MAGNESIUM
Mg
7439-95-4
UN 1869 (chunks)
UN 1418 (powder)

Appearance: Shiny metallic or dull gray metal chips, filings, or gray powder. Magnesium turnings or chunks are usually an indicator of a Grignard reagent. Grignard reactions are addition reactions, where the site of the chlorine atom is the point of addition. Very fine magnesium could be an enhancer for explosives or incendiary devices. Magnesium is available from surplus/camping stores as fire starters.

Odor: None.

Hazards: Magnesium is a flammable solid that will effervesce and give off hydrogen gas in weak acids. Flammable solid.

Incompatibilities: Incompatibility Group 2-A. Chloroform, hydrogen iodide, hydrogen peroxide, and nitric acid.

Used in

Amphetamines:
Synthesis 18, Variation 6
Synthesis 20, Supplementary 8
Synthesis 21, Variations 1 and 2

Chemical Weapons:
Vinylarsine
Methyldichloroarsine
Ethyldichloroarsine

Fireworks:
Colored stars, sparks, strobe formulations, dark and flash effects

Hobbies:
Rocket fuel, Variation 10

Incendiary Devices:
Situation 5
Situation 11

Illegal Drugs:
Cook 3, Variations 1 and 2
Cook 4, Supplementaries 1 and 2

Explosives:
Astrolite G
As an accelerant in almost any explosive

Regulations: Waste Code D001; Hazard Class 4.1; PG III; ERG 138; powder forbidden on passenger planes; cannot be sent by U.S. mail.

Confirmed by: Magnesium Test.

MAGNESIUM CARBONATE
$(MgCO_3)_4Mg(OH)_2 \cdot 5H_2O$
23389-33-5

Appearance: White bulky powder or light friable masses, slowly soluble in water.

Odor: None.

Hazards: None. May act as a skin, eye, or respiratory irritant at high dust levels; avoid breathing dust.

Incompatibilities: Effervesces in acids.

Used in

Explosives:
Lead styphnate

Fireworks

Regulations: Nonhazardous for air, sea, and road freight.

Confirmed by: Magnesium Test.

MAGNESIUM COPPER
MgCu
An alloy manufactured specifically for the fireworks industry.

Used in

Fireworks:
Dark and flash effects

MAGNESIUM OXALATE
$MgC_2O_4 \cdot 2H_2O$
547-66-0
UN 3077

Appearance: White powder that is not soluble in water.

Odor: None.

Hazards: Oxalates are very toxic; however, because magnesium oxalate is so very slightly soluble in water, it acts more like an irritant. It is very irritating to the eyes, causing redness, and causes scaling and itching on the skin.

Incompatibilities: Strong oxidizing agents.

Used in

Fireworks:
Flash delay

Regulations: Environmentally Hazardous Substance, Solid, n.o.s. (Magnesium Oxalate); Hazard Class 9; PG III. Store away from food and beverages. ERG 171.

MAGNESIUM SULFATE
$MgSO_4$
7587-88-9

Appearance: Clear-to-white crystals. The hydrated (contains water) crystals are known as Epsom salts. Epsom salts are suitable for use as a desiccant after driving off the water by heating from 150°F to 200°F. Epsom salts have many uses and are available from drug or hardware stores. Reagent grade will be in plastic jars, probably 500 g or more, or in brown glass jars from a chemical supply house.

Odor: None.

Hazards: None; used as Epson (bath) salts.

Incompatibilities: None.

Used in

Amphetamines:
Synthesis 0.7, Variations 1 and 2
Synthesis 1, Supplementary 5
Synthesis 15, Supplementary 3

Scavenger in extraction solvent to remove water. Not used much, as the percent of yield is not significantly better.

Explosives:
ADBN, ADNB, A-NPNT, BDPF, DANP, DMMD, NENA, NTND, PNT, TBA, TNA, TND, TNEN

Illegal Drugs:
Cook 4, Variations 1 and 2

Tear Agents:
Acrolein
Tear gas

Regulations: None specific.

Confirmed by: Sulfate Test; Magnesium Test.

MALONAMIDE
$CH_2(CONH_2)_2$
108-13-4

Appearance: White powder that is soluble in water. Sold in 100-g poly bottles.

Odor: None reported.

Hazards: NFPA 0, 0, 0. May cause eye and skin irritation.

Incompatibilities: Strong oxidizers.

Used in

Explosives:
TNM

Regulations: No shipping requirements; not considered hazardous waste.

MALONITRILE
$(CN)_2CH_2$
109-77-3
UN 2647

Appearance: White-to-light yellow crystals that are soluble in water.

Odor: None.

Hazards: Hazardous if swallowed; will cause skin irritation.

Incompatibilities: Oxidizing agents, reducing agents, strong acids, and strong bases.

Used in

Tear Agents:
o-Chlorobenzylidene malonitrile, pepper spray

Regulations: Hazard Classes 6.1 and 8; ERG 153.

MALTOSE (Anhydrous)
69-79-4
Also see Sugar.

Used in

Explosives:
MON

MAMATOL®
A baby laxative
Also see Mannitol.

Used in

Amphetamines:
See Chemicals used for final cut

MANDELIC ACID
$C_6H_5CH(OH)COOH$
611-71-2

Appearance: White crystals or powder moderately soluble in water.

Odor: Faint.

Hazards: Harmful if swallowed; irritant. Mandelic acid is used as a medicine.

Incompatibilities: Incompatible with strong bases and strong oxidizing agents.

Used in

Amphetamines:
Synthesis 20, Supplementary 4

Regulations: Nonhazardous for air, sea, and road freight.

MANGANESE(III) ACETATE
(CH3CO2)3Mn·2H2O
19513-05-4

Appearance: Solid. Made with cobalt(II) acetate and excess mananese(II) acetate or manganese(II) acetate salt, manganese oxides, or manganese hydroxides.

Odor: None found.

Hazards: Oxidizer; irritant. This is a mild selective oxidizer.

Incompatibilities: Strong oxidizers.

Used in

Amphetamines:
Synthesis 17

MANGANESE DIOXIDE
MnO_2
1313-13-9

Appearance: Very dark purple crystalline grains, or very fine black powder, that is not water soluble.

Odor: None.

Hazards: Oxidizer. The oxidizing reactions with this material tend to be very slow. Inhalation may cause metal fume fever. Is poorly absorbed; however, may cause abdominal pain. Industrial grades may include lead.

Incompatibilities: Hydrochloric acid; gives off chlorine gas. Easily oxidized materials, such as sulfur, sulfides, phosphides, hypophlophites, chlorates, peroxides, aluminum powder. Mixing with organic materials and heating may cause fires.

Used in

Explosives:
Nitrogen sulfide

Fireworks:
Catalyst

Regulations: Not regulated.

MANGANOUS CHLORIDE
$MnCl_2$

Appearance: Rose-colored deliquescent crystals that are very soluble in water. Sold commercially in drums. Most likely to be found in a glass jar from a laboratory supply house.

Used in

Amphetamines:
Synthesis 14

Used in the tube furnace in place of thorium oxide.

MANNITOL
$C_6H_8(OH)_6$
69-6-8

Appearance: White crystalline very soluble solid. This is considered to be a sugar, although the body cannot absorb it very well. It is therefore used as a sugar substitute for diabetics. It is also considered to be a laxative. It is a sorbitol isomer. Will probably be found as a medical supply or a dietary supplement. Mamatol (*see* above) may very well be this product, translated through a tweaker. I could not find the product Mamatol, but available, it might be the source of mannitol.

Odor: None.

Hazards: Used as a children's laxative and as a chemical used in brain surgery.

Incompatibilities: Strong oxidizers.

Used in

Explosives:
Nitromannitol

Regulations: None specific.

MARIJUANA
8063-14-7

Description: Marijuana grows 3–10 feet tall and has hairy leaves divided into 5 to 7 serrated leaflets; the leaves are often sticky with resin. The plants are distinctly male or female.

Symptoms: Heavy long-term use causes physical problems. Side effects include impaired short-term memory, impaired ability to complete complicated tasks, a sense of time slowing down.

Used in

Illegal Drugs:
Cook 4, Variations, 4, 5 and 6

Regulations: Although regulated heavily by the U.S. government, in many states there are less restrictive statutes. The amount allowable, or degree of criminality for a given amount, varies, but it is usually around 1 oz.

MATCHES
UN 1944

There are two aspects to matches which make them attractive to people intending to break the law. The match heads themselves contain sesquisulfide sulfur compounds to burn, and strike-anywhere matches contain red phosphorus, as do the striker plates on match books and boxes. Each of these components has a value to people manufacturing illicit materials. *Also see* Phosphorus sesquisulfide *and* Red phosphorus.

Hazards: Matches are extremely flammable solids.

Used in

Amphetamines:
Synthesis 1, Variation 2

The striker plates provide a source of phosphorus for a tweeker. Such a person will often spend days scrapin g off the phosphorus from hundreds of books of matches. Look for large numbers of matchbooks with the matches themselves undisturbed.

Explosives:
McGyver explosives; oxidizer–fuel mixtures

Match heads have been used for years. The Unibomber started his career using bombs made from match heads. Match heads in steel pipe can make a formidable bomb.

Regulations: They cannot be shipped on commercial airplanes in bulk. Cannot be shipped on passenger planes but passengers are allowed to carry four matchbooks on their person.

MEAL POWDER

Used in

Explosives:
Potassium chlorate primer, friction primer

Fireworks:
Sparks, glitter

MEK
See Methyl ethyl ketone.

MERCURIC CHLORIDE
$HgCl_2$
7487-94-7
UN 1624

Appearance: Clear-to-white crystals; that are moderately soluble in water; looks like table salt. Most likely found in 1-oz or 1-pint brown glass or brown plastic chemical reagent jars.

Odor: None.

Hazards: Primary hazard is toxic by ingestion and skin absorption. The skin absorption can be greatly enhanced if this salt is in organic solvents. Compare this LD_{50} with any other in this document and try to appreciate how really toxic this material is. Carelessness with this can be lethal. LD_{50} 0.001 g/kg. *Note:* Sodium cyanide has an LD_{50} value of 0.002 g/kg.

Incompatibilities: Strong acids, ammonia, carbonates, metallic salts, alkalis, phosphates, sulfites, sulfates, arsenic antimony, bromides.

Used in

Amphetamines:
Synthesis 6

Mercuric chloride is an initiator in the P-2-P method of making methamphetamine (Synthesis 6). P-2-P is added to alcohol, aluminum, and a small amount of $HgCl_2$ in a reaction vessel. Look for aluminum or a light gray mush. Mercuric chloride may be found dissolved in solvents or in waste sludge. Due to the extreme health and environmental hazard of this poison, any components of Synthesis 6 should be tested for mercury.

Chemical Weapons:
Facility 5

Illegal Drugs:
Cook 4, Supplementary 1

Toxins:
Methyl mercury, toxic chemicals

Regulations: Waste Code D009 (0.2 mg/L); Hazard Class 6.1; PG II; ERG 154.

Confirmed by: Mercury Test.

MERCURIC NITRATE
$Hg(NO_3)_2$
10045-94-0
UN 1625

Appearance: White or slightly yellow deliquescent crystal or powder that is very soluble in water.

Odor: Mild, nitric acid.

Hazards: Poison by inhalation, ingestion, and skin absorption!!!! May cause delayed cyanosis due to the formation of methemoglobin. Oxidizer.

Incompatibilities: Strong reducing agents, combustible materials, most common metals.

Used in

Explosives:
Hg NTA, mercury azide, trimercurychlorate acetylide

Toxins:
Toxic chemicals

Regulations: Hazard Classes 5.1 and 6.1; ERG 141. One of the EPA's TCLP chemicals.

Confirmed by: Mercury Test.

MERCURIC OXIDE
HgO
21908-53-2

Appearance: Red-orange or yellow heavy powder.

Odor: None.

Hazards: Toxic by inhalation, ingestion, or if absorbed through the skin!!!! Has neurological effects; can also affect the liver and the kidney.

Incompatibilities: Strong oxidizers, strong reducers, combustible materials, organic materials, phenols.

Used in

Explosives:
Mercury nitride

Toxins:
Toxic chemicals

Regulations: Hazard Class 6.1. One of the EPA's TCLP chemicals.

Confirmed by: Mercury Test.

MERCUROUS CHLORIDE
HgCl

Appearance: Heavy white powder that is insoluble in water. Darkens upon exposure to air, eventually decomposes into the much more toxic mercuric chloride. *Also see* Mercuric chloride.

Odor: None.

Hazards: Called the "mild" mercury chloride; once used widely in medicinal preparations. Consider it still toxic. *See* Mercuric chloride.

Incompatibilities: Exposure to sunlight will cause this to become the far more toxic mercuric chloride.

Used in

Fireworks:
Chlorine donor

Regulations: Mercury salts are one of the EPA's TCLP chemicals.

Confirmed by: Mercury Test.

MERCURY
Hg
7439-97-6
UN 2809

Appearance: Silver-white, heavy, mobile liquid metal. This metal has been fascinating to people for years, and many people will keep mercury for no reason that they can explain. Can be purchased from hardware stores, in mercury switches found in thermostats, and in quiet switches. Available from dental supply houses; used in amalgam. Also available from chemical supply houses.

Odor: None; not only is it impossible to detect by odor, but most monitoring devices are totally blind to mercury. This can lead to cumulative high exposures over time.

Hazards: Volatile at ordinary temperatures; sublimes constantly despite a vapor pressure of 0.002 mmHg @ 20°C. The vapors are toxic to the liver, kidney, and the central nervous system, and are difficult for the body to excrete. Although inhalation is by far the greatest threat, there is some skin absorption, and despite the fact that mercury is insoluble, ingestion can be a hazard. Note that the Hazard Class for shipping is 8 (Corrosives).

Incompatibilities: Strong acids, sodium thiosulfate, ammonium hydroxide, acetylene, and hydrogen.

Used in

Ammunition:
Exploding bullets

Explosives:
Mercury fulminate

Hobbies:
Gold mining

Illegal Drugs:
Cook 1, Variation 1

Other:
Circuit board recovery, collectible

Mercury has had a fascination for both children and adults since Roman times. It is not uncommon for a person to collect jars of mercury, for no particular reason other than just to have it. This practice can be very dangerous, contaminating an entire house and poisoning the entire family.

Toxins:
Methyl mercury, toxic chemicals

Regulations: Hazard Class 8; PG III; ERG 172. One of the EPA's TCLP chemicals. At present, large amounts of mercury waste must be shipped out of the country for disposal, usually to Canada.

Confirmed by: Mercury Test.

MERCURY FULMINATE
Can be made; *see* Explosives.

Odor: Not found.

Hazards: Explosive; toxic.

Ignition by: Heat, flame, shock, or friction.

Used in

Ammunition:
Load 2, Variation 2
Load 5
Load 6
Load 7

Regulations: Hazard Class 1.1A (wetted with not less than 20% water by weight, or mixtures of alcohol and water. Not allowed in the U.S. mail. One of the EPA's TCLP chemicals.

MESITYLENE
$C_6H_3(CH_3)_3$
108-67-8
Trimethylbenzene

Appearance: Colorless liquid that is insoluble in water.

Odor: Peculiar aromatic.

Hazards: Harmful by inhalation. Vapor pressure 1.9 mmHg @ 20°C. TLV 25 ppm. Harmful by ingestion and a skin irritant. Combustible flash point 46°C.

Incompatibilities: Strong oxidizers.

Used in

Explosives:
TNTPB

Regulations: None specific.

Confirmed by: Determination by iodine.

METHAMPHETAMINE
$C_6H_5CH_2(CHNHCH_3)CH_3$

Appearance: Methamphetamine base is a yellow-to-brown basic liquid. This is the base form. It is not water soluble. Methamphetamine hydrochloride is the water soluble ionic salt, which is a white crystal. As a salt, methamphetamine can have other forms; however, the only other form dealt with in this book is methamphetamine carbonate. There are two forms of methamphetamine, which are isomers (mirror images of each other), *l* and *d* methamphetamine. *d*-Methamphetamine is the one that causes the "desired" effects. *l*-Methamphetamine is used in the Vicks® inhaler; however, for the most part it is considered to have most of the negative effects.

As seen in clandestine labs: May be found as a yellow or white powder, crystals, dissolved in a nonpolar organic solvent, or be in the form of meth oil. Residue may be found in coffee filters, flat cookware, vacuum cleaner hoses, and on mirrors. Methamphetamine has been made into cakes to cook out solvents on the stove in a frying pan.

Odor: Methamphetamine base has a sharp biting odor and taste. The hydrochloride has no odor.

Hazards: A potent central nervous system stimulant that can be smoked, snorted, injected, or taken orally. It increases heart rate, blood pressure, body temperature, and rate of breathing. High doses result in increased irritability and paranoia; withdrawal produces severe depression. Medical dosages rarely exceed 60 mg per day. A classified dose is 5 mg. Binge users may ingest more than 5000 mg per day for several days. A large amount may be toxic by ingestion. LD_{50} 0.005–0.015 g/kg. Mainlining methamphetamine base causes necrosis.

Method of production: Methamphetamine is the intended final product of all "methsyntheses" considered in this book.

Regulations: CDTA 0.

Confirmed by: Methamphetamine Presumptive Drug Test.

METHANAL
See Formaldehyde.

METHANE
CH_4
74-82-8
UN 1971

Appearance: Colorless, odorless gas. May be odorized. Is available as "gas" in most houses. Can be purchased in laboratory bottles or in large cylinders.

Odor: None. Commercial residential gas has typical methyl mercaptan odor.

Hazards: Flammable gas, LEL 5%, UEL 12%.

Incompatibilities: Oxygen, chlorine, bromine.

Used in

Chemical Weapons:
Facility 8

Incendiary Devices:
Situation 0.1

Regulations: Hazard Class 2.1 (natural gas compressed); ERG 115. Methane cylinders should be empty and punctured before disposal.

Monitored by: Combustible gas indicator.

METHANOL (Methyl Alcohol)
CH_3OH
67-56-1
UN 1230

Appearance: Colorless viscosity 1 polar liquid. This material is used for so many formulations, both legal and illegal, that it is not a good key. The most common containers are 1-quart or 1-gallon rectangular hardware store metal tins. Product names include wood alcohol, wood spirit, and HEET; however, it is commonly labeled as methanol. Available from laboratory supply houses in brown or clear 1-quart clear glass 1-gallon or 4-liter chemical reagent bottles or 4-liter plastic bottles.

Odor: Like alcohol, but very mild; most people can not smell. It is not likely to be prevalent.

Hazards: The primary hazard of methanol is as a flammable liquid. Flash point 52°F; LEL 6%; UEL 36% (wide flammable range). Group D atmosphere. Inhalation of the vapors from an evaporation dish could be a significant health hazard. Methanol also passes through the intact skin rapidly, which could be a significant hazard if a large amount of the liquid was spilled on a person. LD_{50} 13 g/kg; TC_{low} 86,000 mg/m; PEL 200 ppm.

Incompatibilities: Incompatibility Groups 3-A and 4-A. Chromium trioxide and nitric acid.

Used in

Amphetamines:
Synthesis 3
Synthesis 6
Synthesis 10
Synthesis 18, Variation 1
Synthesis 21, Variation 2

Chemical Weapons:
Facility 6, Variations 1 and 2

Designer Drugs:
Group 2, Variation 2

Explosives:
ADBN; AN; A-NPNT; DATB; 4,4-DNB; DNFA-P; hexol; hexaditon; HNB; HNF; MGP; NINHT; NDTT; nitroform; picrylchloride; plastic explosives, Variation

7; TA; TNB; TNBCl; TND; TNTPB; Trinitroanisole

Illegal Drugs:
Cook 1, Variation
Cook 1, Supplementary 1
Cook 2, Variation 2
Cook 2, Supplementaries 1 and 2
Cook 3
Cook 5, Variations 1 and 2
Cook 6, Variation 1
Cook 7

Toxins:
Ricin

Regulations: RQ 2270 kg; Waste Codes D001, U154 and F003; Hazard Class 3; PG II; ERG 131.

Confirmed by: Alcohol Test.

METHENAMINE
See Hexamine.

METHOXYBENZENE
See Anisole.

2-METHOXYINDOLE (w)
$CH_3OC_8H_6N$
3189-13-7

Appearance: White-to-yellow scales that are soluble in hot water. Probably from a fragrance supplier, sold as a fragrance made by Epochem®.

Odor: Was not specified, but will be very strong. Indole smells like fecal material.

Hazards: Used as a fragrance.

Incompatibilities: Strong oxidizers.

Used in

Illegal Drugs:
Cook 5

2-METHOXYINDOLE OXALYL CHLORIDE
We could not find a reference to this compound. However, we were able to find

references to both 2-methoxyindole and to oxalyl chloride. See both of those. This is probably a mix of these two compounds. The specific author has done that before in his recipes. Our sources are not necessarily chemists. Apparently, these compounds are still being mixed and tested (by self-administration) for effects, which is well documented on the Web. *See* 2-Methoxyindole *and* Oxalyl chloride.

METHYLAL
See Dimethoxymethane.

METHYL ALCOHOL
See Methanol.

METHYLAMINE
CH_3NH_2
74-89-5
UN 1235 (liquid)
UN1061 (gas)

Appearance: Clear corrosive gas, pH above or near 11. In the laboratory, methylamine is seen in three forms. The gas can come in gas cylinders or be a gas generated from the liquid. The liquid is a 40% solution of methylamine gas bubbled through water and is available through laboratory supply houses and in some industries. It is a solid as methylamine hydrochloride, which can be purchased or manufactured in the lab (Synthesis 19). Methylamine can be made in the laboratory (*see* Synthesis 19). Most likely, methylamine will be purchased from a laboratory supply house as a 40% liquid in brown glass reagent bottles.

Odor: Amine odor, death, ammonia, biting.

Hazards: Primary hazard is as a flammable and corrosive gas. The liquid is corrosive. The gas, as well as the liquid, is an inhalation hazard and a skin irritant. LEL 4.9%; UEL 20.7%; LD_{50} 0.698 g/kg; PEL 10 ppm.

PPE Warning: Methylamine can destroy polycarbonate (the plastic used in most face shields in Level A suits).

Incompatibilities: Incompatibility Group 4-A. Acetic acid, acetic anhydride, chromium trioxide, hydrochloric acid, nitric acid, perchloric acid, and sulfuric acid.

Used in

Amphetamines:
Synthesis 6
Synthesis 7
Synthesis 7A
Synthesis 8
Synthesis 10
Synthesis 18, Variation 6
Synthesis 19
Synthesis 21, Variations 1 and 2

Methylamine is found as a precursor (Synthesis 6), a reagent (Syntheses 6, 7, 8, 9, and 10), an intermediate (Synthesis 21), and as a product (Synthesis 19) in the clandestine methamphetamine laboratory. It can be used to make methylforamide (Synthesis 7).

Illegal Drugs:
Cook 1

Regulations: CDTA (List I) 1 kg; RQ 45.4 kg; Waste Codes D001 and D002; Hazard Classes 2.1 (gas) and 3.8 (liquid); PG II; liquid 132, gas 118.

Confirmed by: GasTec Polytec IV tube stripe 1 turns brown color; pH 11.

2-METHYL-2-AMINO-1,3-PROPANEDIOL (w)

Appearance: A powder.

Hazards: This product is used as a flavoring agent.

Used in

Explosives:
NTDND

4-METHYLAMINOREX
3568-94-3 (there are four isomers)
Also called U4EA, U4Euh, or Euphoria. Is an illegal drug in itself. It can be made in one

step by *dl*-phenylpropanolamine by cyclization with cyanogen bromide, or in several steps using sodium cyanate hydrochloric acid and *dl*-norpseudoephedrine.

Appearance: Large white crystals.

Odor: Not found.

Hazards: Has the same effects as methamphetamine, but is of longer duration. ED_{50} is about 8.8 mg/kg. Stimulant. Strongly associated with pulmonary hypertension.

Incompatibilities: Strong oxidizers.

Used in

Designer Drugs:
Group 4 chemicals to replace P-2-P.

Regulations: Controlled substance analog, with characteristics making it identical to a Schedule 1 substance.

METHYL BENZILATE (w)
$(C_6H_5)_2C(OH)CO_2CH_3$
76-89-1

Appearance: Fine powder.

Odor: Not found.

Hazards: Harmful if swallowed.

Incompatibilities: Strong oxidizers.

Used in

Chemical Weapons:
Facility 13

Designer Drugs:
Group 1, Variations 1 and 3

Regulation: Australian Group chemical, not scheduled.

METHYL BROMIDE
CH_3Br
74-83-5
UN 1062

Appearance: Colorless gas, probably in gas cylinders. May be in large glass ampoules as a liquid. Boiling point 38.4°F at 1 atm. May be in laboratory bottles.

Odor: None, At very high levels may have a chloroform odor.

Hazards: Vapors are very poisonous! Can pass through some protective clothing. Can pass through the skin. Absorption through the skin and eyes may contribute significantly to the overall exposure. PEL 20 ppm ceiling. TLV 1 ppm skin. Affects nervous sysem, lung, kidney, and liver. Flammable gas.

Incompatibilities: Reacts with aluminum and its alloys to form methylated aluminum compounds that are spontaneously flammable in the air. Reacts with zinc, magnesium, tin, and iron surfaces. Avoid acetylene, ammonia, dimethyl sulfoxide, ethylene oxide, oxidizers, and hot metal surfaces.

Used in

Illegal Drugs:
Cook 4, Supplementary 2

Others:
Fumigant

Regulations: Do not attempt to dispose of residual or unused quantities. Return to the supplier. Shipment of containers refilled without owner's consent is a violation of federal regulations. This is a restricted pesticide that was phased out in 2005. RQ 1000 lb. Hazard Class 2.3; shipping label: Poison Gas, Inhalation Hazard. ERG 123.

Monitored by: Methyl bromide indicator tube.

2-METHYL-1-BUTANOL
See Amyl alcohol.

METHYL CHLORIDE
CH_3Cl
74-87-3
UN 1063

Appearance: Colorless gas; a liquid at 0°C. May be seen in large glass ampoules as a liquid. May be in laboratory bottles, which will be in ice baths, as this product is easiest to use in reactions as a liquid.

Odor: Ethereal, sweet taste.

Hazards: Flammable gas, flash point below 32°F (–45°C); LEL 7%; UEL 19%. Toxic by inhalation, irritant, possible carcinogen. Gases may be absorbed through the skin.

Incompatibilities: Natural rubber and neoprene composites. May react violently with interhalogens, magnesium, zinc, potassium, or sodium.

Used as

Chemical Weapons:
Facility 1, Variation 1
Facility 1, Supplementary 1
Facility 1, Cyclosarin, Variation 1 and 2
Facility 1

Regulations: Hazard Class 2.1, Susbsidiaries 6.1 and 3. Flammable gas, forbidden in passenger transportation ERG 115.

Confirmed by: Beilstein Test.

METHYL CYANIDE
See Acetonitrile.

METHYL DIETHANOLAMINE (w)
$CH_3N(C_2H_4OH)_2$

Appearance: Colorless liquid that is miscible in water. Will be in a brown glass bottle.

Odor: Amine-like.

Hazards: Low toxicity; combustible; pH of solution will be near 11.

Incompatibilities: Oxidizers.

Used in

Chemical Weapons:
Facility 4, N1-3

METHYL DISULFIDE
$C_2H_6S_2$
624-92-0
UN 2381

Appearance: Liquid that is insoluble in water.

Odor: Irritating odor.

Hazards: Potentially fatal if inhaled; skin irritation. Symptoms are nausea, headache, drowsiness, dizziness, lung congestion, death.

Incompatibilities: Oxidizers.

Used in

Chemical Weapons:
Thiosarin, thiosoman

Explosives:
RDX

Regulations: Hazard Class 3; PG II: labeling requirements; ERG 130.

METHYLENE BLUE
$C_{16}H_{18}N_3ClS \cdot 3H_2O$

Appearance: Dark green crystals with bronze luster or crystalline powder. Soluble in water to produce a dark blue liquid. Methylene blue is a dye that is used often as a reagent.

Odor: None.

Hazards: Ingestion of small amounts will turn a person's urine green for a short time.

Incompatibilities: Strong oxidizers.

Used in

Fireworks:
Green smoke

Regulations: Not specifically regulated.

METHYLENE CHLORIDE
CH_2Cl_2
75-09-2
UN 1593
Dichloromethane

Appearance: Colorless liquid that separates and sinks in water. This material is used for so many formulations, both legal and illegal, that it is not a good key.

Odor: Chlorinated, sharp.

Hazards: Possible long-term effects from repeated exposures. Possible asphyxiate.

Incompatibilities: Magnesium; explodes with ethylenediamine.

Used in

Amphetamines:
Synthesis 0.7, Variation 1
Synthesis 18, Variation 6
Synthesis 18, Supplementaries 2, 3, and 4

Chemical Weapons:
Facility 1, Variations 1 and 2
Facility 4

Date Rape:
GHB, Variation 2

Explosives:
Process 2, Variations 1, 2, and 3
Process 5
ADBN, ADNB, A-NPNT, BDPF, DANP, DATBA, DINA, DMMD, 4,4-DNB, DNPU, F-TNB, HNS, MNA, NDTT, Nitrocellulose, Nitroform, PEN, PNT, PVN, RDX, TBA, TNB, TNEN, TNT, UDT, UDTNB

Illegal Drugs:
Cook 1, Variation 1

Recreational Drugs:
Bananadine

Tear Agents:
Bromoacetone, mace

Regulations: Hazard Class 6.1; PG III; ERG 180.

Confirmed by: Beilstein Test.

METHYLENE DIFORAMIDE (w)

Used in

Explosives:
MNA

METHYLENE DINITRITRAMINE (w)

Used in

Explosives:
DDD

METHYLENE IODIDE
CH_2I_2
75-11-6

Appearance: Yellow-to-brown very refractive liquid. This is the heaviest organic liquid known. It is slightly soluble in water. Decomposed by moisture and light.

Odor: Irritating.

Hazards: Inhalation, irritant, Harmful if ingested. NFPA Health 2. Irritant to the skin. Flash point 230°F, NFPA Fire 2.

Incompatibilities: Oxidizing agents, alkali metal salts, strong bases. Forms explosive mixtures with alkali metals and their alloys.

Used in

Explosives:
TNA

Regulations: Waste should be incinerated. Not specifically regulated by DOT.

METHYL ETHER
See Dimethyl ether.

METHYL ETHYL KETONE
$CH_3COC_2H_3$
78-93-3
UN 1193

Appearance: Clear viscosity 1 liquid.

Odor: Nail polish, alcohol, to ether. Organic solvent is found in clear glass jugs or plastic bottles. Can be bought from hardware stores in rectangular metal tins.

Hazards: Primary hazard is flammable liquid; flash point 16° F; LEL 1.4%; UEL 11.4%. Group D atmosphere. The vapors are a central nervous system depressant by inhalation. LD_{50} 3.4 g/kg; PEL 200 ppm.

Incompatibilities: Nitric acid, chromium trioxide, perchloric acid, and sulfuric acid.

Used in

Amphetamines:
Can be used as a polar solvent in many reactions.

Explosives:
Methyl ethyl ketone peroxide
Methyl ethyl ketone oxyacetylene

Hobbies:
Plastic explosives, Variation 2

Illegal Drugs:
Cook 9

Regulations: CDTA (List II) 145 kg; RQ 2270 kg; Waste Codes D001, D035 (200 mg/L), U159, and F005; Hazard Class 3; PG II; ERG 127.

Confirmed by: Carbonyl Test.

N-METHYLFORAMIDE (w)

Used in

Amphetamines:
Synthesis 7

METHYL GREEN
$C_{25}H_{30}N_3Cl_2$

Appearance: Paris light green powder, which is very soluble in water, forming a blue-green solution.

Odor: None.

Used in

Explosives:
Diazodinitrophenol

N-METHYLGLUCONAMIDE (w)

Used in

Explosives:
MGP

N-METHYL HYDROXYACETAMIDE (w)

Used in

Explosives:
NMHAN

METHYL IODIDE
CH_3I
74-88-4
UN 2644
Iodomethane

Appearance: A colorless transparent liquid, which turns yellow or brown on exposure to light and is partially soluble in water. It will be in brown glass bottles.

Odor: Ether-like odor..

Hazards: Methyl iodide is a sever vesicant, may produce blisters within a few hours of contact. It is a strong narcotic and anesthetic. Poisonous!!!!

Incompatibilities: Strong oxidizing agents and strong bases.

Used in

Amphetamines:
Used to make methylamine from hexamine in a reaction not covered in this book.

Chemical Weapons:
Facility 1

Toxins:
Methyl mercury

Regulations: Hazard Class 6.1; ERG 151.

METHYL ISOBUTYL KETONE
See 4-Methyl-2-pentanone.

1-METHYL-4-ISOPROPYLIDINE (w)

Used in

Illegal Drugs:
Cook 4, Variation 3

METHYL NITROAMINE (w)

Used in

Explosives:
DMMD

N-METHYL-p-NITROANILINE (w)
$C_7H_8N_2O_2$
100-15-2

Used in

Ammunition:
Load 3

4-METHYL-2-PENTANONE
$(CH_3)C_5H_{12}O$
108-10-1
UN 1245

Appearance: Most commonly known as MIBK. Clear viscosity 2 liquid that is very slightly soluble in water, floating on water with a sharp curved line.

Odor: Minty odor; threshold around 0.5 ppm.

Hazards: Flammable, flash point 64°F. Harmful if inhaled at PEL 100 ppm, or ingested at LD_{50} (various animals) from 1600 to 2020 mg/kg. Irritating to eyes, respiratory tract, and skin. Prolonged exposure to the skin, especially on the hands, can cause distal necropsy, numbing of the fingers, which can become permanent. May form explosive peroxides.

Incompatibilities: Vapors may ignite explosively on contact with potassium *tert*-butoxide. Oxidizing materials, reducing materials, acids, and alkalis.

Used in

Designer Drugs:
Group 2, Variation 2

Regulations: RQ 1000 lb; Hazard Class 3; PG II; ERG 127.

Confirmed by: Carbonyl Test.

METHYLPHENIDATE
$C_{14}H_{19}NO_2$
113-45-1

Appearance: This is the base for Ritalin. Ritalin is actually methylphenidate·HCL, CAS 298-59-9, $C_{14}H_{19}NO_2 \cdot HCl$, the salt of methylphenidate. As such, we were not surprised that we could not find this chemical either in catalogs or as an MSDS. This is typical of chemicals used for designer drugs, which often start with drugs known to have an effect. α-phenyl-2-piperidineacetic acid methyl ester is probably easy to make using piperidineacetic acid and a Grignard reagent such as the one made with bromobenzene. While Ritalin is far more potent than methamphetamine, it has never caught on as a street drug. Modifications of the drug might start off by making the base out of the salt, using products like Ritalin®, Ritalina®, Ritalin LA®, and Rilatine®.

Hazards: This is a stimulant drug.

Used in

Designer Drugs:
Group 4 chemicals to replace P-2-P

Regulations: Ritalin is a Schedule 2 drug.

METHYLPHOSPHINYL DICHLORIDE (w)
676-83-5

Used in

Chemical Weapons:
Prealkylated compound that can be used to make nerve agents.

Regulations: Australian Group Schedule 2B.

METHYLPHOSPHINYL DIFLUORIDE (w)

Used in

Chemical Weapons:
Prealkylated compound that can be used to make nerve agents.

Regulations: Australian Group Schedule 2B.

METHYLPHOSPHONIC ACID
CH_5PO_3
993-13-5

Appearance: White solid that is soluble in water.

Odor: None

Hazards: Corrosive, causes burns. Harmful if ingested or inhaled or comes into contact with the skin. This product is very destructive to the mucous membranes.

Incompatibilities: Strong oxidizers and strong bases.

Used in

Chemical Weapons:
Prealkylated compound that can be used to make nerve agents.

Regulations: No specific regulations; however, the Hazard Class is 8. Australian Group Schedule 2B.

METHYLPHOSPHONIC DICHLORIDE
Agent DC
Cl_2PCH_3O
676-97-1
UN 9206

Appearance: Crystalline deliquescent solid that reacts with water. This is binary Agent DC. This product is available through chemical supply houses in small brown glass jars.

Hazards: Very water reactive. Contact with any alcohol will make this compound into a nerve agent. These are not as hazardous as the compounds formed with Agent DF.

Incompatibilities: Water, alcohol.

Used in

Chemical Weapons:
Facility 1

Regulations: Australian Group Schedule 2B.

METHYLPHOSPHONIC DIFLUORIDE
Fl_2PCH_3O
676-99-3

Appearance: Crystalline deliquescent solid that reacts with water. This is binary Agent DF. This will probably have to be made on site; *see* Chemical weapons, Facility 1.

Hazards: Contact with any alcohol will make this compound into a nerve agent.

Incompatibilities: Contact with isopropyl will immediately make sarin. Water reactive.

Used in

Chemical Weapons:
Facility 1

Regulations: Australian Group Schedule 1B. It is unlikely that the FBI would allow this compound to be moved by anyone except themselves.

METHYLPHOSPHONOTHIO DICHLORIDE (w)
676-98-2

Used in

Chemical Weapons:
Prealkylated compound that can be used to make nerve agents.

Regulations: Australian Group Schedule 2B.

METHYLPHOSPHONOTHIOIC DICHLORIDE (w)

Used in

Chemical Weapons:
Facility 1

METHYL SODIUM (w)
MH_3Na

Appearance: Clear viscosity 1 liquid.
Odor: None.

Hazards: Flammable caustic liquid, which will burn to form sodium oxide or sodium hydroxide caustic solids.

Incompatibilities: Water reactive.

Used in

Chemical Weapons:
Facility 1

Regulations: Although not specifically regulated, probably due to the small amount in commerce, testing should be done, as this material is both caustic and flammable.

Confirmed by: Burning to white caustic crust.

METHYL STYRENE
$C_6H_5(CH_3)C=CH_3$
100-81-1 (*meta*)
611-15-4 (*ortho*)
622-97-9 (*para*)
UN 2618 (vinyltoluene stabilized)

Appearance: Clear liquid that is not soluble in water. This will probably be a mixture of o-, m-, and p-methyl styrene, called vinyltoluene.

Odor: Irritating.

Hazards: Carcinogen. Will polymerize if heated. Irritating to eyes and mucous membranes. Combustible; flash point 114 to129°F, depending on the isomer or the mixture. Harmful if inhaled or ingested. Irritating to the skin and respiratory tract. May cause nosebleeds.

Incompatibilities: Oxidizers and strong acids.

Used in

Amphetamines:
Synthesis 18A

Regulations: Hazard Class 3; PG III; ERG 130P. Waste should be incinerated.

METHYLSULFONYLMETHANE (w)

Used in

Amphetamines:
See Chemicals used for final cut.

METOL®
Trademark for methyl-*para*-aminophenol sulfate. Colorless needles that are soluble in water. Will come in glass bottles.

Hazards: Moderately irritating.

Used in

Hobbies:
Photography, developing

METRIOL
See 2-(Hydroxymethyl)-2-methylpropane-1,3-diol.

MIBK
See 4-Methyl-2-pentanone.

MICROBALLOONS

Used in

Explosives:
Binary explosives V, Supplementary 2

MILK POWDER

Appearance: White powder, which forms a very good suspension in water, partly soluble. *See also* Casein.

Odor: None.

Hazards: This is food.

Incompatibilities: Strong oxidizers.

Used in

Explosives:
Nitrated milk powder

Illegal Drugs:
Used as a cut

MINERAL OIL
CnHn hydrocarbons around C_{16}

Appearance: Clear viscosity 3 oil.

Odor: None.

Hazards: None significant.

Incompatibilities: Strong oxidizers.

Used in

Explosives:
RDX
Binary Agents, tovan

MINERAL SPIRITS
CnHn hydrocarbons around C_8
See Petroleum ether; *also see* Nonpolar liquids that float.

MONKSHOOD
Poisonous plant

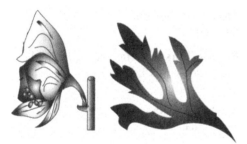

Description: Also called *Friar's hood* and *wolfbane*. High plant with a slim stem and beautiful blue blossoms. Monks hood is one of the most poisonous plants. It is often used for criminal purposes. It grows on wet grassland, stony or rocky slopes, and near forest streams higher than 1200 m.

Where found: Found at higher altitudes in the northern Rocky Mountain and Pacific coast states. The yellow Idaho native is found from Pennsylvania to Georgia.

Deadly parts: The entire plant is poisonous.

Symptoms: The symptoms start rapidly, death in 10 minutes to a few hours. The poison is aconitine, which can be ingested or absorbed through the skin. Burning and tingling; numbness in the tongue, throat, and face; nausea; vomiting; blurred, then dimness of vision; prickling of skin. Paralysis of the respiratory system, low blood pressure, slow and weak pulse, chest pain, giddiness, sweating, and convulsions. Pronounced feeling of cold. Slow paralysis of the heart muscle causes death.

Antidotes: There is no specific antidote.

Field test: Addition of potassium permanganate produces a red crystalline precipitate.

Used in

Toxins:
Aconitum napellus

MONOCHLOROACETIC ACID
See Chloroacetic acid.

MONOGLYME
See Dimethyl glycol.

MONOMETHYLAMINE
See Methylamine.

MORNING GLORY SEEDS

Used in

Illegal Drugs:
Cook 1

MORPHINE
$C_{17}H_{19}NO_3$
57-27-2

Appearance: Probably obtained illegally from a hospital, or stolen from ambulances.

Used in

Illegal Drugs:
Cook 8, Variations 1 and 2

Regulations: Morphine is classified as a Schedule 2 drug under the Controlled

Substances Act. Internationally, morphine is a Schedule 1 drug.

MORPHOLINE
C_4H_9NO
110-91-8
UN 2054

Appearance: Colorless, hygroscopic oily liquid with a pH near 11 that is soluble in water.

Odor: Ammonia- or amine-like.

Hazards: Inhalation hazard can be fatal. Readily absorbed through the skin; skin contact may cause serious burns. Caustic.

Incompatibilities: Strong oxidizers, strong acids, acid chlorides, and acid anhydrides.

Used in

Amphetamines:
Synthesis 20

Explosives:
Binary explosives, Supplementary 1

Regulations: Hazard Class 8; PG II; ERG 132.

MOTHBALLS
See Paradichlorobenzene *or* Naphthalene.

MURIATIC ACID
See Hydrochloric acid.

MUSHROOMS
Although there are many illegal products that involve mushrooms, we have avoided the discussion and inclusion of mushrooms in this document. The identification of mushrooms is a very precise and difficult science/art.
However, some mushrooms were not easy to ignore, and therefore have been included in this document by the following names:

Death cap mushroom
Psilocybin

N

NAPHTHALENE
$C_{10}H_8$
91-20-3
UN 1334

Appearance: White scales, powder, or balls that are insoluble in water. Naphthalene Sublimes at standard temperatures and pressures.

Odor: Strong coal tar odor (mothballs).

Hazards: Inhalation hazard. Sublimes noticeably at room temperature.

Incompatibilities: Strong oxidizers.

Used in

Ammunition:
Smokeless powder

Explosives:
TNN

Incendiary Devices:
Situation 2

Regulations: Hazard Class 4.1; Combustible Solid; ERG 133

NATURAL GAS
See Methane.

NEOPENTYL GLYCOL
$HOCH_2C(CH_3)_2CH_2OH$

Appearance: White crystalline solid that is partially soluble in water. A synthetic elastomer.

Used in

Explosives:
NPF

NEROL
See Citral.

NICKEL METAL
Ni
7440-02-0

Appearance: Silver-white, ductile, malleable, somewhat magnetic metal. Available in sheets, shot (granules) wire, and ingots.

Odor: None.

Hazards: Nickel may be carcinogenic. Some people are allergic to nickel on their skin. Prolonged skin contact may result in nickel itch for those who are allergic.

Incompatibilities: Nickel is very inert.

Used in

Chemical Weapons:
Facility 9, Variations 1 and 2

NEWSPAPER

Used in

Explosives:
Nitrocellulose

Incendiary Devices:
Situation 3, solid accelerants

N-NISOTRODIPHENYLAMINE (w)

Used in

Ammunition:
Load 3

4-NISOTRODIPHENYLAMINE (w)

Hazards: Nisotroamines tend to be carcinogens.

Used in

Ammunition:
Load 3

NITRIC ACID
HNO_3
7697-37-2
UN 2031
UN 2032 (red fuming)

Appearance: Clear-to-slightly yellow liquid, depending upon concentration.

Higher concentrations of nitric acid are associated with fuming, which goes from white (weaker), to yellow, and finally, red (concentrated). The hazard of this compound changes considerably with the concentration. Red fuming nitric acid is the quintessential hazardous material. Nitric acid is sold in clear glass bottles with red lids. It can also be found in large plastic carboys and stainless steel drums. The presence of red or yellow fuming nitric acid and concentrated sulfuric acid or fuming sulfuric acid is a good indicator that explosives are being made.

We have broken nitric acid down by percentages. Even though available in industry (70% nitric acid is the seventh most produced product in the United States), high percentages, 70% or more, must be made by distillation on site; *see* Explosives, Process 2.

Odor: Irritating to chlorine; has been described as sharp green grass.

Hazards: Very strong oxidizer. The liquid is corrosive. The fumes are a dangerous inhalation hazard. The primary hazard from this material is its extreme reactivity. It can explode in contact with many organic compounds. When mixed with other acids and organic compounds, it creates high explosives. On contact with many metals, it releases nitrogen tetroxide gas. LD_{50} 0.43 g/kg; LC_{50} 67 ppm @ 4 hours; PEL 2 ppm (most likely to be nitrogen dioxide in the air); PEL 5 ppm ceiling.

Incompatibilities: Incompatibility Group 6-A. Acetic acid, acetic anhydride, acetone and acetic acid, acetylene, ammonia, sodium cyanide, ethylenediamine, hydrogen peroxide, hydrogen iodide, lithium, sodium, potassium, magnesium, organic material, phosphine, pyridine, sodium hydroxide, sulfuric acid in the presence of many organic materials. Unless you are a knowledgeable chemist, it is unwise to put this acid with anything.

Used in

Amphetamines:
Synthesis 3
Synthesis 14

Used to prepare platinum or palladium catalysts from the metals or salts into forms to be used in hydrogenation synthesis (Synthesis 3). It may be used to prepare thorium salts to make the oxide used in the tube furnace (Synthesis 14). Nitric acid is not a clandestine methamphetamine laboratory chemical. Could be dangerous if used in place of other acids.

Chemical Weapons:
Facility 10, Variations 1 and 2
Chloropicrin

Explosives:

Reagent-grade nitric:
Process 1, Variation 1
ADN, ADBN, copper fulminate, cyclotrimethylene trinitriamine, DNP, HMX, mercury fulminate, mercury nitride, methyl picric acid, nitroguanidine, nitrocellulose, picric acid, urea nitrate, styphnic acid, urea nitrate

70% Nitric:
AN, DINA, DNR, EDDN, HgNTA, lead styphnate, MNA, MNTA, nitro PCB, NTD, NU, nitrated milk powder, RDX, silver acetylide, TCTNB, TNN.

90% Nitric:
A-NPNT, DATAB, DNPU, DNAN, diazodinitrophenol, nitroglycern, DNR, DTN, EGDN, ETN, HNTCAB, hexanitrate, KDN, NTND, PETN, PGDN, picric acid, quebrachitol nitrate, tetraniline, TEX TNEN, TNM

99% Nitric:
DANP, diazodinitrophenol, DITN, DINA, DNFA-P, EDT, isoitolnitrate, methyl picric acid, MNA, MON, MPG, MX, NENA, NINHT, nitrocellulose, nitroform, nitrostarch, NMHAN, PEN, PETN, PNT, PVN, RDX, TAEN, tetryl, TNAD, TNPU, TNT, TNN, TNM

Hobbies:
Photography

Recreational Drugs:
Rush, Variations 1 and 2

Regulations: RQ 454 kg; Waste Codes D001 and D002; Hazard Class 8; PG II and I (red fuming) 8, 5.1; ERG 157; forbidden on passenger planes; cannot be sent by U.S. mail. 70% or greater nitric acid cannot be sold to unlicensed persons.

Confirmed by: Nitric Acid Test.

_m_-NITROANILINE
$NH_2C_6H_4NO_2$
99-09-2
UN 1661

Appearance: Yellow needles or powder that are soluble in water.

Odor: Not found.

Hazards: Toxic by inhalation and skin contact. Containers may explode when heated.

Incompatibilities: Strong acids and strong oxidizers.

Used in

Explosives:
Tetraniline

Regulations: Hazard Class 6.1; ERG 153.

NITROBENZENE
$C_6H_5NO_2$
98-95-3
UN 1662

Appearance: Viscosity 2 to 3 clear-to-yellow/brown but opaque liquid. Most likely found in brown glass 1-quart bottles. Nitrobenzene is slightly soluble in water.

Odor: Strong; volatile oil almond. Odor threshold 0.3 ppm.

Hazards: Poisonous!!! by ingestion or inhalation. Vapor pressure 0.15 mmHg @ 20°C, PEL 1 ppm (skin) ceiling. If absorbed through intact skin, nitrobenzene is synergistic with ethanol, a combination that can be lethal to a person drinking alcoholic beverages after skin contact with nitrobenzene. Skin contact will leave a yellow stain. Nitrobenzene is also a skin irritant, which can cause reddening or even blistering. Nitrobenzene is combustible, flash point 190°F.

Incompatibilities: Strong oxidizers.

Used in

Explosives:
McGyver, styphnic acid

Regulations: Hazard Class 6.1; PG II, label: Poisonous Material; ERG 152. One of the EPA's TCLP chemicals.

NITROCARBOL
See Nitromethane.

NITROCELLULOSE
9004-70-0

Appearance: Depends on what the starting product was, can look like cellophane, cotton, or cloth.

Odor: None.

Hazards: Deflagrates upon ignition.

Incompatibilities: Heat.

Used in

Explosives:
Smokeless powder

Fireworks:
Binder

Regulations: Shipped with less than 25% water or alcohol by weight; Hazard Class 1.1D.

6-NITRO-*ortho*-CRESOL (w)
$NO_2(CH_3)C_6H_3OH$
UN 2446

Appearance: Yellow crystals that are very slightly soluble in water. Made in step 2 of Illegal drugs, Cook 5; by purchase of product can skip to step 3.

Odor: Not found.

Hazards: Toxic by inhalation, skin absorption, and ingestion.

Incompatibilities: Strong oxidizers.

Used in

Illegal Drugs:
Cook 5, Variation 1

Regulations: Label Nitrocresols Solid; Hazard Class 6.1; ERG 153.

NITROETHANE
$CH_3CH_2NO_2$
79-24-3
UN 2842

Appearance: Colorless viscosity 1 liquid. Model airplane fuel is often dyed a pink color and comes in 1-gallon plastic bottles.

Odor: Chlorinated-type odor.

Hazards: Primary hazard is flammable liquid, flash point 82°F; LEL 3.4%; LD_{50}(rat) 1100 mg/kg. PEL 100 ppm.

Incompatibilities: Incompatibility Group 4-A. Sodium hydroxide. In contact with hydrocarbons and heated, it burns explosively (e.g., racing fuels). Alkali salts of this compound can explode when dry.

Used in

Amphetamines:
Synthesis 1, Supplementary 5
Synthesis 15
Synthesis 18, Variation 1

Explosives:
Binary explosives, Supplementaries 3 and 4

Hobbies:
Model airplane fuel

Regulations: CDTA (List I) 2.5 kg; Waste Code D001; Hazard Class 3; PG III; ERG 129.

Confirmed by: Gray flame when ignited.

NITROFORM
9011-05-6

It is not likely that this product could be purchased. We could find no vendor of this product.

    Nitroform is a secondary explosive that is difficult to ignite. It is, however, an ingredient in so many other explosives, it is listed here. Can be made; *see* Explosives, Nitroform

Used in

Explosives:
HNF, HNH-3, NTND, silver nitroform, TBA, TNEN, TNP

Regulations: Hazard Class 4.1.

NITROGEN (Liquid)
N_2
7727-37-9
UN 1977

Appearance: Fuming liquid that will probably be in a dewar or thermos of some type.

Odor: None.

Hazard: Cryogenic. Although nitrogen gas makes up 80% of the atmosphere, the liquid is dangerous upon contact. The NFPA lists this basically inert gas as Health 3. Heat (much lower than body temperature) applied to the containers can cause an explosion. Nitrogen is an asphyxiate.

Incompatibilities: Can explode when heated. Metals, oxidizing agents.

Regulations: Storage subject to OSHA 29 CFR 1910.101. Hazard Class 2.2; label Nitrogen Refrigerated Liquid; ERG 120.

NITROGEN (Gas)
N_2
7727-37-9
UN 1066

Appearance: Clear gas which will be in either a lecture bottle or a large cylinder.

Odor: None.

Hazards: Simple asphyxiate. Note that NFPA Health is listed as 1 as opposed to 3 for the liquid nitrogen. Eighty percent of the atmosphere is nitrogen gas. The hazard is due to quick expansion and it being an odorless, colorless gas.

Incompatibilities: This gas is inert, and is often used to make reactive compounds inert. However, when exposed to EXTREMELY high temperatures in air, it may break down into nitrogen dioxide gas.

Used in

Amphetamines:
Synthesis 21, Variation 2

Chemical Weapons:
Facility 3
Facility 4, Variation 5

Illegal Drugs:
Cook 2, Variation 2
Cook 2, Supplementary 1
Cook 4, Variation 3

Regulations: Hazard Class 2.2; ERG 121.

NITROGEN DIOXIDE GAS
NO_2
10102-44-0
UN 1067

Appearance: Nitrogen dioxide gas is one of the few colored gases. It is a yellow to orange to red gas. May be purchased in lab bottles or in large cylinders. In cylinders, nitrogen dioxide is in equilibrium with nitrogen tetroxide (10544-72-6), with 99% + in the form of nitrogen dioxide. Usually, this gas must be manufactured in the laboratory by adding nitrate salts to concentrated sulfuric acid.

Odor: Lower levels, green grass, to "clean" chlorine odor; slightly higher levels irritating.

Hazards: This is a toxic oxidizing gas. PEL 5 ppm @ 15 min STEL. IDLH 20 ppm. Can cause severe lung damage, eye, and skin burns. Symptoms may be delayed.

Incompatibilities: Explosions may occur on contact with ammonia, boron trichloride, carbon disulfide, cyclohexane, formaldehyde, nitrobenzene, toluene, incompletely halogenated hydrocarbons, propylene, alcohols, and ozone. Incompatible with water, bases, flammable and combustible materials, copper, aluminum.

Used in

Explosives:
Process 1, Variation 1

Regulations: Do not attempt to dispose of residual or unused quantities. Return cyclinder to supplier. Hazard Class 2.3; RQ 10 lb; poison gas, oxidizer, corrosive (inhalation hazard). Forbidden in passenger and air transportation. ERG 124.

Monitored by: Nitrogen dioxide indicator tube.

NITROGEN OXIDE (Nitric Oxide)
NO
10102-43-9
UN 1660

Appearance: Colorless gas that may react in contact with the air or water. The product contains N_2O, NO, N_2O_2, NO_2, N_2O_3, N_2O_4 and N_2O_5. This product will be in a laboratory bottle, or it could be made at the site, which could be dangerous. Nitrogen oxide is also a synonym for *dinitrogen monoxide*.

Odor: Described as grain, green grass, sharp, and clean. Odor threshold near 1 ppm.

Hazards: This is a local acute inhalation hazard and can cause severe damage to the lungs in a very short time. PEL 25 ppm.

Incompatibilities: Air and water. Avoid contact with combustible materials, metals, bases, metal oxides, reducing agents, halocarbons, halogens, metal carbides.

Used in

Hobbies:
Rocket fuel, Variation 1

Regulations: Hazard Classes 2.3, 5.1, and 8. Poison gas. Hazard Zone A. Forbidden in passenger rail cars and air and air cargo. RQ 10 lb; ERG 124.

Monitored by: Nitric oxide indicator tube.

NITROGLYCERIN
$CH_2NO_3CHNO_3CH_2NO_3$
UN 3064, UN 1204, UN 3343, UN 3357, UN 3319

Appearance: Viscous clear to pale yellow and slightly soluble in water. VERY UNSTABLE liquid. Can be made; *see* Explosives, nitroglycerin. There is obviously a source for this product, as it is listed in the ingredients for many ammunition loads. Should be kept under refrigeration.

Hazards: Explosive. Very toxic, is absorbed through the skin immediately, causing dynamite headaches due to the large increase in heart rate, which can be lethal.

Used in

Ammunition:
Loads 2 and 3

Regulations: Must be shipped as a mixture. Desensitized with not less than 40% nonvolatile water-insoluble phlegmastizer by weight. Hazard Class 1.1D. Wetted with not less than 25% alcohol by weight. Depending on the concentration; can be ERG 127 or 113.

NITROGUANADINE
$H_2NC(NH)NHNO_2$

Appearance: Long, thin, flat, flexible, lustrous needles or small, thin, elongated plates. These are the alpha and beta forms of the product. Can be made; *see*

Explosives, Nitroguanidine. If shipped, will be in a 20% or greater water solution.

Hazards: Shock- and heat-sensitive explosive. Toxic.

Used in

Ammunition:
Load 3

Explosives:
NINHT

Regulations: Must be shipped in 20% water or greater. Dry or wetted with less than 20% water by weight; Hazard Classs 1.1D; ERG 113.

NITROMETHANE
CH_3NO_2
75-52-5
UN 1261

Appearance: Colorless oily liquid that is slightly soluble in water. Nitromethane, which is to be used in binary explosives, is often dyed red. Can be made; *see* Explosives, Binary explosives, Variations 5 and 6. Nitromethane is not a very sensitive explosive by itself, but forms explosive compounds with many amines.

Odor: Disagreeable and choking.

Hazards: Flammable, flash point 95 to 112°F, This is deceptive, as nitromethane is very easily ignited. Poisonous!!! by ingestion, Reactive. Less toxic by inhalation. NFPA Health 1, Flammable 3, Reactive 4. PEL 25 ppm.

Incompatibilities: Forms explosive compound on contact with many amines. Nitromethane is corrosive to lead and its alloys, and copper.

Used in

Chemical Weapons:
Facility 10, Variations 3 and 4

Explosives:
Binary explosives, Variations 1, 2, 3, and 4

Hobbies:
Car racing fuels, rocket fuels (liquid), boat racing fuels

Illegal Drugs:
Cook 6, Variation 1

Regulations: Hazard Class 3; PG II; Flammable Liquid; ERG 129.

NITRONIUM TETRAFLUOROBORATE (w)
NO_2BF_4

Appearance: Solid.

Hazard: Corrosive; wear gloves when handling nitronium tetrafluoroborate, which can cause skin burns and throat irritation if inhaled.

Used in

Explosives:
ADN, HNIW, KDN

2-NITROPHENYL AMINE (w)

Used in

Ammunition:
Load 3

NITRO SIL
See Ammonia.

NITROSONIUM TETRAFLUOROBORATE (w)
H_2NOBF_4
14635-75-7

Used in

Explosives:
HNIW

o-NITROTOLUENE
$C_7H_7NO_2$
88-72-2 (*ortho*)
99-99-0 (*para*)
99-08-1 (*meta*)
UN 1664

Appearance: Light yellow liquid that is not soluble in water.

Odor: Not found.

Hazards: Carcinogen. Highly toxic; may be fatal if inhaled, Vapor pressure 0.1 mmHg @ 20°C. Readily absorbed through the skin. Combustible, flash point 106°C (slightly above the boiling point of water).

Incompatibilities: Oxidizing agents, strong bases, sulfuric acid, reducing agents, hydrogen, and sodium.

Used in

Designer Drugs:
Group 3, Supplementary 2

Regulations: Hazard Class 6.1; PG II.

NITROUREA
$NH_2(CO)NHNO_2$
556-89-8

Appearance: White crystalline powder that is slightly soluble in water. This probably will be made in-house, but it can be purchased in steel drums.

Odor: Not found.

Hazards: Unstable secondary explosive material.

Used in

Explosives:
DPT

Regulations: Hazard Class 1.1D. A label. Forbidden on passenger or air transportation. Possession by or sale of nitrourea is a Class C felony.

NITROUS OXIDE
N_2O
10024-97-2

Appearance: Clear gas. Most likely to be found in a laboratory bottle. Also available in pressurized whipped cream containers from the supermarket. Laughing gas. Can be made; *see* Recreational drugs.

Odor: None.

Hazards: Intoxicating; moderate explosion hazard.

Used in

Hobbies:
Rocket fuel

NITROX®
See Sodium nitrite.

NONPOLAR LIQUIDS THAT FLOAT
These can be specified in a formulation; however, the solvent is often just a carrier solvent, or an extraction solvent, in which case other nonpolar solvents (even chlorinated) could be substituted. In many cases a specific nonpolar solvent is not really indicative of the product:

Benzene
Benzine
Charcoal lighter
Coleman® white gasoline
Ether (Sometimes)
Gasoline
Heptane
Hexane
Kerosene
Ligroin
Mineral spirits
Paint thinner
Pentane
Petroleum distillates
Petroleum ether
Petroleum naphtha
Ronson® lighter fluid
Stoddard solvent
Toluene
Xylene
Zippo® lighter fluid

These compounds can often, although not always, be interchangeable. Freon® (nonpolar solvent sinks) and many chlorinated compounds have been used as nonpolar extraction fluids. To determine if the solvent is essential, look to see if its structure is a part of the final product.

Incompatibilities: Ammonium nitrate, chromic acid, hydrogen peroxide, halogens, nitric acid, sodium peroxide.

Used in

Amphetamines:
Cooks to extract methamphetamine from almost all of the cooking processes use nonpolar solvents.

Chemical Weapons:
Extraction methods are used for buzz.

Illegal Drugs:
Cooks can use to extract the final product from the cooking processes listed below:

Cook 1, Variation 2
Cook 1, Supplementary 1
Cook 4, Variation 6
Cook 5, Variation 1

Recreational Drugs:
Bananadine

Regulations: RQ 2270 kg; Waste Code D001; Hazard Class 3; PG II; ERG 128.

Confirmed by: Floats on water; determination by iodine.

d,l-NOREPHEDRINE
$C_9H_{13}NO$
492-41-1

Appearance: White powder.

Odor: None.

Hazards: Harmful by inhalation and ingestion. Absorbed through the skin. Stimulant.

Incompatibilities: Strong oxidizers.

Used in

Designer Drugs:
Group 4, Euphoria, Variation 1

Regulations: Nonhazardous for air, sea, and road freight. Schedule 4 controlled substance.

NORITE

Appearance: A mafic intrusive igneous rock composed largely of the calcium-rich plagioclase labradorite and hypersthene with livine. Norite, essentially a ground-up rock, is indistinquishable from gabbo.

Odor: None.

Hazards: This is inert.

Incompatibilities: This is inert.

Used in

Illegal Drugs:
Cook 1, Variation 1
Cook 6

Regulations: Nonhazardous for air, sea, or road freight.

Confirmed by: Calcium Test.

d,l-NORPSEUDOEPHEDRINE
$C_9H_{13}NO$
492-39-7

Appearance: White powder. Most commonly known as cathine.

Odor: None.

Hazards: Harmful by inhalation and ingestion. Absorbed through the skin. Stimulant.

Incompatibilities: Strong oxidizers.

Used in

Designer Drugs:
Group 4, Euphoria, Variation 2

Regulations: Schedule 4 controlled substance; one of the drugs for which Olympic athletes are tested.

O

OIL ORANGE
See Phenethyl alcohol.

Regulations: No specific regulations.

OIL SCARLET (w)

Used in

Fireworks:
Smoke dye

OLEANDER
Poisonous plant

Description: Oleander can reach up to 5 m. It has a dark green spear-shaped leaves and fragrant white, red, pink, or orange blossoms. The poison contains cardiac glycosides, olendrin, and nerioside. Ingestion of one leaf can be lethal.

Where found: Can grow almost anywhere.

Deadly parts: All parts are poisonous, including the nectar of the flower.

Symptoms: Gastrointestinal and cardiac effects. Nausea and vomiting, excess salivation, abdominal pain, diarrhea that may or may not contain blood, irregular heart rate: racing, then slower, irratic beating. Extremities may become pale and cold. May cause unconsciousness, respiratory paralysis, and death.

Antidotes: Prompt vomiting is encouraged. Atropine can be used cautiously. Treatment for heart symptoms.

Used in

Hobbies:
Gardening

Toxins:
Nerium oleander

OLEIC ACID
$CH_3(CH_2)_7CHCH(CH_2)_7COOH$
112-80-1

Appearance: Colorless to pale yellow or light greenish-yellow liquid that is insoluble in water. Can be replaced with olive oil. Olive oil is a 70% glyceryl ester of oleic acid.

Odor: Slight.

Hazards: Is used as a drug coloring agent, maximum level permitted is 5%. May act as an irritant.

Incompatibilities: Strong oxidizing agents. Aluminum.

Used in

Explosives:
Smokeless powder

Regulations: Nonhazardous for air, sea, and road freight.

OLEUM
$xSO_3 \cdot H_2O$
8014-95-7
UN 1831 (less than 30%)
UN 1830 (51% or greater)

Appearance: Fuming sulfuric acid. Oleum is a thick, yellow jelly-like hygoscopic material. *Also see* Sulfuric acid, 20% fuming *and* 30% fuming. Made in Explosives, Process 3 and Process 3, Supplementaries 1, 2 and 3. SO_3 (a hydrolyzer) + H_2O = H_2SO_4. H_2SO_4 itself does not become an acid until at least 7% water is added. The hydrogen in acid is the disassociated hydrogen from water. Pyrosulfuric acid, $H_2S_2O_7$ is a solid (described as black) and the easiest form to ship.

Odor: May be biting, depending on concentration. The vapor pressure will depend on the concentration of sulfur trioxide. *See* Sulfur trioxide.

Hazards: Boils immediately and explosively in water. Oleum will quickly dehydrate the skin (can make skin into charcoal). Called toxic by ingestion (a real understatement); incidents of suicide by drinking these compounds include descriptions of quick, painful deaths accompanied by destruction of the body. Carcinogen if inhaled.

Incompatibilities: Water!!!! Incompatible with organic materials, powdered metals, bases, halides.

Used in

Chemical Weapons:
Facility 6, Variation 1

Regulations: Waste Code D002; Hazard Class 8; PG I; ERG 137.

OLIVE OIL
8001-25-0
See Oleic acid.

OLIVETOL
(OH)2C6H4C5H11
500-66-3

Appearance: 5-Pentyl resorcinol is a colorless crystal. This is difficult to obtain, as it is a scheduled precursor. Can be made; see Illegal drugs, Cook 4, Supplementary 1.

Odor: Not found.

Hazards: Harmful if ingested.

Incompatibilities: Strong oxidizers.

Used in

Illegal Drugs:
Cook 4, Variations 1, 2, and 3

Regulations: This is a listed precursor.

OPIUM POPPY

Location: These poppies are grown widely as ornamentals in various colors. Although technically illegal, many seed companies and nurseries grow and sell live plants and seeds in many variations (many do so without knowing).

Hazards: While eating poppy seeds will cause a person to fail an opium drug test (even just lemon poppyseed muffins), we could find no physical effects from eating the seeds. *Also see* Lettuce.

Used in

Illegal Drugs:
Cook 8, Supplementaries 1 and 2

Hobbies:
Gardening

Regulations: Possession of any part of *Papaver somniferum* other than the seed is illegal, as this is a Schedule 2 controlled substance. Poppyseeds are used in cooking; this is legal, which explains why if you have to take random drugs tests, you are advised to stay away from poppyseed cake. Poppy seed teas can be calming.

ORTHO-CHLOROSTYRENE
See o-Chlorostyrene.

OXALIC ACID
HOOCH2CH2OOH
144-62-7

Appearance: White powder.

Odor: None.

Hazards: Poison. Found in some foods, including spinach and rhubarb.

Incompatibilities: Oxidizers. May be destructive to mercury or silver.

Used in

Hobbies:
Photography

Other:
Used as a silver and copper cleaner

Regulations: Hazard Class 6.1; otherwise, not specifically regulated.

OXALINE
$C_3H_4N_2$

Appearance: A white crystalline organic base. Made by the action of ammonia on glyoxal.

Used in

Explosives:
Plastic explosives, Variation 2

OXALYL CHLORIDE
$(COCl)_2$
79-37-8
UN 2922

Appearance: Colorless liquid that is decomposed by water. Can be made with oxalic acid and phosphorus pentachloride. Important: *also see* 2-Methoxyindole oxalyl chloride.

Odor: Pungent odor; lachrymator.

Hazards: Toxic by inhalation (may be fatal by inhalation) and ingestion. Causes severe burns; extremely destructive to the mucous membranes. When heated, breaks down to carbon monoxide. This product has been used as a military poison gas.

Incompatibilities: Reacts violently with water, liberating toxic gas. Bases, alcohols, steel, oxidizing agents, and alkalis.

Used in

Designer Drugs:
Group 3, Variation 1

Illegal Drugs:
Cook 5, Variation 1
Cook 7

Regulations: Hazard Class 8; PG II; label; Corrosive Liquid Toxic, n.o.s.; ERG 154.

OXONE
See Sodium peroxide.

OXY ACETYLENE

Appearance: Will probably be found as a welding or cutting apparatus, with two tanks, one oxygen, the other acetylene. The acetylene tank will be very squat. Both are gases; the oxygen has no odor, whereas the acetylene will have a mild garlic metallic odor. The gases are available at any welding supply house. *See* Oxygen; also *see* Acetylene.

Odor: Slight metallic garlic.

Hazards: The acetylene has a LEL of 2% and an UEL of 100%, which makes it an extremely flammable gas. Oxygen will ignite grease, and when confined can be explosive. Together these gases are **extremely** flammable.

Incompatibilities: These gases mixed have been known to explode in the presence of grease, especially on the valving.

Used in

Explosives:
Methy ethyl ketone peroxide and oxyacetylene

Regulations: *See* Acetylene; *also see* Oxygen.

OXYGEN
O_2
7782-44-7
UN 1072

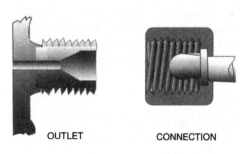

OUTLET   CONNECTION

Appearance: Colorless gas. Cylinders with stainless steel valves.

Odor: None.

Hazards: Primary hazard is as an oxidizer. Although oxygen does not itself burn, it enhances the combustibility of anything with which it comes in contact. Entry is not allowed into confined spaces where the measured level of oxygen is greater than 23.5%.

Incompatibilities: Organic compounds, especially grease and oil. Hydrogen.

Used in

Amphetamines:
Oxygen may be an artifact of using acetylene gas for a chemical blanket. Medical oxygen bottles have been used to carry ammonia.

Explosives:
Process 3, Supplementary 1
Process 3, Variations 1 and 2

Chemical Weapons:
Facility 8, Variation 2

Regulations: Hazard Classes 2.2 and 5.1; ERG 122.

Monitored by: Oxygen meter. Normal atmosphere is 21% oxygen; 23.5 is considered enriched.

OXYGEN-PRODUCING FERTILIZER
Ammonium nitrate is usually the oxidizer in many oxidizing fertilizers. However, other oxidizers can be and have been used (we have even found nitroform). If the fertilizer produces enough oxygen to enhance a closed space, it can be used as part of an explosive. If the label states that the fertilizer is oxygen producing, then consider it as a possible component for a bomb. Fertilizer bombs in empty metal cylinders were utilized in many abortion clinic bombings.

Used in

Explosives:
McGyver bombs (most often used to bomb abortion clinics)

OZOKERITE

Appearance: White or yellow tasteless waxy cakes. A mixture of hydrocarbons of complex composition purified by treatment with concentrated sulfuric acid and filtered through bone carbon. Earth wax, mineral wax, cerosin, and cerin. Not water soluble.

Odor: Slightly waxy.

Hazards: This is a substitute for bees wax, which is resistant to oxidizers. Has been used for waxed paper.

Incompatibilities: Fairly inert.

Used in

Explosives:
Plastic explosives, Variation 7

OZONE
O_3
10028-15-6

Appearance: Clear gas. There are many ways to generate this gas, including devices that use ultraviolet light to generate ozone continuously for use as ozonaters for home or laboratory or municipal water systems. It is possible to see ozone in a laboratory bottle, but not likely. See the regulations.

Odor: Sharp, clean. Ozone is the smell after a large electrical storm.

Hazards: Very strong oxidizer that is very toxic by inhalation; PEL 1 ppm.

Incompatibilities: May decompose spontaneously and violently to oxygen. Ozone even at low pressures is potentially explosive.

Used in

Amphetamines:
Synthesis 20, Supplementary 5

Regulations: This compound can only be shipped under special regulations.

Monitored by: Ozone indicator tubes.

P

P-2-P
See 1-Phenyl-2-propanone.

PAINT STRIPPER
For antique wooden furniture. The stripper sold under the name Geocities®, manufactured by Specialty Wood Products, is γ-butyrolactone. *See* γ-Butyrolactone.

PAINT THINNER
C8 hydrocarbons
See Nonpolar liquids that float.

PALLADIUM
Pd
7440-05-3
UN 3089 (palladium powder)

Appearance: Silvery metal. May be found as a catalyst on charcoal (called *Pearlman's catalyst*).

Odor: None.

Hazards: Depends on the physical state of the metal. As a metal, it is inert. As a powder, it is a flammable solid, depending on how fine the powder is. NFPA lists this as Health 1, Fire 1, Reactivity 0. Fine powders can be irritants to the skin and mucous membranes.

Incompatibilities: Sodium borohydride. The powder should be stored away from heat.

Used in

Amphetamines:
Synthesis 3
Synthesis 4

Palladium is a catalyst in Syntheses 3 and 4 and could be used in other synthesis or almost anywhere platinum is used. It may be in the form of the metal, bromide, chloride, or the oxide. It might be found as coatings for metal beads (from a catalytic converter), on charcoal, or on barium sulfate (Synthesis 4).

Regulations: These regulations apply to the powdered material, not to metal or foils. Hazard Class 4.1; label Metal Powder, Flammable, n.o.s. (Palladium Powder); PG III; ERG 170.

PALLADIUM BLACK (w)
PdO
1314-88-05

Appearance: Dark black to greenish-black powder. Brown glass chemical reagent jars in storage or in reaction vessels in P-2-P laboratories. There is no commercial product that contains this compound.

Odor: None.

Hazards: Minimal. NFPA Health 1.

Used in

Amphetamines:
Synthesis 3

PALLADIUM CHLORIDE
$PdCl_2$
7647-10-1

Appearance: Black-brown deliquescent powder or crystals that are soluble in water. Sold as Enplate® activator 440 and Niklad® 262.

Odor: None.

Hazards: May be a sensitizer. Irritating to the eyes and respiratory system.

Incompatibilities: Strong oxidizing agents.

Used in

Hobbies:
Photography

PALLADIUM ON BARIUM SULFATE (5%)

Used in

Hobbies:
Photography

Regulations: Nonhazardous for air, sea, and road freight.

PALLADIUM ON CHARCOAL (5%)

Used in

Designer Drugs:
Group 3 Supplementary 2

Hobbies:
Photography

para-DICHLOROBENZENE
$C_6H_4Cl_2$
106-46-7
UN 1592

Appearance: Clear waxy cakes; or granular chunks.

Odor: Strong, urinal cakes; mothballs.

Hazards: Sublimes; vapor pressure 0.6 mmHg @ 20°C. Flash point 65°C. Harmful if inhaled or swallowed.

Used in

Fireworks:
Fuel deoxidizer

Regulations: Hazard Class 6.1; PG III.

PARAFFIN OILS
C18 hydrocarbon and better mix
See Nonpolar liquids that float.

Used in

Fireworks:
Whistle formula

Regulations: None specific.

PARAFORMALDEHYDE
$(CH_2O)_n$
3052-58-94
UN 2213

Appearance: White crystalline powder that is slowly soluble in water. *Also see* Formaldehyde. Used to fumigate mattresses.

Odor: Formaldehyde.

Hazards: Irritant, moderately toxic by ingestion; combustible, flash point 160°F.

Incompatibilities: Strong oxidizers.

Used in

Ammunition:
Load 4

Explosives:
ANPNT

Regulations: Hazard Class 4.1; ERG 133.

PARAQUAT
$[CH_3(C_3H_4N)_2CH_3]·2CH_3SO_4$
1910-42-5
UN 2811

Appearance: A yellow solid that is very soluble in water. It will always be seen as a liquid solution. Although this chemical is no longer commercially available in the United States, considerable amounts were sold and are still in storage. Paraquat is still used extensively by commercial practitioners. Because paraquat is so highly toxic, in the United States it is marked with a blue dye and a chemical that gives it a sharp odor. However, foreign brands such as Gramoxine®, which are available, are not so marked.

Odor: None, but if sold in the United States, sharp.

Hazards: Poison by ingestion. Unlikely to be absorbed through the intact skin, but will pass through cuts and rashes at very high levels. Can be inhaled in the lungs as a smoke. Wet paper stock and waste are flammable and considered a fire risk.

Incompatibilities: Strong oxidizers.

Used in

Toxins:
Toxic chemicals

Regulations: Paraquat is classified as "Restricted Use," which means that it can be used only by people who are licensed applicators. Paraquat is a restricted waste.

PARLON
$(C_6H_6Cl_4)_n$

Appearance: Chlorinated polymer.

Used in

Fireworks:
Chlorine donor, strobe formulations

Regulations: Chlorinated compounds should be incinerated.

PEARLY GATES SEEDS
See also Heavenly blue seeds.

Used in

Illegal Drugs:
Cook 2

PENACOYL ALCOHOL (w)
$C_5H_{11}OH$
464-07-3

Appearance: Viscosity 2 clear liquid.

Odor: Not found.

Hazards: Combustible.

Incompatibilities: Oxidizers.

Used in

Chemical Weapons:
Facility 1

Regulations: Hazard Class 3; Australian Group Schedule 2B.

PENAZONE
Probably a product name for pancurium bromide paverine chlorohydrate.

Used in

Hobbies:
Photography

PENTABORANE
Industrial chemical weapon
B_5H_9
56-38-2
UN 1380

Appearance: Colorless liquid that hydrolyzes slowly in water. Probably will be in metal cylinders. This is not an off-the-shelf item, even from compressed gas houses or Laboratory supply houses. Since this is a closely watched commodity, it is known that several canisters are unaccounted for at this time.

IF YOU ENCOUNTER THIS MATERIAL, THE SITUATION IS VERY DANGEROUS! ANY EXPOSURE CAN KILL YOU!

Odor: Pungent, detected below the IDLH; some describe it as burned chocolate.

Hazards: Although pure pentaborane is NOT flammable, ANY contamination makes it likely to ignite spontaneously in air. This is HIGHLY toxic by ingestion or inhalation. This is an explosion risk. A lethal dose is in the ppb range.

Incompatibilities: Air.

Regulations: Hazard class 4.2; PG I; forbidden in passenger transportation. ERG 135.

PENTAERYTHRITOL
$C(CH_2OH)_4$
115-77-5

Appearance: White crystalline powder that is soluble in water. Sold in drums and sacks.

Odor: Not found.

Hazards: Combustible, flash point 240°F, low toxicity, LD_{50}(rat) 19,500 mg/kg. Limit exposure to dust.

Incompatibilities: Strong acids, strong oxidizing agents, acid chlorides, and acid anhydrides.

Used in

Explosives:
PEN, PETN

Regulations: Nonhazardous for air, sea, and road freight.

PENTANE
C_5H_{12}
109-66-0
UN1256

Appearance: Colorless viscosity 1 liquid. *See* Nonpolar liquids that float. Pentane is most likely to be found in brown 4-liter laboratory supply house glass jugs. It is not an over-the-counter commercial product. This compound would be used as an extraction solvent. A person using this compound is probably also associated with a legitimate laboratory or is a college student new to the clandestine laboratory scene, as the cost of this purified hydrocarbon is over $70 a gallon and it will work no more effectively than Coleman® white gas.

Odor: Mild thinner.

Hazards: Flammable, flash point <40°F; LEL 1.5%; UEL 7.8%. Intoxication by inhalation; absorbed through the skin; most likely contact problem is drying of the skin, which can cause cracking and infection. Pentane is similar to petroleum ether, hexane, and Coleman white gas. It is also similar in many ways to paint thinner, petroleum distillates, petroleum naphtha, Stoddard solvent, and mineral spirits.

Incompatibilities: Chromium trioxide.

Used in

Amphetamines:
Pentane is a solvent for methamphetamine base extraction. The waste product will probably have a pH of 14, due to residual meth base and sodium hydroxide from the extraction process.

Chemical Weapons:
Specified for vinylarsine

Regulations: Waste Code D001; Hazard Class 3, PG I and II; ERG 128.

Confirmed by: Determination by iodine.

PERCHLORIC ACID
$HClO_4$
7601-90-3
UN 1873 and UN 1802

Appearance: Colorless viscosity 1 liquid. Fuming perchloric acid is very dangerous. It will come from a laboratory supply house in a clear glass bottle or jug. There is no commercial over-the-counter source for this compound.

Odor: Acid irritating.

Hazards: The primary hazard is as a very strong oxidizer. The liquid is the ultimate corrosive. It forms explosives upon reacting with metal salts and organic materials. The visible fumes and even invisible fumes are an inhalation hazard. The primary hazard from this material is its extreme reactivity. The hazard of this compound changes considerably with the concentration. It can explode in contact with many organic compounds, or when mixed with other acids, salts, metals, and organic compounds, it creates high explosives. Perchloric acid forms salts which become high explosives when dry. Fifty percent solutions should be treated with great caution; solutions greater than 72% should be handled only after consultation with a person familiar with the hazards of this incredibly dangerous material. Unknown concentrations should be treated as worst case.

Incompatibilities: Incompatibility Group 6-A. Forms explosive compounds on contact with acetic acid, acetic anhydride, acetone and acetic acid, acetylene, ammonia, sodium cyanide, ethylenediamine, hydrogen peroxide, hydrogen iodide, lithium, sodium, potassium, magnesium, and organic materials. As a rule, any oxidizable material that is treated with perchloric acid will become a low-order explosive material. Phosphine, pyridine, sodium hydroxide, any metal or metal salt, can react violently with perchloric acid. Unless you are a knowledgeable chemist, it is unwise to put this acid with anything. Does not play well with others.

Used in

Amphetamines:
Synthesis 4

Explosives:
BDC, SATT

Regulations: Waste Codes D001 and D002; DOT (>50%); Hazard Class 8; PG I (<50%) Class 5.1; PG II, Labels 5.1,8; ERG 140/143; forbidden on passenger planes; cannot be sent by U.S. mail.

Confirmed by: Perchlorate Test.

PERCHLOROBENZENE
See Hexachlorobenzene.

PERCHLOROFLUORIDE (w)
$ClFO_3$
7616-94-6
UN 3083

Appearance: Colorless gas. Will be in a special lecture bottle.

Odor: Sweet odor that resembles gasoline and kerosene.

Hazards: Toxic. Powerful oxidizer.

Incompatibilities: Organic materials, reducing agents.

Used in

Explosives:
PCB

Regulations: Hazard Classes 2.3 and 5.1; ERG 124.

PESTICIDES, Arsenic
Industrial chemical weapon

PESTICIDES, Carbamates
Industrial chemical weapon

PESTICIDES, Fumigants
Industrial chemical weapon

PETRO-AG®
See Sodium diethylnaphthalene sulfuric acid.

PETROLEUM DISTILLATES
Mixed hydrocarbons with formulas near C_5H_{12}. *See* Nonpolar liquids that float.

PETROLEUM ETHER
UN 1271

Mixed hydrocarbons with formulas near C_5H_{12}. *See* Nonpolar liquids that float.

PETROLEUM NAPHTHA
Mixed hydrocarbons with formulas near C_8H_{18}. *See* Nonpolar liquids that float.

Appearance: Colorless viscosity 1 liquid. *Also see* Nonpolar liquids that float. Supplied from laboratory supply houses or hardware stores in 1- and 5-gallon metal containers.

Odor: Mild thinner.

Hazards: Flammable; flash point < 0°F; LEL 1.1%; UEL 5.9%; Group D atmosphere. Intoxication by inhalation, absorbed through the skin; however, most likely contact problem is drying of the skin, which can cause cracking and infection. Note that this is pretty much like pentane, hexane, Coleman white gas, and even to a degree, paint thinner, petroleum distillates, petroleum naphtha, Stoddard solvent, and mineral spirits.

Incompatibilities: Incompatibility Group 6-B. Chromium trioxide.

Used in

Amphetamines:
Petroleum naphtha is a solvent for methamphetamine base extraction. *See* Extractions.

Chemical Weapons:
Facility 1

Explosives:
Nitro-PCB, NPF

Illegal Drugs:
Cocaine, LSD, psilocin, dimethyltryptamine

Regulations: Waste Code D001; Hazard Class 3; PG I and II; ERG 128.

Confirmed by: Determination by iodine.

PEYOTE CACTUS

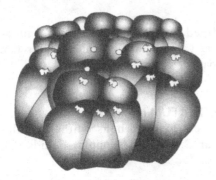

Appearance: Peyote is a small, spineless, leafless, globe-shaped cactus. The mescal buttons of the peyote cacti provide a source of hallucinatory alkaloid. The top part of the cacti rarely rises above an inch or so above the soil. Peyote is a native of the Chihuahan Desert and is annually harvested by the millions.

Effects: Noticeable psychoactive effects can usually last up to 10 hours. Effects can be different during each use, due to varying potency, the amount ingested, and the user's expectations, mood, and surroundings. Some experience sensations that are enjoyable; others can include terrifying thoughts and anxiety, fears of insanity, death, or losing control. Some users experience "flashbacks" or hallucinogen persisting perception disorder (HPPD), which are reoccurrences of hallucinations long after ingesting the drug.

Used in

Illegal Drugs:
Cook 6, Variations 2 and 3

Hobbies:
Gardening

Regulations: The DEA has listed this as a Schedule 1 drug.

PHENETHYL BROMIDE
C_8H_9Br
303-130-8

Appearance: Colorless-to-yellow liquid that is insoluble in water, and sinks.

Odor: None found.

Hazards: Harmful if swallowed; eye irritant.

Incompatibilities: Strong oxidizers.

Used in

Designer Drugs:
Group 2, Variation 2

Regulations: Nonhazardous for air, sea, and road freight.

PHENETHYL TOSYLATE (w)
$C_6H_5(C_2H_5)C_7H_8O_3$

Appearance: White solid. Tosylate = *p*-toluenesulfonate; there is probably another synonym that has not been found.

Odor: Not found.

Hazards: Irritant.

Incompatibilities: Strong oxidizers.

Used in

Designer Drugs:
Group 2, Variation 2

Regulations: Nonhazardous for air, sea, or road freight.

PHENMETRAZINE
$C_{11}H_{15}NO$
134-49-6

Appearance: Solid white powder. Like many of the precursors for designer drugs, this is a drug itself, and not easily available. It was sold as Preludin an anorectic. This has been removed from the market, and phenmetrazine is rarely prescribed at this time.

Odor: Not found.

Hazards: Considered very addictive and as more potent than methamphetamine.

Incompatibilities: Strong oxidizers.

Used in

Designer Drugs:
Group 4 chemicals to replace P-2-P

Regulations: Schedule 2 chemical.

PHENOL
C_6H_5OH
108-59-2
UN 1671

Appearance: White deliquescent crystals that appear like snow.

Odor: Strong hospital odor; the smell of chloroseptic, camphophonique, and white library paste used in kindergartens.

Hazards: Phenol is very corrosive upon skin contact. In larger amounts it will pass through the skin, and can stop the heart. Vapor pressure 1 mmHg @ 40°C. Flash point 175°F. Although it will burn and would be an inhalation hazard, the flash point and vapor pressure preclude those hazards under standard conditions.

Incompatibilities: Fuming nitric acid; other oxidizers. Explosive!

Used in

Chemical Weapons:
Facility 10, Variation 1

Explosives:
DNP, picric acid

Regulations: Hazard Classes 6.1, 4.1, and 8; ERG 153.

PHENOLPHTHALEIN
$C_{20}H_{14}O_4$
77-09-8

Appearance: White or faintly yellowish-white very fine crystalline powder that does not dissolve in water and is unstable in air.

Odor: None.

Hazards: Carcinogen. Was once used as a laxative; is also used as a truth serum. Ingestion, nausea, vomiting, and diarrhea; skin and eye irritant. NFPA Health 1.

Incompatibilities: Strong oxidizing agents.

Used in

Illegal Drugs:
Cook 2, Variation 8

Regulations: In isopropyl solutions, UN 1219, Hazard Class 3; PG II.

PHENYL ACETIC ACID
$C_6H_5CH_2COOH$
103-82-2

Appearance: Shiny white plate crystals. If it is not in the reaction vessel, it will probably be found in storage in brown glass chemical reagent jars. Phenylacetic acid cannot be purchased as a commercial over-the-counter product. Phenylacetic acid can be synthesized from benzyl cyanide (Synthesis 20).

Odor: Sweet floral; strong honey-like; cat urine.

Hazards: Combustible.

Incompatibilities: Incompatibility Group 6-B. Chromium trioxide.

Used in

Amphetamines:
Synthesis 11
Synthesis 12
Synthesis 14
Synthesis 20, Supplementaries 1 to 8

Phenylacetic acid is a reagent/precursor in three syntheses to manufacture P-2-P. It is used with acetic anhydride and sodium acetate (Synthesis 11). A biker method using phenylacetic acid with lead acetate to make P-2-P is not uncommon (Synthesis 12). Acetic acid catalyzed by thorium oxide in a tube furnace to make P-2-P is not common (Synthesis 14). Phenylacetic acid is made in Synthesis 20. Phenylacetic acid has a distinctive odor which will cling to the facility and to all people who are in the facility for a long time. Organic chemistry provides a set of building blocks, and the number of ways that these building blocks can be attached to one another is endless. Phenylacetic acid is an important compound in organic chemistry. As such, there are an endless number of ways that this key component in the manufacture of P-2-P can be made. We decided not to embark upon the slippery slope of assigning each a synthesis number in this book, as it is obvious that such a course could lead to an endless number of syntheses, which would defeat our intention of cross-referencing reactions in this book for ease of identification. For completeness and to help in understanding what is occurring should one of these unusual methods be encountered, listed next are several more syntheses that can be used to make phenylacetic acid: Benzene and chloroacetic acid in the presence of ferric chloride and heat. Benzoylformic acid in the presence of hydrazine hydrate and KOH. Styrene and morpholine in the presence of sulfur makes phenylacetothiomorpholine, which with 50% sulfuric acid will make phenylacetic acid. Mandelic acid, hydriodic acid, and red phosphorus. Allyl benzene, ozone, hydrogen peroxide, and sulfuric acid. Phenyl ethyl alcohol and dinitrogen tetroxide, toluene, potassium metal, sodium peroxide, *n*-heptane, and dry ice. Acetophenone, sodium hydroxide, hydroxylamine, hydrogen chloride and sulfur benzyl chloride, and magnesium (Grignard reagent) ether reflux with carbon dioxide gas.

Regulations: CDTA (List I) 1 kg.

PHENYL ACETONE
See 1-Phenyl-2-propanone.

PHENYLACYL CHLORIDE
See Chloroacetopherone.

1-PHENYLBUTANONE (w)
The similarity of this compound and 1 phenyl-2-propanone is very obvious. It appears that this compound is already unavailable, as we could find no body that sold it nor any information about it on the Web. *See* 1-Phenyl-2-propanone.

Used in

Designer Drugs:
Group 4 chemicals to replace P-2-P

2-PHENYLCYCLOHEXYLAMINE (w)

Used in

Designer Drugs:
Group 4 chemicals to replace P-2-P

2-PHENYLCYCLOPYLAMINE (w)

Used in

Designer Drugs:
Group 4 chemicals to replace P-2-P

m-PHENYLENEAMINE (w)
$C_6H_4(NH_2)_2$
2526-57-63

Appearance: Colorless needles that are water-soluble and will become red on standing in the air.

Odor: Not found.

Hazards: Strong irritant to the skin, toxic by ingestion and inhalation. TLV 0.1 mg/m^3.

Incompatibilities: Strong acids and oxidizing agents.

Used in

Explosives:
DATB

Regulations: None specific.

PHENYL ETHYL ALCOHOL
$C_6H_5CH_2CH_2OH$
60-12-8

Appearance: Colorless liquid. Will be in tins or glass bottles. Called *orange oil*.

Odor: Floral odor.

Hazards: This is a component of many essential oils. Harmful by inhalation, Vapor pressure 1 mmHg @ 58 °C ingestion and skin absorption.

Incompatibilities: Strong oxidizing agents and strong acids.

Used in

Amphetamines:
Synthesis 20, Supplementary 5

Fireworks:
Smoke dye

Regulations: None specific.

PHENYL ETHYL CHLORIDE
$ClC_2H_4C_6H_5$
622-24-2

Appearance: Colorless clear liquid that is very slightly soluble in water.

Odor: Not found.

Hazards: Combustible, flash point 60°C. The toxicology of this compound has not been fully investigated. Ingestion, nausea, vomiting. and diarrhea; skin irritation; causes respiratory tract irritation.

Incompatibilities: Strong oxidizing agents and strong bases.

Used in

Designer Drugs:
Group 2, Variation 1

Regulations: Not specifically regulated.

PHENYLHYDRAZINE
$C_6H_5NHNH_2$
100-63-0
UN 2572

Appearance: Colorless-to-pale yellow liquid that is moderately soluble in water.

Odor: None found.

Hazards: Corrosive; Toxic by ingestion, inhalation, or if absorbed through the skin. Suspected carcinogen. TLV 5 ppm.

Incompatibilities: Strong oxidizers, metal oxides.

Used in

Designer Drugs:
Group 3, Supplementary 1

Regulations: Hazard Class 6.1; PG II; ERG 152.

PHENYLISOPROPYLAMINE (w)
Looking at the structure of this compound, it is not surprising that it could replace P-2-P in the synthesis of methamphetamine. That would account for why it is not available in catalogs or on the Web. *See* 1-Phenyl-2-propanone.

Used in

Designer Drugs:
Group 4 chemicals to replace P-2-P

d,l-PHENYLPROPANOLAMINE
See Cathine, *d,l*-Norephedrine, *d,l*-Norpseudoephedrine, *and* Catha edulis.

1-PHENYL-2-PROPANONE
$C_6H_5COCH_3$
103-79-7

Appearance: Amber crystal. The density of 1.00025 makes its reactions in water unpredictable. P-2-P is always seen in liquid form (we are not certain what puts it into solution). As seen in clandestine labs: rust (yellow red) in reaction vessel; a yellow liquid (street name, "pee" or "P") in a small reaction vessel (distillation method), or orange liquid in jar or bottle (street name, "red oil"), a product of the pyridine neutralization method. It is unlikely that this will be in a chemical reagent bottle, as it is virtually impossible to buy P-2-P in that form at this time.

Odor: Pungent fruity perfume-like odor; honey; cat urine. A chemical odor typical of clandestine labs, even in the red phosphorus/hydriodic acid labs, where there is no P-2-P. Persons familiar with P-2-P labs claim the smell of a red phosphorus/hydriodic acid lab is the same.

Hazards: Combustible. Intoxication by inhalation; LD_{50} 0.54 g/kg.

Incompatibilities: Incompatibility Group 4-A. Chromium trioxide.

Used in

Amphetamines:
Synthesis 6
Synthesis 7
Synthesis 8
Synthesis 9
Synthesis 10
Synthesis 11
Synthesis 12
Synthesis 13
Synthesis 14
Synthesis 15
Synthesis 16
Synthesis 17
Synthesis 18
Synthesis 18A

Regulations: Schedule 2 item. This is a banned product and will not be found as a product.

Confirmed by: Ketone Test 3 using 2,4-dinitrophenol.

PHENYLPROPYL CHLORIDE
See 1-Phenyl-3-chloropropane.

PHENYLUREA (w)
$C_7H_8N_2O$
64-10-8
UN 2767

Appearance: Off-white cream crystals.

Odor: Not found.

Hazards: Toxic.

Incompatibilities: Strong oxidizing agents.

Used in

Explosives:
DNPU

Regulations: ERG 151.

PHLOROGLUCINOL
$C_6H_3(OH)_3 \cdot H_2O$
6099-90-7
1,3,5-Trihydroxybenzene

Appearance: White or yellowish crystals that are soluble in water.

Odor: None.

Hazards: The amount of tissue damage depends on the length of contact. Very toxic by skin contact; permeates the skin. Inhalation, severe overexposure can result in lung damage. Eye contact will result in corneal damage or blindness. NFPA Health 3.

Incompatibilities: Strong oxidizers and strong bases.

Used in

Explosives:
Lead TNP, TPG

Regulations: Not a DOT-controlled compound; however, should still be considered Hazard Classes 8 and 6.1, n.o.s.

PHOSGENE
Industrial chemical weapon
$COCl_2$
75-44-5
UN 1076

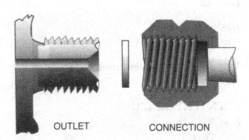

OUTLET   CONNECTION

Appearance: Colorless gas. This rarely encountered gas is produced accidentally by high energy in contact with chlorinated solvents. Can be generated in a laboratory; *see* Facility 3. Phosgene can be purchased from compressed gas supply houses.

Odor: Sharp grassy odor.

Hazards: Extremely toxic by inhalation; LC_{50} 3.2 g/M3; PEL 0.1 ppm.

Used in

Chemical Weapons:
Facility 7, Variation 2
Agent CX, phosgene oxime

Regulations: Hazard Classes 2.3 and 8. Cylinder must be empty before disposal. ERG 125.

PHOSPHINE
PH_3
7803-51-2
UN 2199

Appearance: Colorless gas. Probably created by accident during the synthesis of a compound containing phosphorus under acid or basic conditions. If purchased, will be in a laboratory bottle.

Odor: Described as dead fish and dead cats; very strong, unpleasant odor.

Hazards: Extremely toxic by inhalation; LD_{50} (rat) 11 ppm; PEL 0.3 ppm; STEL 1 ppm; IDLH 50 ppm.

Used in

Amphetamines:
Synthesis 1

Unintentionally produced. As seen in clandestine labs: This is the "kitchens of death" gas, which has been encountered often. It is produced in a reaction vessel during the manufacture of hydriodic acid by the action of heat and hydrogen on red phosphorus (Synthesis 1). This gas is said to be a major killer of cooks. It is generated when red phosphorus comes into contact with caustics and/or acids, especially in contact with metal, generating hydrogen gas. Using hypophosphorous acid and iodine appears to generate considerable amounts of phosphine (Synthesis 1, Variation 4). Heat as part of the cooking process increases production of this gas. It can also be generated by phosphorus in a process under a hydrogen chemical blanket.

Regulations: Hazard Classes 2.1 and 2.3; ERG 119.

PHOSPHORIC ACID
H_3PO_4
7664-83-2
UN 1805

Appearance: Colorless viscosity 2 liquid. A full bottle will be very heavy. Sold in clear glass 1-gallon chemical reagent bottles with white lids. Also found in commercial tile cleaners. We have found no illegal uses for phosphoric acid.

Odor: Acidic. Phosphoric acid is not very volatile, and an odor should not be noted.

Hazards: The primary hazard is skin or eye contact. It does not cause pain on initial contact, which can lead to severe burns later. This acid has very low volatility and will not be in the air except in extremely peculiar circumstances. LD_{50} 1.5 g/kg.

Incompatibilities: Incompatibility Group 6-B. Magnesium, potassium, sodium, lithium, sodium hydroxide, and ammonia.

Used in

Amphetamines:
There is some indication that this is brought into the lab under the notion that it could replace red phosphorus, in the same manner as hypophosphorous acid.

Synthesis 4

If hydrochloride gas is bubbled through phosphoric acid, it can be used in place of perchloric acid in one of the many variations of Synthesis 4.

Regulations: RQ 2270 kg; Waste Code D002; Hazard Class 8; PG III; ERG 154.

Confirmed by: Phosphate Test.

PHOSPHORUS (Red)
P
7723-14-6
UN 1338

Appearance: Red phosphorus is an orange-red to dark red-purple granular pill or powder. Purchased from laboratory supply houses in brown glass jars with white lids.

Odor: None.

Hazards: Primary hazard is as a flammable solid. However, under many conditions this material can form phosphine gas. The white smoke that comes off phosphorus as it burns is phosphorus pentoxide, which is an acid anhydride. If there is flame in the absence of air, red phosphorous can burn to white phosphorus, which is pyrophoric (air reactive). LD_{50} 4.41 g/kg.

Incompatibilities: Alkalis, oxygen, and reducing agents.

Used in

Amphetamines:
Synthesis 1
Synthesis 20, Supplementary 4

Red phosphorus is a catalyst in the production of methamphetamine from ephedrine using hydriodic acid (Synthesis 1). Red phosphorus is not used up in the reaction and is often reclaimed after the reaction for reuse. It is also a catalyst in the synthesis of hydriodic acid from iodine and water under hydrogen. In "Bevis and Butthead" labs, the source of red phosphorus may be the striker plates from matchbooks. This will be a fine reddish-brown to purple powder which is not pure red phosphorus. May be seen on ground glass such as the necks of reaction vessels, where the friction of inserting connectors can ignite the phosphorus. A red powder or crust will be found on bed sheets spread over 5-gallon plastic pails in Mexican national laboratories, and the red powder or crust will be found on coffee or Whatman filters. The strong red color of this catalyst sometimes looks almost like blood.

Ammunition

Explosives:
McGyver; Armstrong powder

Regulations: RQ 0.454 kg; Waste Code D001; Hazard Class 4.1; PG III; ERG 133.

Confirmed by: Ignition Test.

PHOSPHORUS OXYCHLORIDE
$POCl_3$
10025-87-3
UN 1810

Appearance: Colorless fuming liquid. Found in clear 1-quart or 1-gallon glass jugs.

Odor: Irritating.

Hazards: Primary hazard is strong irritant to eyes, skin, and lungs. Corrosive acid. LD_{50} 0.38 g/kg; LC_{50} 48 ppm for 4 hours.

Incompatibilities: Incompatibility Group 3-B. Magnesium, potassium, sodium, lithium, sodium hydroxide, and ammonia.

Used in

Amphetamines:
Synthesis 3

Chemical Weapons:
Facility 1
Facility 2
Facility 3

Explosives:
NPF

Regulations: RQ 454 kg; Waste Code D002; Hazard Class 8; PG II; Labels 8 and 6.1, Poison Inhalation Hazard, Zone B; ERG 137; forbidden on passenger planes; cannot be sent by U.S. mail. Australian Group Schedule 3B.

Confirmed by: PolyTec IV indicator tube.

PHOSPHORUS PENTABROMIDE
PBr_5
7789-69-7
UN 2691

Appearance: Yellow crystalline mass that is soluble in water.

Odor: Not found.

Hazards: Water reactive, forms strong acid, corrosive and toxic.

Incompatibilities: Reacts with water and bases.

Used in

Amphetamines:
Synthesis 3

Regulations: Waste Code D002; Hazard Class 8; PG II; Labels 8; ERG 137; forbidden on passenger planes; cannot be sent by U.S. mail.

PHOSPHORUS PENTACHLORIDE

PCl_5
10026-13-8
UN 1806

Appearance: Slightly yellow crystalline mass. Available from a laboratory supply house only, will be in brown glass jars, probably sealed with wax or tape.

Odor: Irritating.

Hazards: Primary hazard is water reactive. Flammable, corrosive to the eyes and skin.

Incompatibilities: Reacts strongly with water.

Used in

Amphetamines:
Synthesis 3

Chemical Weapons:
Facility 1
Facility 3

Regulations: Waste Code D002; Hazard Class 8; PG II; ERG 137; forbidden on passenger planes; cannot be sent by U.S. mail.

PHOSPHORUS PENTASULFIDE

$P2S_5$
1314-80-3
UN 1340

Appearance: Light yellow crystalline masses. Decomposed by water, forming phosphoric acid and hydrogen sulfide gas.

Odor: Peculiar odor.

Hazards: Ignites by friction. Flammable. Toxic by inhalation and ingestion. Contact with water or acids liberates H_2S. *See* Hydrogen sulfide.

Incompatibilities: Heat (releases) phosphorus pentoxide, and sulfur dioxide, Water releases H_2S.

Used in

Chemical Weapons:
Facility 3

Regulations: Hazard Class 4.3; label Dangerous When Wet, Flammable Solid, Phosphorus pentasulfide free from white or yellow phosphorus. Forbidden in passenger transportation. Requires export permits from the DTCC. ERG 139.

PHOSPHORUS PENTOXIDE

P_2O_5 (P_4O_{10})
1314-56-3
UN 1807

Appearance: Dry, fine, white powder that reacts by generating heat with water. Will be tightly sealed in brown glass jars with either wax or tape.

Odor: None; however, if moist, may be biting.

Hazards: Classified as an acid; it is a hydrolyzer, the anhydrous form of phosphoric acid. This material is very water reactive, becoming very hot when wet. Corrosive; causes burns. Eye contact may lead to serious permanant damage. Visible vapors released when wet are phosphoric acid.

Incompatibilities: Moisture. Reacts violently with water, alcohols, metals, sodium potassium, ammonia, oxidizing agents, HF, peroxides, magnesium, and strong bases.

Used in

Explosives:
EDT, HMX, nitrocellulose, Solex

Regulations: Hazard Class 8; PG II; Transport Citatory 2; ERG 137.

PHOSPHORUS SESQUISULFIDE
P_4S_3
1314-85-8

Appearance: Yellow crystalline mass that is soluble in water. Most likely to be found as match heads (a mixture of 25% phosphorus sesquisulfide and 75% potassium perchlorate is used for the heads of strike-anywhere matches. Also called phosphorus trisulfide, tetraphosphours trisulfide, or phosphorus sulfide.

Odor: Not found.

Hazards: Ignites by friction; irritant; toxic.

Used in

Explosives:
Oxidizer-fuel mixtures

Regulations: Hazard Class 4.1; flammable solid; forbidden on planes, except three books of matches for personal use.

PHOSPHORUS TRIBOMIDE
PBr_3
7789-60-8
UN 1808

Appearance: Slightly yellow crystalline mass. Available from a laboratory supply house only, will be in brown glass jars, probably sealed with wax or tape.

Odor: Irritating.

Hazards: Corrosive; inhalation hazard; primary hazard is water reactive.

Incompatibilities: Reacts violently with water, sodium, and potassium. Incompatible with strong bases, acids, alcohols, strong oxidizing agents, and organic materials.

Used in

Amphetamines:
Synthesis 3

Regulations: Waste Code D002; Hazard Class 8; PG II; ERG 137; forbidden on passenger planes; cannot be sent by U.S. mail.

PHOSPHOROUS TRICHLORIDE
PCl_3
7719-12-2
UN 1809

Appearance: Colorless, clear, fuming liquid, hydrolyzes in water.

Odor: Biting.

Hazards: Corrosive, water reactive.

Incompatibilities: Reacts slowly, but violently, with water.

Used in

Chemical Weapons:
Facility 1, Supplementary 1
Facility 1, Cyclosarin
Facility 1, Thiosarin
Facility 1, Soman
Facility 2, Variation 2
Facility 3

Regulations: RQ 454 kg; Waste Code D002; Hazard Class 8; PG II; Labels 8 and 6.1; Poison Inhalation Hazard, Zone B; ERG 137; forbidden on passenger planes; cannot be sent by U.S. mail. Australian Group Schedule 3B.

Confirmed by: PolyTec IV tube.

PHTHALATE
$C_6H_4(COOH)_2$salt

Appearance: Depends on which salt of phthalic acid; colorless crystals.

Used in

Explosives:
PTN

PHTHALOCYANINE BLUE

Appearance: Blue dye.

Used in

Fireworks:
Blue smoke

PICRIC ACID
$C_6H_2OH(NO_2)_3$
1918-02-1
UN 1336 (wetted)

Appearance: Picric acid is a brilliant yellow crystal, often in a liquid (aqueous) form. Can be purchased from chemical supply houses. Can be made; *see* Facility 10.

Odor: None.

Hazards: Can be explosive by shock or friction when dry or when contaminated.

Incompatibilities: Metal salts, especially alkali metal bases, and heavy metal salts such as lead, ammonia, and metals.

Used in

Chemical Weapons:
Facility 10

Explosives:
Ammonium picramate, Ammonium picrate, DDNP, lead picrate, sodium picrate.

Regulations: Wet or dry with less than 10% water; label Explosive A, forbidden in passenger planes, greater than 10% water; flammable solid.

Confirmed by: Picric Acid Test.

PICRYL CHLORIDE
Can be made; *see* Explosives, Picryl chloride.

Hazards: Explosive.

Used in

Explosives:
Hexaditon, HNBP, TNTPB, trinitroanisole

Regulations: Explosive label.

PIMPINELLA ANISUM
See Anise.

PINACOLONE (w)
$(CH_3)_3CCOCH_3$
75-97-8

Appearance: Liquid.

Odor: Not found.

Hazards: Health 2, Fire 1, Reactivity 0.

Incompatibilities: Strong oxidizers.

Regulations: Australian Group chemical, not scheduled. Requires export permits from the DTCC.

PINACOYL ALCOHOL (w)
See Penacoyl alcohol.

PIPERIDINE
$CH_2CH_2CH_2CH_2CH_2NH$ (cyclic)
110-89-4
UN 2401

Appearance: Colorless liquid that is soluble in water.

Odor: Pepper.

Hazards: May be fatal if inhaled. Vapor pressure 40 mmHg @ 20°C. Strong irritant may cause burns on skin. Contact with eyes may lead to permanent damage. Toxic by ingestion. Highly flammable, flash point 16°C.

Incompatibilities: Strong oxidizers, acids, organic acids, and water.

Used in

Illegal Drugs:
Cook 3, Variatons 1 and 2

Tear Agents:
o-Chlorobenzylidene malonitrile

Regulations: Hazard Class 8; labels 8 and 3; PG I; ERG 132.

PIPERIDONE (w)
$NC_5H_{10}=O$ (cyclic)
We were not able to find this compound. This has been true of many of the

compounds that the DEA has listed as possible precursors for designer drugs, indicating that they have already been greatly restricted or the DEA chemists prefer to use the most obscure of synonyms. Piperidone appears as a component for many other compounds, such as methylpiperidone, which chemically could easily be reduced to piperidone. From its structure we assume that it is a liquid, flammable, and has a pH near 11.

Used in

Designer Drugs:
Group 2, Variation 2

N-(4-PIPERIDYL)PROPIONANILIDE (w)

We were not able to find this compound. This has been true of many of the compounds that the DEA has listed as possible precursors for designer drugs, indicating that they have already been greatly restricted.

Used in

Designer Drugs:
Group 2, Variation 1

PIPERONAL

$C_6H_3(CH_2)OOCHO$ (bicyclic)
120-57-0
Similar to and used like safrole. *See* Safrole.

Used in

Amphetamines:
Synthesis 18, Variation 5

PIPERONYL ALCOHOL

$C_3H_7C_6H_2(OCH_2)CH_2OC_2H_4OC_2H_4OH$
495-76-1

Appearance: White powder.

Odor: None.

Hazards: The toxicological aspects of this compound have not been investigated. Ingestion may irritate the digestive tract. Inhalation may irritate the respiratory system, Vapor pressure not found, but flash point is >230°F.

Incompatibilities: Oxidizing agents.

Used in

Amphetamines:
Synthesis 18, Variation 6

Regulations: There are no specific regulations.

PIPES and TUBES

Appearance: Plastic tubes, cardboard tubes, galvanized pipes, small metal tubes.

Used in

Amphetamines:
Synthesis 22

Plastic pipe has been used to transport liquefied ammonia gas.

Explosives:
Dynamite McGyver explosives

Galvanized pipe bombs are the most common types of explosives in the United States. Small-diameter metal tubes have been used to make blasting caps.

Illegal Drugs:
Cook 4, Supplementary 3

PIPE TOBACCO

Used in

Toxins:
Nicotine

PLATINUM

Pt
7440-06-4

Appearance: Very silvery nontarnishing metal. Most likely a chemical reagent type of foil (small square metal). Could use junk platinum from a jewelry store.

Odor: None.

Hazards: None, used in jewelry.

Incompatibilities: Incompatibility Group 2-A.

Used in

Amphetamines:
Synthesis 3

Platinum is a metal catalyst that replaces palladium black or Raney nickel in the ephedrine-to-methamphetamine hydrogenation method using thionyl chloride and chloroform.

Explosives:
Process 3, Supplementary 1
Process 3, Variation 1

Regulations: None specific unless is finely divided powder; then is considered Hazard Class 4.1.

PLATINUM OXIDE
PtO

Appearance: Black powder.

Odor: None.

Used in

Amphetamines:
Synthesis 10

PLUTONIUM-238
Radioactive material
Pu

Appearance: Can be made by irradiation of neptunium-237 or as a by-product of nuclear reactors. Since 1993 plutonium-238 has been purchased from Russia. Plutonium was widely distributed by the U.S. government at one time, in order to start an industry based on this metal, which was created as a by-product in reactors. In small pieces this metal is very pyrophoric.

Emits: This is a powerful alpha source.

Half-life: 86 years.

Applications: Atomic bomb making, batteries. Although plutonium is not considered to be something generally used, and therefore not generally available, considerable amounts of this element are not accounted for.

Used in

Dirty Bombs

Regulations: Radioactive elements are highly regulated and have specific requirements for both transportation and disposal.

POLONIUM
Radioactive material
Po

Appearance: Will be in a well-marked antistatic brush. Inside the brush will be a sealed container containing the polonium. The restrictions on this metal are so strong that there will be only a few microcuries of polonium. Polonium has a half-life of 138 days, which means that the brush will not work after 6 or 7 months. Other possible sources are the oil drilling industry and instrumentation calibration.

Hazards: Polonium is a strong alpha emitter. As such, outside the body it is relatively harmless; however, inside the body it is very toxic.

Used in

Hobbies:
Photography, dusting negatives; electronics, dusting circuits

Toxins:
Polonium

Regulations: The small amount of polonium available in the brushes sold today is due to a ruling by the pre-ERDA agency in 1972. It might be easier to wait 6 months and ship this as lead.

Confirmed by: Most radiation meters do not detect alpha particles. An alpha detection meter will detect this product as a strong source.

POLYBUTADIENE
This is a product of Nippon Zeon Co.

Used in

Explosives:
C_4

Hobbies:
Rocket fuel, Variation 2

POLYESTER ADIPATE

Used in

Ammunition:
Load 3

POLYETHYLENE

Used in

Explosives:
PETN

POLYETHYLENE GLYCOL
$H(OCH_2CH_2)_nOH$

Appearance: Clear colorless viscous liquids, to waxy solids that are soluble or miscible with water. As they become more polymerized, they become more like crystalline powder or flakes.

Odor: odorless.

Hazards: Combustible, flash point 270°C. Nontoxic.

Incompatibilities: Strong oxidizers.

Used in

Hobbies:
Rocket fuel, Variation 5

Regulations: Nonhazardous for air, sea, or road freight.

POLYMERIC FOAM

Used in

Explosives:
Binary explosives, Supplementary 2
Plastic explosives, Variation 3

POLYPROPYLENE GLYCOL
$H[OCH(CH_3)CH_2)CH2]_nOH$
25322-69-4

Appearance: Viscous colorless liquid; the viscosity depends on the average molecular weight of the sample.

Odor: None found.

Hazards: MAY be harmful by inhalation; ingestion LD_{50} (rat) 56 mL/kg, and through skin absorption. Vapor pressure less than 0.01 mmHg @ 20°C.

Used in

Explosives:
Plastic explosives, Variation 1
Plastic explosives, Variation 4

POLYSTYRENE
$(C_6H_5C=CH_3)_n$
9003-53-6

Appearance: Viscosity 6 liquid with yellow-brown to pearlescent overtones. Comes in 1-pint, 1-quart and 1-gallon metal tins from any hardware store as boat resin. Can also come as white powder or beads, or as a clear solid (polymerized).

Odor: Boat resin.

Hazards: Flammable; flash point will depend on how much styrene monomer remains in the mixture. The solid forms will more likely be combustible. The liquid will polymerize. Styrene is a strong irritant, and the plastics can be irritating.

Incompatibilities: Strong oxidizing agents.

Used in

Explosives:
Plastic explosives, Variation 2

Hobbies:
Casting resins; boating

Regulations: The liquid resins can be flammable or combustible; Hazard Class 3; the solids are nonhazardous for air, sea. and road freight.

POLY(TETRA FLUOROETHYLENE) RESIN

Used in

Explosives:
Plastic explosives, Variation 6

POLYURETHANE

Formulations for polyurethane can vary greatly. Ingredients such as TDI or 4,4'-methylene bis-(2-chloroaniline) can have a significant impact on hazard.

Used in

Explosives:
Plastic explosives, Variation 3

POLY(VINYL ALCOHOL)

Appearance: Powdery off-white water-soluble plastic powder.

Odor: None.

Hazards: Harmful if ingested, inhaled, or on contact with skin.

Used in

Explosives:
PVN

Regulations: Nothing specific.

POLY(VINYL CHLORIDE)
$(CH_2CHCl)_n$
9002-86-2

Appearance: Poly(vinyl chloride) can come in a resin which looks like fine glass beads that are insoluble in water. This would probably be purchased from the manufacturer in 35-lb fiberboard drums.

Odor: None.

Hazards: Irritating to eyes, skin and respiratory system.

Incompatibilities: This is plastic; PVC plastic is fairly inert.

Used in

Fireworks:
Fuel additive, chlorine donor

Regulations: None specific to this material. This compound is highly investigated and regulated at its source to ensure that the very minimum of vinyl chloride remains in the final product.

POPPERS
See Rush.

PORCELAIN GRAINS

Used in

Fireworks:
Sparks

PORTABLE COOKER®
See Hexamine.

POTASSIUM
K
7440-09-7
UN 2257

Appearance: Potassium metal is a shiny, slightly purple, silvery metal when fresh. Should be found submerged in an organic solvent or in a sealed metal container. The potassium will probably be covered with an oxide and look like a brie cheese.

Odor: None.

Hazards: Primary hazard is water reactive. Potassium is the most reactive of the common alkali metals (cesium is more reactive). Potassium will quickly break into a purple flame when exposed to water. If there is sufficient oxide, it explodes due to encapsulation by the more rigid salt. The resulting solution is very caustic. Left in the open air, it degenerates slowly to potassium hydroxide; however, the core of the caustic remains potentially explosive. Free potassium may be found in a sludge. Potassium quickly forms a white-yellow crust of potassium oxide, even when it is submerged in an organic solvent.

Incompatibilities: Air, chloroform, chromium trioxide and heat, hydrogen, iodine, nitric acid, and water (reacts violently).

Used in

Amphetamines:
Potassium has not been found in clandestine methamphetamine laboratories at this time. It is an easy substitute for lithium and/or sodium. *See* Sodium *and* Lithium.

Synthesis 20, Supplementary 6

Illegal Drugs:
Cook 5, Variation 1

Regulations: Waste Codes D001 and D003; Hazard Class 4.3; PG I; Dangerous When Wet; ERG 138; forbidden on passenger planes; cannot be sent by U.S. mail.

Confirmed by: Violent reactions with water; lavender Flame Test.

POTASSIUM ALUMINUM

Used in

Hobbies:
Photography

POTASSIUM BIFLUORIDE
KHF_2
7789-29-9
UN 1811

Appearance: Colorless crystals, soluble in water, decomposed by heat.

Odor: Not found, but probably contains some hydrofluoric acid. *See* Hydrofluoric acid.

Hazards: Highly toxic and corrosive to tissue permeates the skin. *See* Hydrofluoric acid.

Incompatibilities: Once in contact with any moisture will become hydrofluoric acid. *See* Hydrofluoric acid.

Containers: Cannot be contained in glass!!

Used in

Chemical Weapons:
Facility 1

Regulations: Hazard Classes 6.1 and 8; ERG 154; Australian Group chemical, not scheduled.

POTASSIUM BISULFIDE
KHS_2
1312-73-8

Appearance: Solid that is soluble and hydrolyzes in water. It is a poorly defined mixture of potassium sulfide, potassium polysulfide, potassium thiosulfate, and potassium bisulfide.

Odor: Not found.

Hazards: Sulfides are toxic. Becomes corrosive, pH near 14 in water.

Incompatibilities: Strong acids; release H_2S gas.

Used in

Chemical Weapons:
Choking agents

Hobbies:
Metal working

POTASSIUM BROMIDE
KBr
7758-02-3

Appearance: White crystalline granules or powder, somewhat hydroscopic, soluble in water.

Odor: Not specified, but has a bitter saline taste.

Hazards: Moderately toxic if swallowed or inhaled.

Incompatibilities: Strong acids (produces bromine gas), oxidizers.

Used in

Hobbies:
Photography

Regulations: None specific.

Confirmed by: Bromide Test.

POTASSIUM CARBONATE
$K_2CO_3 \cdot 1\frac{1}{2} H_2O$
584-08-7

Appearance: Colorless or white small crystals (granular). It is not hygroscopic but is very soluble in water. The aqueous solution is very caustic.

Odor: None.

Hazards: Solutions are caustic pH 11.6. Irritating by inhalation; can cause skin burns; ingestion is corrosive to mouth, throat, and GI tract; can cause circulatory collapse. NFPA Health 2.

Incompatibilities: Acids cause effervescence. Incompatible with magnesium and chlorine trifluoride.

Used in

Amphetamines:
Synthesis 18, Supplementary 2

Explosives:
DINA, KNF, KDN, nitroform, yellow powder

Hobbies:
Photography

Illegal Drugs:
Cook 3, Variaton 1
Cook 5, Variation 1

Recreational Drugs:
Rush, Variation 1

Regulations: Not specifically regulated.

Confirmed by: Carbon dioxide indicator tube.

POTASSIUM CHLORATE
$KClO_3$
3811-04-9
UN 1485

Appearance: White crystalline material. Can be bought through laboratory supply houses. Can also be purchased at hardware or welding supply stores as Solidox®, which comes in an aluminum can containing six gray sticks, which must be broken up.

Odor: None.

Hazards: Oxidizer; releases oxygen when heated.

Incompatibilities: Metals, organic compounds, and acids.

Used in

Ammunitions:
Load 7
Load 8
Rim-fire primer mix

Incendiary Devices:
Situation 0.4
Situation 10

Explosives:
McGyver plastic explosive, lead nitroanilate, potassium chlorate primer

Fireworks:
Whistle formula, salutes, smoke dye formulation

Hobbies:
Rocket fuel, Variation 1

Regulations: Hazard Class 5.1; PG II; Oxidizer label; ERG 140.

POTASSIUM CHLORIDE
KCl
7447-40-7

Appearance: Depending on where it has been mined, potassium chloride can be pink-red, or if it has been purified at all, it will look pretty much like table salt. The purer the product, the more soluble it becomes in water . The majority of the potassium chloride produced is used for making fertilizer, where it may be found in 100-lb paper sacks. For poisoning people it is most likely to be seen as a square plastic jar from a chemical supply house.

Odor: None.

Hazards: This is Gatorade. At very high levels inside of the body, can stop the heart.

Incompatibilities: Inert.

Used in

Hobbies:
Photography

Illegal Drugs:
Cook 2, Supplementary 3

Toxins:
Toxic chemicals

Regulations: None specific.

Confirmed by: Chloride Test.

POTASSIUM CHROMATE
K_2CrO_4
7789-00-6
UN 3805

Appearance: Yellow crystals that are soluble in water. Will come from a laboratory supply house in glass jars or cardboard containers.

Odor: None.

Hazards: Chromates are considered to be carcinogens. Toxic by ingestion and inhalation. Oxidizer.

Incompatibilities: Organic compounds, sulfur, and aluminum.

Used in

Fireworks:
Catalyst, coating

Regulations: Hazard Classes 5.1 and 8; Oxidizing Solid Corrosive, n.o.s. PG III; RQ 375 lb. Chromates are on the TCLP and can be no more than 1% of the waste.

Confirmed by: Chromate Test.

POTASSIUM CITRATE
$K_3C_6H_5O_7 \cdot H_2O$
886-84-2

Appearance: Colorless or white deliquescent crystals or powder that is soluble in water. This is sold in 100-lb fiber drums.

Odor: None.

Hazards: Low toxicity.

Incompatibilities: Oxidizers.

Used in

Hobbies:
Photography

Regulations: Not specifically regulated.

POTASSIUM CYANIDE
KCN
151-50-8
UN 1680

Appearance: White deliquescent granular powder or fused pieces. It is gradually decomposed by the carbon dioxide in the atmosphere.

Odor: HCN; at low levels a little like chlorine, at higher levels like burned almonds.

Hazards: Poisonous!!! Ingestion and inhalation hazards. Although not likely to be absorbed through the skin, it can pass through cuts.

Incompatibilities: Acids.

Used in

Chemical Weapons:
Facility 2
Facility 13
Facility 14

Designer Drugs:
Group 1, Variation 3
Group 4, Euphoria, Variation 2

Hobbies:
Gold mining, entomology

Illegal Drugs:
Cook 3, Variation 2
Cook 9
Other, circuit board gold recovery

Toxins:
Toxic chemicals

Regulations: Hazard Class 6.1; Poison B; Shipping Poison label; ERG 157; Australian Group chemical, not scheduled.

Confirmed by: Cyanide Test.

POTASSIUM DICHROMATE
$KHCr_2O_7$
778-50-9

Appearance: Crystalline bright orange granular water-soluble solid. Probably in glass jars from a laboratory supply house; however, available in 100-lb bags and 400-lb drums.

Odor: None

Hazards: Oxidizer; poison; environmental hazard; suspected carcinogen.

Incompatibilities: Organic compounds, sulfur, and strong reducing agents.

Used in

Amphetamines:
Synthesis 0.7, variation 1

Fireworks

Hobbies:
Photography, ceramics

Regulations: Hazard Classes 5.1 and 8; Oxidizing Solid Corrosive, n.o.s. Chromates are on the TCLP and can be no more than 1% of the waste.

Confirmed by: Chromate Test.

POTASSIUM DINITROPHENATE (w)

Used in

Fireworks:
Whistle formula

POTASSIUM FERROCYANIDE
$K_4Fe(CN)_6 \cdot 3H_2O$
13943-58-3

Appearance: Yellow, soft, slightly florescent crystals that are soluble in water.

Odor: None.

Hazards: The compound has relatively low toxicity; however, upon heating it releases hydrogen cyanide gas. *See* Hydrogen cyanide.

Incompatibilities: Contact with concentrated nitric acid can release hydrogen cyanide gas.

Used in

Chemical Weapons:
Facility 8

Hobbies:
Photography, lithography

Regulations: Not specifically regulated.

Confirmed by: Cyanide Test.

POTASSIUM FLUORIDE
KF
7664-39-3
UN 1812

Appearance: White deliquescent crystalline powder. Water soluble.

Odor: Not found.

Hazards: Corrosive; etches glass. Toxic. Strong skin irritant.

Incompatibilities: Acids will make this compound into hydrofluoric acid.

Used in

Amphetamines:
Synthesis 18, Supplementary 3

Chemical Weapons:
Facility 1

Regulations: Solid requires no label; liquid requires corrosive label; ERG 154; Australian Group chemical, not scheduled. Requires export permits from the DTCC.

Confirmed by: Fluoride Test.

POTASSIUM FORMATE
HCO$_2$K
590-29-4
This compound is used to deice airports and in oil drilling muds.

Used in

Hobbies:
Photography

POTASSIUM HEXACHLOROPALLADATE (w)

Used in

Hobbies:
Photography

POTASSIUM HYDROXIDE
NaOH
1310-58-3
UN 1823 (solid)
UN 1824 (liquid)

Appearance: White deliquescent crystals or pellets which begin to take up water immediately when exposed to the air. The pellets usually appear to be wet. Can also be a clear liquid. This product is most likely obtained from laboratory supply houses in square white plastic 500- or 1000-g jars.

Odor: Stringent, but can be detected only if you are too close. The odor is much like detergent if you can detect it.

Hazards: The primary hazard of potassium hydroxide is that it is a strong corrosive. This is a dangerous skin-contact compound. There is no pain upon contact, but pain will occur after considerable damage to the eyes or skin has been completed. This product is very difficult to wash off, especially out of the eyes. LD$_{50}$ 0.5 g/kg in 10% solution.

Incompatibilities: Incompatibility Group 1-A. All acids, acetaldehyde, and chromium trioxide. Will become very hot (to boiling) in water.

Used in

Amphetamines:
Synthesis 4
Synthesis 5
Synthesis 18, Supplementary 4
Synthesis 20, Supplementary 1

Used in conjunction with hydrazine in Synthesis 5. May be used in Synthesis 4 to salt the end product, as potassium perchlorate is insoluble. Using sodium hydroxide would remove perchlorates at the same time as it was producing meth oil.

Chemical Weapons:
Facility 13

Explosives:
TA, KDN

Hobbies:
Photography

Illegal Drugs:
Cook 2, Variations 2 and 3
Cook 2, Supplementaries 1 and 2
Cook 6, Variation 1

Regulations: RQ 454 kg; Waste Code D002; Hazard Class 8; PG (liquid) II and III (solid) I; ERG 129; forbidden on passenger planes; cannot be sent by U.S. mail.

Confirmed by: pH Test.

POTASSIUM IODIDE
KI
7681-11-0

Appearance: White crystalline powder that is soluble in water.

Odor: None.

Hazards: Low toxicity, supplement up to 0.01% in table salt. Although large amounts may cause problems by ingestion, this compound is considered to be nonhazardous according to directive 67/548/EEC.

Incompatibilities: Strong reducing agents, acids, steel, aluminum, alkali metals, brass, magnesium, zinc, cadmium, copper, tin, and nickel.

Used in

Designer Drugs:
Group 2, Variation 1

Explosives:
A-NPNT, 4,4-DNB

Hobbies:
Photography

Regulations: Nonhazardous for air, sea, and road freight.

Confirmed by: Iodide Test.

POTASSIUM METABISULFITE
$K_2S_2O_5$
16731-55-8

Appearance: White granules or powder that is soluble in water and oxidizes in air. Will be in brown glass jars.

Odor: None.

Hazards: Low toxicity. Used as a food preservative. Eye and respiratory irritant.

Incompatibilities: Strong acids.

Used in

Hobbies:
Photography

Regulations: Not specifically regulated.

POTASSIUM NITRATE
KNO_3
7757-79-1
UN 1486

Appearance: White granular solid soluble in water. Can be bought from chemical supply houses or from drugstores as saltpeter.

Odor: None.

Hazards: Oxidizer. Harmful if swallowed.

Incompatibilities: Strong reducing agents, organic compounds, combustible materials.

Used in

Ammunition:
Load 1
Load 3
Load 6
Load 3, Variation 1

Chemical Weapons:
Facility 10, Supplementary 1

Explosives:
Process 2, Variation 1
Black powder, F-TNB, hexaditon, NDTT, picryl chloride, RDX, TNN, trinitroanisole, UDTNB, yellow powder

Fireworks:
Black powder, flash powder, rockets, sparks

Hobbies:
Photography
Rocket fuels, Variation 7

Regulations: Hazard Class 5.1; PG III; Oxidizer label; ERG 140.

POTASSIUM NITRITE
KNO_2
7758-09-0
UN 1488

Appearance: White granular solid soluble in water.

Odor: None.

Hazards: Oxidizer. Toxic if swallowed. Causes cyanosis. Eye, skin, and respiratory irritant.

Incompatibilities: Acids will release nitrogen tetroxide gas. *See* Dinitrogen tetroxide. Incompatible with strong reducing agents, combustible materials, cyanides, and ammonium salts.

Used in

Explosives:
DDNP, KNF, nitroform

Regulations: Hazard Class 5.1; PG II; Oxidizer label, ERG 140. Considered to be very damaging to the environment.

POTASSIUM OXALATE
$K_2C_2O_4 \cdot H_2O$
6487-48-5
UN 2811

Appearance: Colorless transparent crystals that are soluble in water.

Odor: None.

Hazards: Toxic by ingestion and inhalation.

Incompatibilities: Strong oxidizers.

Used in

Hobbies:
Photography stain remover

Regulations: Hazard Class 6.1; PG II; label Toxic Solid Organic n.o.s. (Potassium Oxalate). ERG 154.

POTASSIUM PERCHLORATE
$KClO_4$
7778-74-7
UN 1489

Appearance: Colorless crystals or white crystalline powder that are soluble in water.

Odor: None.

Hazards: Oxidizer. Will ignite organic materials. Decomposed by concussion, and organic material. Harmful if ingested, inhaled, or absorbed through the skin. Destructive to mucous membranes. May cause aplastic anemia and other blood disorders.

Incompatibilities: Organic materials and reducing agents.

Used in

Fireworks:
Sparks, colored sparks, whistle formula, salutes, strobe formations

Regulations: Hazard Class 5.1; Transport Category 2; PG II; Oxidizer label; ERG 140.

Confirmed by: Perchlorate Test.

POTASSIUM PERMANGANATE
$KMnO_4$
7711-64-7
UN 1490

Appearance: Dark purple water-soluble granular material.

Odor: None.

Hazards: Very strong oxidizer.

Incapatibilities: Organic compounds; many other oxidizers, such as hydrogen peroxide and reducing agents.

Used in

Amphetamines:
Synthesis 0.7, Variation 2

Explosives:
DNAT, nitrogen trichloride, TNA, TND

Regulations: Hazard Class 5.1; Oxidizer label; ERG 140.

POTASSIUM PHOSPHATE
$K_1H_2PO_4$, K_2HPO_4, K_3PO_4
7778-53-2 (tribasic)
7758-11-4 (dibasic)
7758-77-0 (monobasic)

Appearance: White powder or crystals that are soluble in water. Unfortunately, the source being an underground publication, the potassium phosphate was not specified. Most of the information here applies to all three; however, we have emphasized the tribasic form, as it is the most hazardous. Mono- and dibasic are not as caustic, and therefore do not have shipping requirements, as does the tribasic.

Odor: None.

Hazards: This applies mostly to the tribasic. Skin and eye irritant.

Incompatibilities: Strong oxidizers.

Used in

Illegal Drugs:
Cook 2, Supplementary 3

Regulations (applies mostly to the tribasic): Hazard Class 8; PG II.

POTASSIUM PYROSULFITE
See Potassium metabisulfite.

POTASSIUM SODIUM TARTRATE
$KNaC_4H_4O_6 \cdot 4H_2O$
6381-59-5

Appearance: Colorless transparent efflorescent crystals or white powder that is soluble in water. Also called *Rochelle salts*.

Odor: None.

Hazards: Nontoxic. Nonhazardous according to Directive 67/548/EEC.

Incompatibilities: Strong oxidizing agents.

Used in

Amphetamines:
Synthesis 18, Variation 5

Regulations: Nonhazardous for air, sea, and road freight.

POTASSIUM SULFATE
K_2SO_4
778-80-5

Appearance: Colorless or white hard crystals or powder that are soluble in water.

Odor: None.

Hazards: This is a food additive. Inhalation may cause a nuisance dust effect; ingestion may cause gastrointestinal disturbances.

Incompatibilities: Aluminum magnesium.

Used in

Ammunition:
Load 3
Load 4, Variation 1

Regulations: None specific.

Confirmed by: Sulfate Test.

POTASSIUM TETRACHLOROPALLADATE(II) (w)

Used in

Hobbies:
Photography

POTATO CHIP BAG

Used in

Incendiary Devices:
Situation 7

POTATOES

Appearance: Brown, lumpy.

Hazards: High carborhydrates.

Used in

Hobbies:
Making vodka

Illegal Drugs:
Cook 5, Supplementaries 1 and 2

POTATO SPIRIT OIL
See Isoamyl alcohol.

PRELL® detergent

Used in

Incendiary Devices:
Situation 1

PROCAINE
$C_6H_4NH_2COOCH_2N(C_2H_5)_2 \cdot HCl$
51-05-8
UN 2811

Appearance: Small colorless crystals or white crystalline powder that is soluble in water.

Odor: None.

Hazards: Used as a local anesthetic. Is a skin irritant, eye irritant, and inhalation hazard. Severe over-exposure can result in death. NFPA Health 2.

Incompatibilities: Strong oxidizers.

Used in

Amphetamines:
See Chemicals used for final cut.

Regulations: Hazard Class 6.1; PG III; label Toxic Solid, Organic n.o.s.

PROPANE
C_3H_8
74-98-6
UN 1978

Appearance: Colorless gas. Found as 400-g soldering torches and 5- or 10- gallon barbecue tanks. Small 2- oz propane torches are used to smoke methamphetamine in glass pipes. REMEMBER: Ammonia is transported in the same type of tank. Look for green or blue corrosion if ammonia is suspected.

Odor: None; may be odorized.

Hazards: Primary hazard is flammability. Group D atmosphere.

Incompatibilities: Oxygen gas.

Used in

Amphetamines:
This is an accessory to the cook. Primary uses will be heating.

Explosives:
Binary explosives, Variation 5

Incendiary Devices:
Situation 0.1

Regulations: Hazard Class 2.1; label Liquefied Petroleum Gas; ERG 115.

Monitored by: Combustible gas indicator.

PROPANE MATCH
See Propane.

2-PROPANOL
See Isopropyl alcohol.

PROPIONITRILE
C_2H_5CN
107-12-0
UN 2404

Appearance: Colorless liquid that is soluble in water.

Odor: Not found.

Hazards: Toxic by ingestion inhalation and skin absorption. Flammable liquid, flash point 6°C.

Incompatibilities: Strong oxidizing agents, strong bases, strong acids, and strong reducing agents.

Used in

Explosives:
RDX

Regulations: Hazard Class 3; PG II; ERG 131; can be shipped by air; Cyanide Solutions, n.o.s.; Poison label.

Confirmed by: Cyanide Gas Test.

PROPIONYL CHLORIDE
C_3H_6OCl
79-03-8
UN 1815

Appearance: Clear liquid that reacts with water.

Odor: Lachrymator.

Hazards: Flammable, flash point 52°F. Inhalation, causes burns to respiratory tract; ingestion, causes gastrointestinal tract burns; eyes, causes burns, lachrymator.

Incompatibilities: Water, strong bases, and alcohols.

Used in

Designer Drugs:
Group 2, Variation 2

Regulations: Hazard Class 3; PG II; ERG 132.

PROPYLENE CHLOROHYDRIN
$CH_2ClCHOHCH_3$
127-00-4
UN 2611

Appearance: Colorless liquid that is soluble in water. Most likely found in bottles from a laboratory supply house. Sold in drums.

Odor: Mild.

Hazards: Moderate fire hazard, flash point 125°F. Toxic by ingestion and inhalation. Toxicology has not been fully investigated.

Incompatibilities: Strong oxidizers.

Used in

Chemical Weapons:
Facility 2
Facility 4

Regulations: Hazard Classes 6.1 and 3; PG II; Transport Category 2; ERG 131.

PROPYLENE GLYCOL
$CH_3CHOHCH_2OH$
57-55-6

Appearance: Colorless viscosity 2 hygroscopic liquid that is miscible in water.

Odor: None.

Hazards: Inhalation, no effects reported; ingestion, relatively nontoxic, large amounts may adversely affect people with kidney problems, LD_{50} (rat) 20 g/kg; eyes, may cause transitory stinging. Flash point 210°F.

Incompatibilities: Strong oxidizing agents.

Used in

Chemical Weapons:
Facility13

Explosives:
PGDN

Hobbies:
Photography

Regulations: No specific regulations.

Confirmed by: Alcohol Test.

PSEUDOEPHEDRINE
$C_6H_5COCH_2NH_2$

Appearance: May be found as tablets, powder, or as the active ingredient in some nasal inhalers. As seen in clandestine labs: White, can be dyed blue, and becomes yellow upon standing. May be found dissolved in alcohol or water. Sudafed®.

Odor: None; may smell musty on standing.

Hazards: A large amount may be toxic by ingestion.

Used in

Amphetamines:
Pseudoephedrine is a precursor in Syntheses 1, 2, 3, 4, and 5 of methamphetamine by hydrogenation. It is not considered as efficient as ephedrine.

Regulations: CDTA (List I) 1 kg.

Confirmed by: Ephedrine Tests; NIC Test.

PSILOCYBE MYCALLIAL
Psilocybe cubensis
Psilocybe caerulipes
Psilocybe mexicana

Description: There are many genus of this mushroom that contain psilocybin. The San Ysidro (cubensis) is most common. Usually found growing on or near cow dung in pastures during warm rainy periods from February to November. Will turn purple upon bruising, but turning purple is not specific to this mushroom. *Psilocybe caerulipes* is found in the summer as a solitary mushroom on decomposing logs and debris of hardwood trees. *Psilocybe cyanescens* is found in autumn scattered, grouped, or clustered in woods, on earth, among leaves and twigs and decomposing wood in the northwestern United States. *Psilocybe mexicana* is found from May to October isolated or sparse at altitudes above 4500 feet, in limestone regions, among mosses and herbs, mostly in Mexico.

Effects: Psilocybin content is about 5 mg/g of dried mushrooms. Recreational dose is approximately 2 to 5 g of dried mushroom. This causes a pleasant, even ecstatic effect, including a deep sense of connection to others, confusion, hilarity, and a general feeling of connection to nature and the universe.

Used in

Illegal Drugs:
Cook 5, Variation 2

Recreational Drugs:
Shrooms

Regulations: Listed as Schedule 1 drugs under the United Nations 1971 Convention on Psychotropic Substances. Also listed by the DEA as a Schedule 1 drug. In many states there is ambiguity about the legal status of possession of psilocybe mushrooms and the spores. Title 21, Section I, of the *United States Code* (1970 edition) makes possession illegal. In Florida it is legal to have the mushroom as long as you have collected it; and in New Mexico it is legal to use it to manufacture psilocin as long as it is for personal use.

Confirmed by: Psilocin Test.

PVC
See Poly (vinyl chloride).

PYRIDINE
C5H5N
110-86-1
UN 1282

Appearance: Coffee-colored viscosity 2 liquid. Must be purchased from a laboratory supply house.

Odor: Like rotten soy sauce; unpleasant.

Hazards: Flammable, flash point 19°C. Caustic; inhalation, ingestion, and skin-contact hazard exposures may cause irreversible effects to the liver, kidney, and central nervous system. PEL 15 ppm.

Incompatibilities: Strong acids and strong oxidizing agents.

Used in

Amphetamines:
Synthesis 11
Synthesis 18, Variation 6
Synthesis 18, Supplementary 3

Pyridine is a reagent used to neutralize P-2-P (Synthesis 11). This method of neutralization is used in place of distillation. The final product is called red oil, a viscous orange liquid.

Chemical Weapons:
Facility 2
Facility 14

Explosives:
NPF

Tear Agents:
o-Chlorobenzylidene malonitrile

Regulations: Hazard Class 3; ERG 129.

Confirmed by: Determination by iodine.

PYROCATECHOL
See Catechol.

PYROGALLIC ACID
$C_6H_6O_3$
87-66-1

Appearance: White lustrous crystalline phenol-like material that is soluble in water.

Hazards: Toxic; however, it is used for topical skin preparations to treat certain diseases.

Used in

Hobbies:
Photography

PYROGALLOL
$C_6H_3(OH_3)$

Appearance: White lustrous crystals that turn gray on standing in light and that are is soluble in water. This is pyrogallic acid heated with three times it weight in water. *See* Pyrogallic acid.

Used in

Hobbies:
Photography

PYROSULFURIC ACID
See Oleum.

PYRROLIDINE
C_4H_9N
123-75-1
UN 1922

Appearance: Colorless-to-pale yellow liquid that is soluble in water.

Odor: Penetrating amine-like odor.

Hazards: Harmful upon ingestion; causes burns to the skin and eyes.

Incompatibilities: Strong oxidizers.

Used in

Designer Drugs:
Group 3, Supplementary 2

Regulations: Hazard Class 3.2; ERG 132.

Q

QAT
See Catha edulis.

QUEBRACHITOL
$C_7H_{14}O_6$
642-38-6

Appearance: A sweet crystalline compound that comes from quebracho bark. May be sold at health food stores. Available from chemical supply houses.

Hazards: More like sugar than anything else.

Used in

Explosives:
Isoitol nitrate, quebrachitol nitrate

3-QUINUCLIDINOL
$HC(HOHCH_2-)(CH_2CH_2-)_2N$
6238-13-7

Used in

Chemical Weapons:
Facility 13

Designer Drugs:
Variations 1, 2, and 3

Regulations: Australian Group Schedule 2B. Requires export permits from the DTCC.

R

RADIATOR FLUID
See Ethylene glycol *and* Propylene glycol.

RADIUM-226
Radioactive material
Ra

Appearance: Most likely seen as radium bromide or radium chloride, a pellet or solution encased within a ceramic outhousing. This is one of the few radioactive elements considered for dirty bombs that is naturally occurring.

Emits: Alpha and gamma rays.

Half-life: 1603 years.

Applications: Industrial applications; has been used to make glow-in-the-dark exit signs.

Used in

Dirty Bombs

Regulations: Radioactive elements are highly regulated and have specific requirements for both transportation and disposal. Exemption Quantity (EQ) 10,000 Bq.

RANEY NICKEL
Ni_2H
7440-02-3

Appearance: Grayish-black powder or cubic crystals. New catalog reagent may be stored under argon. Raney nickel contains hydrogen and several percent aluminum. Aluminum appears to be the reactive portion of the compound. Stored under water or alcohol in a chemical reagent brown glass jar. There is no commercial over-the-counter product containing Raney nickel.

Odor: None.

Hazards: Primary hazard is pyrophoric (air reactive). Raney nickel will burst into flame spontaneously upon contact with air. The smoke is toxic by inhalation. Carcinogen. Since Raney nickel will probably be used in the same reaction that uses phosphorus pentachloride, phosphorus oxychloride, or thionyl chloride, this is a very dangerous laboratory.

Incompatibilities: Incompatibility Group 2-A. Bursts into flame spontaneously on contact with air.

Used in

Amphetamines:
Synthesis 3

Raney nickel is a replacement catalyst for the second step of Synthesis 3, using thionyl chloride and chloroform to hydrogenate ephedrine to methamphetamine.

Designer Drugs:
Group 3, Supplementary 2

Illegal Drugs:
Cook 1, Variation 1
Cocaine

Regulations: Waste Code D003; Flammable Solid label; forbidden on passenger vehicles and planes with less than 40% water.

RAT NIP
See Red phosphorus.

RB-810
We could find no reference for this material. It is a commercial rocket fuel.

Used in

Hobbies:
Rocket fuel, Variation 4

RED DEVIL® DRAIN CLEANER
See Sodium hydroxide.

RED GUM
See Accroides (red) gum.

RED PHOSPHORUS
See Phosphorus (red).

RED TAR

Found in

Amphetamines:
Red tar is the result of failed P-2-P reactions.

RESORCINOL
$C_6H_4(OH)_2$
108-46-3
UN 2876

Appearance: White needle-like crystals that can become pink on standing in sunlight. Slightly soluble in water.

Odor: Slightly peculiar.

Hazards: Has been used medicinally.

Incompatibilities: Acetanilid, albumin, alkalis, camphor, ferric salts, menthol, spirit nitrous ether, urethane.

Used in

Explosives:
Styphnic acid

Hobbies:
Photography

RESORCINOL DIACETATE (w)
$(CH_2CO_2)C_8H_4$
108-58-7

Hazards: Irritant.

Used in

Explosives:
DNR

RHODAMINE B
$C_{28}H_{31}ClN_2O_3$
509-34-2

Appearance: Green crystals or reddish-violet powder that is very soluble in water. Used as a red dye.

Odor: None.

Used in

Fireworks:
Red smoke

RHODIUM(III) CHLORIDE (w)
$RhCl_3$

Appearance: Brownish-red powder, not soluble in water.

Used in

Hobbies:
Photography

RHODODENDRON
Poisonous plant

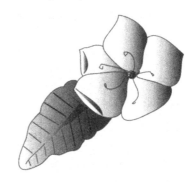

Description: Large shrub, flowers are bell-shaped and usually odorless. Azaleas, native rhododendrons, are fragrant.

Where found: All over the United States.

Deadly parts: All parts are poisonous. Symptoms: About 6 hours after ingestion. Nausea, irritation, drooling, vomiting, increased tear formation, paralysis, slowing of pulse, lowering of blood pressure, diarrhea, seizure, coma, and death. The poison is carbohydrate andromedotoxin.

Used in

Hobbies:
Gardening

Toxins:
Rhododendron ponticum

ROCHELLE SALTS
See Potassium sodium tartrate.

RODINAL

Appearance: This is a commercial product based on a very old formula that many companies produce. It is an aqueous solution of paraminophenol hydrochloride, potassium metabisulfite, and sodium hydroxide. It might be diluted with water.

Hazards: Sodium hydroxide is approximately one-third of the formula, so this is a caustic solution. *See* Sodium hydroxide.

Incompatibilities: Strong acids.

Used in

Hobbies:
Photography, developer

Regulations: RQ 454 kg; Waste Code D002; DOT Class 8, PG (liquid) II and III; ERG 129; forbidden on passenger planes; cannot be sent by U.S. mail.

ROHYPNOL
Date rape
Commercially available drug with prescription.

ROMAN WORMWOOD
See Artemisia pontica.

RONSON® LIGHTER FLUID
Mixed hydrocarbons with average around C_5. *See* Nonpolar liquids that float.

ROSIN

Appearance: Derived from pine trees.

Hazards: Harmless.

Incompatibilities: Strong oxidizers.

Used in

Ammunition:
Load 3

Regulations: None specific.

RUBBING ALCOHOL
See Isopropyl alcohol.

RUSH
Butyl nitrite, a recreational drug used to prolong orgasm during sex. Used mostly by homosexuals. Butyl nitrite replaces amyl nitrite in products for sale. *Also see* Amyl nitrite. The Fake® bottle is a competitor; we can only assume that the product inside is the same.

Used in

Recreational Drugs:
A person buying Rush® probably is not making it.

Regulations: It is illegal to sell amyl nitrite if the person buying it is going to inhale it; amyl nitrite is considered to be a medicine and therefore cannot be sold without prescription. It is therefore sold as a room odorizer or as a tape deck head cleaner.

RUTHENIUM(III) CHLORIDE (w)
$RuCl_3$

Appearance: Black deliquescent solid that is not soluble in water and decomposes in hot water.

Used in

Hobbies:
Photography

RYE GRAIN

Used in

Hobbies:
Brewing beer

Illegal Drugs:
Cook 5, Supplementary 3

RYE GRASS

Used in

Illegal Drugs:
Cook 2

S

SAFROLE
$C_{10}H_{10}O_2$
945-59-7

Appearance: Colorless-to-slightly yellow viscosity 2 liquid that is insoluble in water.

Odor: Sassafras odor, reminiscent of root beer.

Hazards: Carcinogen.

Incompatibilities: Strong oxidizers.

Used in

Amphetamines:
Synthesis 17
Synthesis 18

Used in place of P-2-P in methamphetamine synthesis to make Ecstasy.

Hobbies:
Brewing root beer

Regulation: Safrole is listed as a carcinogen. The FDA has banned the use of safrole in food.

SALICYLIC ACID

Used in

Explosives:
Lead nitroanilate

SALINE SOLUTION
Saline is a solution of sodium chloride in sterile water, used commonly for intravenous infusion, rinsing contact lenses, and nasal irrigation or jala neti. It can be obtained from medical supply houses, on the Web, or made at home.

Used in

Illegal Drugs:
Cook 2, Variation 8

SALT
See Sodium chloride.

SALT CAKE
See Sodium sulfate.

SALT PETER
See Potassium nitrate.

SAND
Sand will probably be in brown paper bags that come from a hardware supply house. The coarseness of the sand is not of concern here.

Used in

Recreational Drugs:
Sand is used to control and stabilize the temperature while heating ammonium nitrate in a reaction vessel in order to make nitrous oxide (laughing gas).

SARAN RESIN
$(C_3H_2Cl_{12})_n$

Appearance: Plastic resin.

Used in

Fireworks:
Chlorine donor

SASSAFRAS OIL

Appearance: 2% volatile oil, which is available from health food stores.

Odor: Mild aromatic.

Hazards: Used as a flavoring agent in foods.

Used in

Amphetamines:
Synthesis 18, Supplementary 1

Distilled to make safrol. *See* Safrole.

Hobbies:
Brewing root beer

Regulation: The FDA outlawed sassafras for food preparation because it contains safrole, a carcinogen.

SAWDUST

Used in

Fireworks:
Anticaking

SELENIOUS ACID
See Hydrogen selenide.

SELENIUM
Se
7782-49-2
UN 2658

Appearance: A nonmetallic element of the sulfur group. Selenium exists in several forms. It melts at 220°C, boils at 690°C. Selenium burns with a bright blue flame to form a red smoke. Insoluble in water.

Hazards: May be fatal if inhaled, swallowed or absorbed through the skin. Stomach pains, vomiting, diarrhea, coughing, and chest pains.

Used in

Chemical Weapons:
The oxide smoke is very toxic. It is used as a metal fume choking agent.

Toxins:
Chemical poisons
Difficult to detect poison.

Regulations: One of the EPA's TCLP chemicals; DOT 6.1; PG III.

SELENIUM DIOXIDE
SeO$_2$
7446-08-4
UN 3283

Appearance: Yellowish white-to-reddish powder or crystals.

Odor: Not found.

Hazards: Dangerously toxic. Fatal if inhaled, ingested LD$_{50}$ (rat) 68 mg/kg, or absorbed through the skin. Vesicant. A gram in contact with the skin may be fatal.

Incompatibilities: Organic materials, strong acids, ammonia, nitric acid, and halogen acids.

Used in

Hobbies:
Photography

Regulations: Hazard class 6.1; PG III; Transport Category 3. One of the EPA's TCLP chemicals.

SHELLAC

Appearance: Derived from the excrement of insects that feed on resiniferous trees in India. Also said to be the shell of the lac bug from Malaysia, dissolved in acetone.

Hazards: For some can be an allergen.

Used in

Fireworks:
Binder, stars

SHOCK®
See Lithium hypochlorite.

SILICA
112945-52-5

Appearance: Glassy beads, designed to melt as the fireworks heat up.

Hazards: Inert.

Used in

Fireworks

SILICA GEL
See Silicic acid.

SILICIC ACID
H$_2$SiO$_3$

Appearance: Precipitated silica. Silicic acid is a white amorphous powder insoluble in water or most acids. Silica gel is a precipitated silicic acid in the form of lustrous granules.

Odor: None.

Hazards: May cause irritation if swallowed. Carcinogen.

Incompatibilities: Avoid iron and oxidizing materials.

Used in

Chemical Weapons:
Facility 1, Variations 1 and 2
Facility 2, Variation 2

Explosives:
DDD

Regulations: None specific.

SILICONE

Used in

Fireworks

SILVER
Ag
7440-22-4

Appearance: Lustrous soft white metal.

Odor: None.

Hazards: The effects of silver are all chronic. The most noticeable is bluing of skin after many years of repeated exposure.

Incompatibilities: Most incompatible compounds are only degrading the silver; these include oxalic acid, ozone, hydrogen sulfide, and tartaric acid. Very finely divided silver is flammable.

Used in

Explosives:
Silver acetylide
Silver fulminate, Variation 2

Regulations: None specific; however, finely divided silver powder is Hazard Class 4.1. Flammable metal.

SILVER CHROMATE
$AgCrO_4$
7784-01-2

Appearance: Brown crystals or powder.

Odor: None.

Hazards: Harmful by ingestion.

Incompatibilities: Inert.

Used in

Hobbies:
Photography

Regulations: No specific DOT regulations. One of the EPA's TCLP chemicals.

SILVER IODIDE
AgI
7783-96-21

Appearance: White-to-yellow crystalline powder that is soluble in water. The material is light senstive.

Odor: None.

Hazards: May be harmful if swallowed.

Incompatibilities: Strong oxidizers.

Used in

Explosives:
Silver KNF

Hobbies:
Photography

Regulations: One of the EPA's TCLP chemicals.

SILVER NITRATE
$AgNO_3$
7761-88-8
UN 1493

Appearance: Colorless, transparent, large crystals or small white crystals. Usually found as an aqueous solution of silver nitrate.

Odor: None.

Hazards: Oxidizer, poisonous, corrosive.

Incompatibilities: Organic materials, alkali hydroxides, hydrazine, carbides, magnesium powder, antimony salts, arsenites, bromides, creosote, and ammonia.

Used in

Explosives:
Diaminesilver chlorate, silver azide, silver NENA, TNEN

Hobbies:
Photography

Regulations: Hazard Class 5.1; PG II; ERG 140. One of the EPA's TCLP chemicals.

SILVER OXIDE
AgO
20667-12-3
UN 3085

Appearance: Brownish-black heavy powder.

Odor: None

Hazards: Oxidizer. This is the least irritating of the silver preparations.

Incompatibilities: Organic materials, most common metals, ammonia, and magnesium.

Used in

Explosives:
HNH-3, silver nitride

Hobbies:
Photography

Regulations: Hazard Class 5.1; PG II; ERG 140; Oxidizing Solid, Corrosive n.o.s. One of the EPA's TCLP chemicals.

SILVER PERCHLORATE
Ag ClO$_4$
7783-93-9

Appearance: White deliquescent crystals that are very soluble in water.

Odor: None.

Hazards: Chronic ingestion may lead to bluing of the skin. Irritating® the eyes and skin.

Incompatibilities: Reducing agents and organic materials.

Used in

Explosives:
SATP

Regulations: Hazard Class 5.1; PG II; ERG 140; Oxidizing Solid, Corrosive n.o.s. One of the EPA's TCLP chemicals.

SILVER(I) SELENIDE (w)
Ag$_2$Se
1302-09-6

Appearance: Cubic thin gray plates.

Odor: Not found.

Hazards: Poison! May cause contact dermatitis.

Used in

Hobbies:
Photography, toning

Regulations: Hazard Class 6.1; Toxic Solid n.o.s. One of the EPA's TCLP Chemicals.

SIMICARBAZIDE·HCl
CO(NHNH$_2$)$_2$·HCl

Appearance: Salt of carbohydrazide. Colorless.

Used in

Explosives:
NTO

SLAKED LIME
See Calcium hydroxide.

SNAPPERS
See Rush.

SOAPSTONE
See Talc.

Used in

Ammunition:
Load 2, Variation 2

SODIUM
Na
7440-23-5
UN 1428

Appearance: Sodium metal is a shiny silvery metal when fresh. Should be found submerged in an organic solvent or in a sealed metal container. Sodium quickly forms a white yellow crust of sodium oxide even when it is submerged in an organic solvent. It may look like brie cheese or like a hardened sponge.

Odor: None.

Hazards: Pure sodium is an incredibly hazardous material. It will react violently in the presence of water, producing hydrogen and igniting. The resulting solution is sodium hydroxide. In air, sodium will form sodium oxide and peroxide on the surface.

Incompatibilities: Incompatibility Group 2-A. Air, chloroform, chromium trioxide and heat, hydrogen, iodine, nitric acid, and water.

Used in

Amphetamines:
Synthesis 2
Synthesis 9
Synthesis 16

Sodium metal is used as a catalyst in the hydrogenation of ephedrine to produce methamphetamine using the ammonia/alkali metal (Nazi) method (Synthesis 2). Sodium metal is used with ethanol in the production of methamphetamine using P-2-P and methylamine (Synthesis 9). Sodium metal is used to make sodium ethoxide (Synthesis 16). May be found in powder (ephedrine) in small plastic bags. During reaction sodium metal will be found in a blue solution in an open glass container that smells like ammonia (Synthesis 2).

Designer Drugs:
Group 1, Variation 1

Explosives:
Sodium azide

Fireworks:
Color

Illegal Drugs:
Cook 1, Variation 1
Cook 4, supplementary 2

Regulations: RQ 4.54 kg; Waste Codes D001 and D003; Hazard Class 4.3; PG I; label Dangerous When Wet; ERG 138;

forbidden on passenger planes; cannot be sent by U.S. mail.

SODIUM ACETATE (Anhydrous)
NaCH$_3$COO
27-09-3

Appearance: White crystals. It will probably come from a laboratory supply house in a square white plastic jar or in chemical reagent brown glass jars.

Odor: Vinegar.

Hazards: Prolonged contact with skin will cause irritation.

Incompatibilities: Ammonia.

Used in

Amphetamines:
Syntesis 11

Sodium acetate is one of the three ingredients of the most common method of producing P-2-P (Synthesis 11). It will be in a reaction vessel with a yellow-red liquid and a reflux column and have the odor of cat urine (due to phenyltacetic acid).

Hobbies:
Photography

Confirmed by: Acetate Test; sodium yellow Flame Test.

SODIUM ARSENITE
NaAsO$_2$
7784-46-5
UN 2027 (solid)
UN 1686 (aqueous solution)

Appearance: White or grayish-white powder, somewhat hygroscopic, absorbs CO$_2$ from the air, freely soluble in water.

Odor: None

Hazards: Very poisonous.

Used in

Chemical Weapons:
Arsenicals

Toxins:
Toxic chemicals

Regulations: Hazard Class 6.1; PG II and II (solutions), PG I (solid); Poison label; ERG 154 (solution) and 151 (solid).

SODIUM ASCORBATE
$NaC_6H_8O_6$
134-03-2

Appearance: White-to-off-white solid. Salt of vitamin C.

Odor: None.

Hazards: Believed to be harmful to health.

Incompatibilities: Strong oxidizing agents.

Used in

Hobbies:
Photography

Regulations: Nonhazardous air, sea, and road freight.

SODIUM AZIDE
NaN_3
26628-22-8
UN 1687

Appearance: Sodium azide is available from laboratory supply houses. A method to make this material is listed in Explosives Process 1. It is an explosive (used to power air bags in cars); used in the manufacture of many explosives. It is also used as a sterilizer in hospitals.

Odor: None.

Hazards: May be fatal by ingestion or inhalation. Is absorbed through the skin. May explode from heat, ignition, shock, and/or friction.

Incompatibilities: Strong oxidizers, mineral acids, water, halogen acids, halogen compounds, barium carbonate, bromine, carbon disulfide, mercury, dimethyl sulfate, common metals (especially brass, copper, lead, silver); and strong acids. Does not get along well with others.

Used in

Explosives:
ADNB, ADNBF, ammonium nitride, azidoethyl, BDPF, copper azide, DANP, DIANP, lead azide, mercury azide, silver azide, TADA, tetrazide

Illegal Drugs:
Cook 9

Regulations: Hazard Class 6.1; PG II, very toxic in the environment; ERG 153.

SODIUM BENZOATE
$C_7H_5O_2Na$
532-32-1

Appearance: White granules or crystalline powder that is soluble in water.

Odor: None.

Hazards: Used as preservative for food products. May irritate skin.

Incompatibilities: Oxidizers.

Used in

Explosives:
HNS

Fireworks:
Whistle formula

Regulations: No specific regulations.

SODIUM BICARBONATE
See Bicarbonate of soda.

SODIUM BIFLUORIDE
NaF,HF
1333-83-1
UN 2439

Appearance: White crystalline powder that is soluble in water.

Odor: Not found.

Hazards: Becomes hydrofluoric acid when wet. Very hazardous in case of ingestion. Strong irritant by skin contact, eye, and inhalation. NFPA Health 3.

Incompatibilities: When wet becomes hydrofluoric acid. *See* Hydrofluoric acid.

Used in

Chemical Weapons:
Synthesis 1

Regulations: Hazard Class 8; corrosive solid; PG II; ERG 154. Australian Group chemical, no schedule. Requires export permits from the DTCC.

SODIUM BISULFATE
$NaHSO_4$
7681-38-1 (anhydrous)
UN 3260

Appearance: Colorless crystals or white fused lumps that are soluble in water.

Odor: Not found.

Hazards: Aqueous solution is strongly acid (normal sulfuric acid). Strong irritant to tissues. Inhalation damage to mucous membranes and upper respiratory tract. Ingestion, fatal severe burns of the mouth, throat, and stomach. Skin contact burns. Eye contact can result in permanent damage.

Incompatibilities: Strong bases, calcium hypochlorite, and sodium carbonate.

Used in

Tear Agents:
Tear gas

Regulations: Although not specifically regulated, exhibits one or more characteristics of a hazardous waste, and may require analysis to determine proper disposal. Hazard Class 8; label Corrosive Solid, Acidic, Inorganic, n.o.s. PG III; ERG 154.

Confirmed by: Sulfate Test.

SODIUM BISULFITE
$NaHSO_3$
7631-90-5

Appearance: White crystalline powder. This product can be obtained from swimming pool supply stores and is used industrially as a water purifier.

Odor: None.

Hazards: The primary hazard is as a slightly corrosive solid, slightly toxic by ingestion. LD_{50} (rat) 2000 mg/kg. Harmful by inhalation or ingestion.

Incompatibilities: Strong oxidizers and acids.

Used in

Amphetamines:
Synthesis 0.7, Variation 2
Synthesis 1, Supplementary 5
Synthesis 13

Sodium bisulfite enhances the reaction between benzene and chloroacetone to produce P-2-P (Synthesis 13). Sodium bisulfite is used to reclaim P-2-P from all types of P-2-P wastes. The use of this compound to reclaim P-2-P has led to the storage of these wastes.

Explosives:
4,4-DNB, TND

Hobbies:
Photography

Illegal Drugs:
Cook 9

Toxin:
Ricin

Regulations: No specific regulations.

Confirmed by: Releases SO_2 when heated; GasTec Polytec IV indicator tube stripe 4 turns faint yellow to yellow.

SODIUM BROMATE
$NaBrO_3$
7789-38-0
UN 1494

Appearance: Colorless, odorless crystals, white granules, or crystalline powder that are soluble in water.

Odor: None.

Hazards: Oxidizer. Ingestion causes irritation and nausea, vomiting, and diarrhea. Inhalation may cause irritation. Causes irritation to eyes and skin.

Incompatibilities: Organic matter may ignite.

Used in

Date Rape:
GHB, Variation 2

Hobbies:
Gold mining

Regulations: Hazard Class 5.1; PG II; ERG 141.

Confirmed by: Bromide Test.

SODIUM BROMIDE
NaBr
7647-15-6

Appearance: White crystals, granules, or powder, absorbs moisture from the air but is not deliquescent.

Odor: None.

Hazards: Used as a nerve sedative. Consumption of large amounts may cause nausea, vomiting, and abdominal pain.

Used in

Amphetamines:
Syntheis 16

Waste of Synthesis 16 if bromobenzene is used.

Explosives:
Process 7

Hobbies:
Photography; gold mining

Regulations: None specified.

Confirmed by: Bromide Test.

SODIUM BOROHYDRIDE
NaBH$_4$
16940-66-2
UN 1426

Appearance: Flammable white to gray powder reacts, generating heat in water. A water-soluble reducing agent. Can be purchased as Hydrofin™, or through laboratory supply houses.

Odor: None.

Hazards: Strong reducing agent. Danger, CORROSIVE; causes burns to any area of contact. Flammable solid, dangerous when wet. Produces hydrogen gas when wet. NFPA Heath 3, Reactivity 2

Fire: Do not extinguish with water.

Storage: Breaks down slowly to hydrogen and sodium oxide.

Incompatibilities: Oxidizers, water, acids, transition metal catalysts, and palladium.

Used in

Amphetamines:
Synthesis 3
Synthesis 10

Designer Drugs:
Group 2, Variation 2

Explosives:
TNB

Regulations: Must be sent to an RCRA approved waste facility. Hazard Class 4.3; ERG 138.

Confirmed by: Boron Test.

SODIUM CARBONATE
Na$_2$CO$_3$
497-19-8

Appearance: Soda ash. White hygroscopic powder that absorbs water from the air.

Odor: None.

Hazards: Solutions are alkaline (pH 11.6). Inhalation irritating; can cause skin burns.

Incompatibilities: Effervesces in acid.

Used in

Amphetamines:
Synthesis 3

Similar to and used like bicarbonate of soda. It is not water reactive but it is still acid reactive. It is slightly more basic than the bicarbonate and is used in meth labs to get a pH of 11. *See* Bicarbonate of soda.

Chemical Weapons:
Facility 1, Supplementary 5
Facility 2, Variation 2
Facility 4, N1-1 and 2
Facility 4, N2-1and 2
Facility 4, N3

Designer Drugs:
Group 2, Variation 1

Explosives:
Dynamite, DATB, DNFA-P, EGDN, ETN, hexanitrate, MNA, MON, MNTA, PGDN, PETN, TNN, TNT

Hobbies:
Photography

Illegal Drugs:
Cook 1, Variation 1
Cook 6, Variation 1
Cook 8, Variation 2
Cook 9

Regulations: Not specifically regulated.

Confirmed by: Carbon dioxide indicator tube.

SODIUM CHLORATE
NaClO$_3$
7775-09-9
UN 1495

Appearance: Colorless crystals or white granules that are soluble in water.

Odor: None

Hazards: Oxidizer. Inhalation of powder may be irritating.

Incompatibilities: Keep out of contact with organic matter or other combustible substances.

Used in

Explosives:
Diaminesilver chlorate, trimercury chlorate acetylide

Fireworks

Regulations: Hazard Class 5.1; PG II; ERG 140.

SODIUM CHLORIDE
NaCl
7647-14-5

Appearance: White crystals that are soluble in water. Sodium chloride is table salt. May also be found as rock salt.

Odor: None.

Hazards: This is table salt. Tastes good. Salt is the classic example for toxicology. If a person has no salt, he or she will die. Salt certainly makes potato chips more tasty, but if one eats too many, hypertension may result. Finally, a handful of salt will kill a person. Dose makes the poison. TL$_{low}$ 12.36 g/kg.

Incompatibilities: Strong sulfuric acid will release copious amounts of hydrogen chloride gas. See amphetamines below.

Used in

Amphetamines:
Synthesis 5
Synthesis 18, Supplementary 4
Synthesis 21, Variation 2

Hydrogen chloride gas generator, producing hydrogen chloride gas for salting-out methamphetamine hydrochloride. Most likely found in sulfuric acid in 5-gallon plastic gasoline containers, modified 5-gallon propane tanks, or pop bottles under slight pressure. Can be used to make salt water for Synthesis 5. Synthesis 18, Supplementary 4

Chemical Weapons:
Facility 2

Tabun, buzz (can make hydrogen chloride with salt and H$_2$SO$_4$)

Designer Drugs:
Group 2, Variation 2

Explosives:
Process 5, Variation 1
Nitroglycerin, Variation 2
Nitromannitol, Variation 2

Illegal Drugs:
Cook 5, Variation 1
Cook 7

Regulations: Not specifically regulated.

SODIUM CYANIDE
NaCN
143-33-0
UN 1689

Appearance: White crystalline powder. Will be in storage in brown glass chemical reagent jars. There is no easily obtainable commercial source for this material.

Odor: At low levels has a chlorine odor. About 1 in 10 people can smell cyanide at 5 ppb. At this level it is reported to smell like chlorine unless the person has smelled cyanide before. Most people smell cyanide as an unpleasant bitter almond odor at about 5 ppm. The IDLH of 60 ppm is much higher than the odor threshold for most people.

Hazards: The primary hazard is as an extremely toxic solid. Reacts with acids to generate hydrogen cyanide gas. This is very toxic by ingestion. It can pass through intact skin. Hydrogen cyanide gas can pass through the skin in sufficient amounts to be lethal to persons wearing an SCBA at levels approaching 4000 ppm. Corrosive; LD_{50} 0.002 g/kg; LC_{low} 120 mg/m^3 @ 1 hour.

Incompatibilities: Incompatibility Group 5-A. All acids.

Used in

Amphetamines:
Synthesis 16

Chemical Weapons:
Facility 2, Variation 2
Facility 8, Variations 1,2, and 4

Designer Drugs:
Group 4, Euphoria, Variation 2

Hobbies:
Gold mining; photography; old glass/silver nitrate fixer (from the Civil War)

Toxins:
Chemical poisons

Regulations: Poison B; Hazard Class 6.1; PG I; ERG 157. Australian Group chemical, not scheduled. Requires export permits from the DTCC.

Confirmed by: Cyanide Test.

SODIUM CYANOBOROHYDRATE (w)
25895-60-7
An additive to SAE 70 motor oil.

Used in

Amphetamines:
Synthesis 8

Confirmed by: Boron Test, green flame.

SODIUM DICHROMATE
Na$_2$Cr$_2$O$_7$·2H$_2$O
10588-01-9

Appearance: Red or red-orange deliquescent crystals that are soluble in water. Sold in brown glass jars from a laboratory supply house, and in fiber drums.

Odor: None.

Hazards: Oxidizer; strong irritant; suspected carcinogen.

Incompatibilities: Strong acids; strong reducing agents; organic and combustible materials.

Used in

Amphetamines:
Synthesis 0.7, Variation 1

Regulations: Disposal of chrome waste is highly regulated. Chrome is an EPA TCLP chemical. Hazard Class 5.1.

SODIUM DIETHYLNAPHTHALENE SULFURIC ACID (w)

Used in

Fireworks

SODIUM DITHIONITE
$Na_2S_2O_4$
775-14-6
UN 1384

Appearance: White crystalline powder that reacts in water.

Odor: Not found.

Hazards: Harmful if swallowed; eye and skin irritant.

Incompatibilities: Strong acids, oxidizing agents, and water.

Used in

Designer Drugs:
Group 3, Supplementary 2

Regulations: Hazard Class 4.2; PG II; ERG 135.

SODIUM DODECYL SULFATE POLY-ACRYLAMIDE GEL

Used in

Toxins:
Clostridium botulinum

SODIUM FLUORIDE
NaF
7681-49-4
UN 1690

Appearance: Colorless crystals; often crushed to a white powder that is soluble in water.

Odor: None.

Hazards: Poisonous!!! May be fatal if ingested, inhaled, or absorbed through the skin. Becomes HF when wet, corrosive. NFPA Health 3.

Incompatibilities: Acids and water will make this into hydrofluoric acid. Will eat glass, reacts with silicates.

Used in

Chemical Weapons:
Facility 1, Variation 1
Facility 1, Cyclosarin

Regulations: Hazard Class 6.1; Corrosive label, PG III. ERG 154. Australian Group chemical, not scheduled. Requires export permits from the DTCC.

SODIUM HEXAFLUOROSILICATE
Na_2SiF_6
16893-85-9

Appearance: White granular powder that is insoluble in water.

Odor: Not found.

Hazards: Ingestion, abdominal cramps, vomiting, inhalation, burning sensation, cough, sore throat, skin and eye redness. TLV as F 2.5 mg/m^3.

Incompatibilities: Reacts with concentrated acids to produce hydrofluoric acid and HF gas.

Regulations: Hazard Class 6.1; PG III. Australian Group chemical, not scheduled. Requires export permits from the DTCC. Keep separate from food and feed-stuffs.

SODIUM HYDROSULFITE
See Sodium dithionite.

SODIUM HYDROXIDE
NaOH
1310-73-2
UN 1823 (solid)
UN 1824 (liquid)

Appearance: White deliquescent crystals or pellets that begin to take up water immediately when exposed to the air. The pellets usually appear to be wet. This product can be obtained from laboratory supply houses in white plastic 500-g or 1000-g jars. It is easily purchased in household products, so is most likely to be found in commercial product containers such as lye, Drano® crystals, Red Devil® lye, tile and grout cleaners, and spray oven cleaners. 100-lb multilayer paper bags are commercially available. The waste is often a gel-like fluffy sludge that contains water and a solvent as well as NaOH. The solvent coating can make the initial pH test 7. Some manipulation will bring the pH up to 14.

Odor: Stringent, but can be detected only if you are too close. The odor is much like detergent if you can detect it.

Hazards: The primary hazard of sodium hydroxide is as a strong corrosive. This is a dangerous skin-contact compound. There is no pain upon contact, but pain will occur after considerable damage to the eyes or skin has been completed. This product is difficult to wash off, especially from the eyes. LD_{50} 0.5 g/kg in 10% solution.

Incompatibilities: Incompatibility Group 1-A. All acids, acetaldehyde, and chromium trioxide. Will become hot to boiling in water.

Used in

Amphetamines:
Synthesis 0.7, Variations 1 and 2
Synthesis 0.2
Synthesis 1
Synthesis 1, Supplementary 5
Synthesis 2, Variations 1 and 2
Synthesis 18
Synthesis 18, Variation 5
Synthesis 20, Supplementary 7

pH adjustment prior to methamphetamine base extraction in most amphetamine synthesis, especially Syntheses 1 and 2. Used to produce sodium metal by electrolysis. May be found in a skillet hooked up to a car battery or battery charger, or as a white crust with gouged-out parts that may contain bits of sodium metal. *See* Synthesis 2, Supplement 1, *and* Synthesis 0.7, Variations 1 and 2. Looks like puffs of popcorn in meth oil. May also be found in buckets of urine to change the pH. See Synthesis 0.2 and Synthesis 18, Supplementaries 1 and 2.

Chemical Weapons:
Facility 1, Supplementary 1
Facility 8, Variation 4
Facility 10, Variation 3
Facility 13

Date Rape:
GHB, Variation 1

Designer Drugs:
Group 1 Variation 1
Group 2, Variation 2

Explosives:
Process 6, Variations 1 and 2
Binary explosives, Variation 5
Sodium Azide, Variations 1 and 2, copper acetylide, DAAT, DDNP, DNAT, DNP, hexol, HNB, HNS, HgNTA, lead picrate, lead azide, lead styphnate, methylpicric acid, nitrocellulose, NENA, picric acid, RDX, sodium picramate, TNA, TND

Hobbies:
Photography

Illegal Drugs:
Cook 1, Supplementary 1
Cook 2, Variations 5 and 8
Cook 4, Variations 1 and 2
Cook 4, Supplementary 2
Cook 5, Variation 1
Cook 6, Variations 1 and 2
Cook 7

Recreational Drugs:
Rush, Variation 1

Regulations: RQ 454 kg; Waste Code D002; Hazard Class 8; PG (liquid) II and III, (solid) I; ERG 129; forbidden on passenger planes; cannot be sent by U.S. mail.

Confirmed by: pH Test of 14.

SODIUM HYPOCHLORITE
NaClO
7681-52-9

Appearance: Most likely to be seen as household bleach (5% sodium hypochlorite solution). Sold at grocery stores in 1-gallon plastic bottles.

Odor: Chlorine.

Hazards: Oxidizing, mildly corrosive. Solid sodium hypochlorite is difficult to obtain and is considered to be an explosive. The aqueous solution is most likely to be encountered, and the hazard is associated with the percent sodium hypochlorite.

Incompatibilities: Ammonia; acids.

Used in

Chemical Weapons:
Facility 10

Explosives:
Astrolite, Supplementary 3

Illegal Drugs:
Cook 2, Supplementary 3

Incendiary Devices:
Situation 1

Regulations: 5% or less not regulated significantly; above 5% it is regulated as Hazard Class 8; PG III; ERG 154.

Confirmed by: Oxidizer Test.

SODIUM METABISULFITE
$Na_2S_2O_3$
7681-57-4
Sodium pyrosulfite

Appearance: White crystals or powder that is very soluble in water.

Odor: Slight odor of sulfur dioxide.

Hazards: Aqueous solutions are acid. Skin contact may cause burns. Toxic by inhalation. This chemical is used as a food preservative, however is harmful by ingestion. Solutions are acidic. NFPA Health 3, Reactivity 1.

Incompatibilities: Water, acids, and oxidizing agents.

Used in

Hobbies:
Photography

Regulations: Not specifically regulated; however, this compound does have aspects that are hazardous and it should be tested to determine the hazard prior to shipment or disposal.

SODIUM METABORATE
$NaBO_2 \cdot H_2O$
15293-77-3

Appearance: White lumps or powder.

Odor: Not found.

Hazards: Eye, skin, and respiratory irritant.

Incompatibilities: Strong oxidizing agents.

Used in

Hobbies:
Photography

Regulations: Not specifically regulated.

SODIUM MONOFLUOROACETATE
See Compound 1080.

SODIUM NITRATE
$NaNO_3$
7631-99-4
UN 1498

Appearance: Colorless, transparent crystals and white granules, which deliquesce in moist air. Dissolves in water. Is sold as a fertilizer; can be purchased from a nursery or garden shop. Fertilizer is not pure.

Odor: None.

Hazards: Oxidizer. Harmful if swallowed; eye and respiratory irritant.

Used in

Explosives:
Process 2

Fireworks:
Color

Regulations: Hazard Class 5.1; PG III; ERG 140.

SODIUM NITRITE
$NaNO_2$
7632-00-0
UN 1500

Appearance: White or slightly yellow hygroscopic granules, rods, or powder that are soluble in water.

Odor: None.

Hazards: Oxidizer. Toxic if swallowed, severe eye irritant. Respiratory and skin iurritant.

Incompatibilities: Reducing agents, strong oxidizing agents, acids (produces NOx gas), finely powdered metals, organic and combustible materials.

Used in

Amphetamines:
Synthesis 18, Supplementary 5

Designer Drugs:
Group 3, Variation 1

Explosives:
Binary explosives, Variation 5
CDNTA, CNTA, diazodinitrophenol, DNP, lead TNP, methyl picric acid, MNTA, NDTT, NINHT, nitroform, NTA, picric acid, RDX, styphnic acid, tetracene tetrazene, TNAD

Illegal Drugs:
Cook 2, Variation 6
Cook 4, Supplementaries 1 and 2
Cook 5, Variation 1

Regulations: Hazard Class 5.1; PG III; ERG 140.

Confirmed by: Oxidizer Test.

SODIUM OXALATE
$Na_2C_2O_4$
62-76-0

Appearance: White crystalline powder that is moderately soluble in water.

Odor: None.

Hazards: Toxic by ingestion.

Incompatibilities: Strong oxidizers.

Used in

Fireworks:
Color, flash delay

Regulations: Nonhazardous for air, sea, and road transport.

SODIUM PERCHLORATE
$NaClO_4$
7601-89-0
UN 1502

Appearance: White deliquesce crystals that are very soluble in water. Can be bought as compressed sticks (very impure) from hardware stores for cutting torches (Solidox®).

Odor: None.

Hazards: Oxidizer. Harmful if inhaled or swallowed.

Incompatibilities: Organic and combustible materials, powdered metals, acids, and reducing agents.

Used in

Explosives:
Binary explosives, Variation 7

Regulations: Hazard Class 5.1; PG II; label Toxic Solid n.o.s.

SODIUM PEROXIDE
Na_2O_2
1313-60-6
UN 1504

Appearance: Yellowish-white granular powder. Absorbs water and CO_2 from the air. Freely soluble in water.

Odor: Not found.

Hazards: Strong oxidizer. In water forms sodium hydroxide, and hydrogen peroxide. Corrosive, causes burns on skin, respiratory tract, or if ingested, lethal.

Incompatibilities: Any oxidizable substances can react explosively with acetic acid, acetic anhydride, benzaldehyde, carbon disulfide, glycerol, ethylene glycol, ethyl acetate, and methanol.

Used in

Explosives:
Benzoyl peroxide

Regulations: Hazard Class 5.1; PG I; ERG 144.

SODIUM PHOSPHATE
$Na_2HPO_4 \cdot 7H_2O$
13472-35-0 (monobasic)
7758-79-2 (dibasic)
10101-89-0 (tribasic)

Appearance: White crystals or granular powder.

Odor: None.

Hazards: Monobasic is used as a laxative. May be harmful if ingested in quantity. Tribasic is caustic.

Incompatibilities: Strong acids.

Used in

Hobbies:
Photography

Regulations: Not specifically regulated.

SODIUM PICRAMATE
$NaOC_6H_2(NO_2)_2NH_2$
831-52-7
UN 1349

Appearance: Yellow salt that is water soluble. Sodium picramate is made in Explosives.

Hazards: Explosive. Toxic by ingestion and skin contact.

Used in

Explosives:
Diazodinitrophenol

Regulations: Dry or wet with 20% water or more, Flammable Solids label. With less than 20% water, not acceptable in passenger vehicles or planes; ERG 113.

SODIUM SILICATE
Na_2SiO_3, $Na_6Si_2O_7$, and $Na_2Si_3O_7$
1344-09-8

Appearance: Colorless-to-white or grayish-white crystal-like pieces or lumps that are insoluble in cold water.

Odor: None.

Hazards: The aqueous solution is strongly alkaline. May cause burns through skin or eye contact; harmful by ingestion; very destructive of mucous membranes.

Incompatibilities: Acids, most metals, organic materials.

Used in

Fireworks:
Whistle formula

Regulations: No specific regulations; however, there are many hazardous aspects to this compound, and it should be tested prior to disposal.

SODIUM SULFATE
Na_2SO_4
7757-82-6

Appearance: Known as salt cake. The thirty-ninth highest-volume chemical produced in the United States. White hygroscopic powder.

Odor: None .

Hazards: Not hazardous according to Directive 67/548/EEC.

Incompatibilities: Strong acids, aluminum magnesium, and strong bases.

Used in

Amphetamines:
Synthesis 18, Supplementary 3

Chemical Weapons:
Facility 1

Designer Drugs:
Group 1, Variation 1

Explosives:
AS-20, DANP EGDN, F-TNB, isoitol nitrate, NDTT, TNEN, TNM, TNP, UDTNB

Illegal Drugs:
Cook 2, Variation 7
Cook 4, Supplementary 2
Cook 5, Supplementary 1
Cook 6, Variation 1
Cook 7
Cook 8, Variation 2

Toxins:
Ricin

Regulations: Nonhazardous for air, sea, and road freight.

SODIUM SULFIDE
$Na_2S·9H_2O$
1313-82-2
UN 1385

Appearance: White, beige, or yellow flakes, may be in granules. Deliquescent. *Also see* Hydrogen sulfide gas.

Odor: Rotten eggs.

Hazards: Forms caustic solutions, will produce toxic hydrogen sulfide gas when acidified. Harmful by inhalation, ingestion, and skin contact. Eye, skin, and respiratory irritant.

Incompatibilities: Strong acids will form hydrogen sulfide gas.

Used in

Chemical Weapons:
Facility 4, Variations 1 and 2
Mustard

Hobbies:
Photography

Illegal Drugs:
Cook 4, Supplementary 2

Regulations: Greater than 30% water, ERG 135; less than 30% water, ERG 153; less than 30% water, ERG 135. Australian Group chemical, not scheduled. Requires export permits from the DTCC.

SODIUM SULFITE
$Na_2SO_3·7H_2O$
7757-83-7

Appearance: White, small crystals or powder that is somewhat soluble in water.

Hazards: Sodium sulfite is quite stable. Harmful by ingestion, inhalation, and skin contact. Irritant.

Incompatibilities: Strong acids.

Used in

Chemical Weapons:
Facility 4

Some mustard formulas list sulfite, not sulfide. This may be an error.

Explosives:
ADNB, DIANP, DNFA-P, DPT, TAEN

Hobbies:
Photography; used in place of hypo for fixing prints

Illegal Drugs:
Cook 5, Variation 1

Regulations: Nonhazardous for air, sea, and road freight.

SODIUM THIOCYANATE
NaSCN
540-72-7

Appearance: Colorless hygroscopic deliquescent crystals or white powder that is soluble in water. Degraded by light. This product will be in brown glass jars from a chemical supply house.

Odor: None.

Hazards: Moderately toxic; 15–30 g by ingestion can be lethal. Irritant to the skin and eyes. Inhalation may cause irritation to respiratory tract.

Incompatibilities: Oxidizers: specifically chlorates, nitrates, peroxides, and strong acids.

Used in

Hobbies:
Photography

Regulations: Not specifically regulated.

Confirmed by: Thiocyanate Test.

SODIUM THIOSULFATE
$Na_2S_2O_3$
7772-98-7

Appearance: White translucent crystals or powder. Usually purchased in the form of photographic chemicals. Also sold as Antichlor®.

Odor: None.

Hazards: None.

Incompatibilities: In concentrated sulfuric acid may generate hydrogen sulfide gas, which is poisonous.

Used in

Amphetamines:
Not commonly seen. It can be used as a drying agent to improve yield. Sodium thiosulfate is a scavenger that removes water and chlorides from solutions. Most important it can remove iodine from the final product.

Hobbies:
Photography

Regulations: Nonhazardous by air, sea, and road freight.

Confirmed by: Yellow precipitate in 3 N hydrochloric acid.

SODIUM TUNGSTATE
$Na_2WO_4 \cdot H_2O$
53125-86-3

Appearance: Colorless crystals or white crystalline powder that effervesces in dry air. Soluble in water.

Hazards: Low toxicity may act as an irritant.

Incompatibilities: Inert.

Used in

Hobbies:
Photography

Regulations: Nonhazard for air, sea, or road freight.

SOLIDOX®
See Potassium chlorate.

SORBITOL
$C_6H_{14}O_8$
50-70-4

Appearance: White sweet crystals soluble in water. This is a polyol.

Odor: None; may be sweet.

Hazards: Occurs naturally in cherries, plums, pears, and apples. An additive in cosmetic creams.

Incompatibilities: Strong oxidizers.

Used in

Explosives:
Hexanitrate

Regulations: FDA GRAS (Generally Regarded as Safe).

STANNIC CHLORIDE
Choking agent (metal fume)
$SnCl_4$
7646-78-8
UN 1827

Appearance: Colorless fuming caustic liquid; soluble in water with the evolution of considerable heat. Like arsenic trichloride, this is one of the few liquid salts.

Odor: Irritating.

Hazards: Corrosive irritant may produce burns. May be lethal on inhalation or ingestion. Permeates the skin. Can cause severe damage to the eyes.

Incompatibilities: Very reactive with metals, incompatible with water alkalis.

Regulations: Hazard Class 8; PG III ERG 137.

STANNOUS CHLORIDE
$SnCl_2 \cdot 2H_2O$
10025-69-1
772-99-8 (anhydrous)
UN 1759

Appearance: White crystalline powder that is soluble in water.

Odor: Not found.

Hazards: Harmful if swallowed, inhaled, or absorbed through the skin. Irritant may cause burns on prolonged exposure.

Used in

Designer Drugs:
Group 3, Supplementary 2

Regulations: Hazard Class 8; Corrosive Solid, n.o.s. ERG 154.

STARCHs
9008-84-9
Carbohydrate polymer
See Cornstarch.

Appearance: White powder.

Odor: None.

Hazards: This is food.

Incompatibilities: Strong oxidizers.

Used in

Ammunition

Explosives

Fireworks

STATIC MASTER® (AntiStatic Brush)
See Polonium.

STEARIC ACID
$CH_3(CH_2)_{16}COOH$
57-11-4

Appearance: Greasy white leaflets. Stearic acid is animal fat; it is a fatty acid; lard. Used in industry and medicine, stearic compounds are often found in shampoos and soaps. Candles are a mixture of stearic acid and paraffin.

Odor: None.

Hazards: Stearic acid is an 18-carbon fatty acid or animal fat; mostly lard.

Incompatibilities: Strong oxidizers, reducing agents and bases.

Used in

Fireworks:
Fuel deoxidizer

Regulations: Nonhazardous for air, sea, and road freight.

STEEL BALLS
Steel balls are specified for grinding up black powder.

Used in

Ammunition:
Load 8

STEEL FILINGS

Used in

Fireworks

STEEL WOOL

Used in

Recreational Drugs:
Nitrous oxide

STERNO®

Appearance: Small metal can containing wax and methanol.

Hazards: Flammable solid.

Used in

Incendiary Devices:
Situation 0.5

STODDARD SOLVENT
Mixed hydrocarbons with formulas near C_8H_{18}. *See* Nonpolar liquids that float.

STREUNEX
See Benzene hexachloride.

STRONTIUM-90
Radioactive material
Sr

Appearance: Strontium is a 2B metal appearing much like calcium. Sr90 is produced in nuclear fission.

Emits: Gamma rays.

Half-life: 25 years.

Applications: Used as a tracer in medical and agricultural studies. It has been used as a power source in long-life lightweight power supplies, in electron tubes, and for the treatment of eye diseases and bone cancer.

Used in

Dirty Bombs

Regulations: Radioactive elements are highly regulated and have specific requirements for both transportation and disposal.

STRONTIUM CHLORIDE
$SrCl_2 \cdot 6H_2O$
10025-70-4

Appearance: Colorless or white crystals that are soluble in water.

Odor: None.

Hazards: May be harmful if swallowed.

Incompatibilities: Strong oxidizers.

Used in

Fireworks:
Color

Regulations: Nonhazardous for air, sea, and road freight.

STRONTIUM HYDROXIDE
$Sr(OH)_2 \cdot 2H_2O$
18480-07-4

Appearance: Colorless deliquescent crystals or white powder that absorbs CO_2 from the air, forming strontium carbonate. Slightly soluble in water.

Odor: None.

Hazards: The solution is very alkaline. Severe eye and respiratory irritant. Harmful if swallowed.

Incompatibilities: Inert.

Used in

Fireworks:
Color

Regulations: None specified; however, there are hazardous aspects to this compound, and some testing should be done prior to disposal.

STRONTIUM NITRATE
$Sr(NO_3)_2$
10042-76-9
UN 1507

Appearance: White granules or powder soluble in water. This is the main component of road flares.

Odor: None.

Hazards: Eye, skin, and respiratory irritant.

Incompatibilities: Sulfur (would make it burn like a road flare), strong reducing agents. Combustible materials.

Used in

Fireworks:
Color star

Regulations: Hazard Class 5.1; PG III; ERG 140.

STRONTIUM NITRITE (w)

Used in

Fireworks:
Color star

STRONTIUM OXALATE (w)
$SrC_2O_4 \cdot H_2O$

Appearance: White crystalline powder.

Odor: None.

Hazards: Toxic.

Incompatibilities: Strong oxidizers.

Used in

Fireworks:
Color

STRONTIUM OXIDE
SrO
1314-11-0

Appearance: White-to-grayish-white porous caustic mass. Strontium oxide and nitride remains after road flares burn. The most important use for this material is inside television tubes.

Odor: None.

Hazards: When treated with water, forms the hydroxide with the evolution of considerable heat.

Incompatibilities: Inert.

Used in

Fireworks:
Color

Hobbies:
Ceramics

STRONTIUM SULFATE
$SrSO_4$
7759-02-6

Appearance: White crystalline powder.

Odor: None.

Hazards: Slightly hazardous in case of skin contact, inhalation, or ingestion. NFPA Health 1.

Incompatibilities: None found.

Used in

Fireworks:
Color

Regulations: No specific regulations.

STRYCHNINE
$C_{21}H_{22}O_2N_2$
57-24-9
UN 1692

Appearance: White needles that are insoluble in water. Used as a pesticide for killing small vertebrates or rodents. Unfortunately, this has become the favorite poison of dog poisoners. This is an alkaloid poison. Very painful death.

Odor: None.

Hazards: Violent poison! $LD_{50} = 1$ mg/kg. Strychnine produces some of the best known, most dramatic, terrifying, and painful symptoms imaginable. In 10 to 20 minutes there are continuous spasms, increasing in intensity and frequency until the backbone arches. Death comes from asphyxiation caused by paralysis of the brain's breathing centers.

Incompatibilities: Strong oxidizers.

Used in

Toxins:
Toxic chemicals

Regulations: Hazard Class 6.1; Poison label; ERG 151.

STYPHNIC ACID (w)

Appearance: Yellowish crystalline graduals. Probably has to be made; *see* Styphnic acid under explosives.

Hazards: Explosive.

Used in

Ammunition:
Load 7
Rim-fire primer mix

Explosives:
Barium styphnate, lead styphnate

STYRENE
$CH_6C_2H_3$
100-42-5
UN 3129

Appearance: Colorless-to-yellowish, very refractive, oily liquid.

Odor: Strong plastic (boat resin) irritating odor.

Hazards: Polymerization and oxidation with the formation of peroxides. Will cause nosebleeds at about 100 ppm in the air. Extremely flammable, flash point 31°F. NFPA Health 2, Fire 3, Reactivity 2. PEL 100 ppm.

Incompatibilities: Strong oxidizers; with small amounts will polymerize; with large amounts will burst into flame.

Used in

Amphetamines:
Synthesis 20, Supplementary 3

Regulations: Hazard Class 3 (inhibited or stabilized); ERG 128P.

SUCROSE
See Sugar.

SUGAR
$C_6H_6O_6$

Appearance: Probably will be found in commercial grocery store paper sacks or small cardboard boxes. Some of the formulas require powdered sugar.

Odor: Sweet.

Hazards: If you have to look this up, you are definitely in the wrong business!

Incompatibilities: Strong oxidizers.

Used in

Ammunition:
McGyver gunpowder

Explosives:
McGyver explosives

Illegal Drugs:
Cook 2, Supplementary 3

Incendiary Devices:
Situation 10
Situation 11

Regulations: None specified.

Confirmed by: Sugar Test.

SUGAR OF LEAD
See Lead acetate.

SULFAMIC ACID
$H_2N·SO_2·OH$
5329-14-6
UN 2967

Appearance: White or colorless nonhygroscopic crystals. Slowly hydrolyzes in water, forming ammonium bisulfate. Has been used for flame-proofing fabrics and wood.

Odor: Not found.

Hazards: Toxic by ingestion. Becomes corrosive, irritating.

Incompatibilities: Considered to be stable.

Used in

Explosives:
KDN

Hobbies:
Photography

Regulations: Hazard Class 8; PG III; ERG 154.

SULFIDE SALTS
See Hydrogen sulfide.

SULFUR
S
7704-34-9
UN 1350

Appearance: Brimstone. Amorphous colloidal, or crystalline yellow, practically odorless and tasteless, usually in clumps, rolls, or powders. Sulfur is available from chemical supply houses and from nurseries as flowers of sulfur. Insoluble in water.

Odor: Slight sulfur.

Hazards: Flammable solid; burns to form sulfur dioxide gas.

Incompatibilities: Can form easy to ignite, or ignite by friction, mixtures with nitrates, chlorates, perchlorates, and other oxidizers. Can form ignitable mixtures with most metals (fine). Chlorine, fluorine, and hydrogen.

Used in

Ammphetamines:
Synthesis 20, Supplementaries 3 and 7

Ammunition:
Load 5
Load 9

Chemical Weapons:
Facility 3
Facility 4, Variation 5
Facility 4, N1-2

Explosives:
Black powder, DDNP, nitrogen sulfide, McGyver

Fireworks:
Black powder, rockets, fuel deoxidizer, sparks, glitter, salutes, strobe formulations

Hobbies:
Rocket fuels, Variation 7

Regulations: ERG 133.

SULFUR CHLORIDE
See Sulfur monochloride.

SULFUR DICHLORIDE
SCl_2
10545-99-0
UN 1828

Appearance: Red-brown fuming liquid that decomposes in water.

Odor: Chlorine.

Hazards: Toxic by inhalation, irritant to the lungs.

Incompatibilities: Strong bases.

Used in

Chemical Weapons:
Facility 4

Regulation: Corrosive label; forbidden on passenger vehicles; ERG 137; Australian Group Schedule 3B.

SULFUR DIOXIDE
Industrial chemical weapon
SO_2
7446-09-5
UN 1079

Appearance: Colorless nonflammable gas; will probably be found in a lab bottle.

Odor: Pungent; characteristically irritating effects on the upper respiratory system.

Hazards: An oxidizing and reducing agent. Lethal by inhalation, irritating. PEL 2 ppm.

Used in

Explosives:
Process 3, Variations 1 and 2
Process 3, Supplementary 1

Regulations: Hazard Class 2.3; ERG 125. Forbidden in passenger vehicles, planes, and rail.

SULFUR MONOCHLORIDE
SCl
10025-67-9
UN 1828

Appearance: Light amber-to-yellowish-red, fuming oily liquid.

Odor: Penetrating odor; causes tears and effects breathing.

Hazards: Inhalation.

Used in

Chemical Weapons:
Facility 4

Regulations: Corrosive label, forbidden in passenger vehicles; ERG 137; Australian Group Schedule 3B.

SULFUR TRIOXIDE (Anhydrous)
SO$_3$
7446-11-9
UN 1829

Appearance: Colorless liquid or crystals that sublimes easily, very hygroscopic; long prisms when freshly sublimed; fuming strongly in air. Boils at 45°C. Must be made on site; *see* Explosives, Process 4.

Odor: Irritant; pungent.

Hazards: Highly toxic, strong irritant, oxidizing agent, corrosive, reacts with water, evolving heat, sometimes with almost explosive violence becoming sulfuric acid.

Incompatibilities: Water, powdered metals, bases, cyanides. Does not get along well with others.

Used in

Chemical Weapons:
Corrosive smoke

Explosives:
Process 3, Supplementary 1

Illegal Drugs:
Cook 2

Regulation: Hazard Classes 4.3 and 8; forbidden in passenger vehicles, planes, and rail. ERG 137.

SULFURIC ACID
H$_2$SO$_4$
7664-93-9
UN 1830, UN 1832, UN 2796

Appearance: Colorless-to-brown oily liquid. Often found as purchased in 1-gallon chemical reagent bottles with yellow lids. Commercially available over-the-counter products include Instant Power Guaranteed Liquid Drain Cleaner® and Liquid Fire® (drain openers) and Rooto Root Killer®. Newly purchased unused automobile batteries contain usable sulfuric acid, which can be concentrated. The presence of very concentrated sulfuric acid and concentrated nitric acid is a very strong indicator of the manufacture of explosives. *See* Explosives process 3 to make 30% sulfuric acid (fuming sulfuric acid).

Odor: Irritating but rarely detected, due to low vapor pressure.

Hazards: Primary hazard is corrosive. Concentrated sulfuric acid is reactive enough to boil water, even the water in skin.

Sulfuric acid reacts with organic acids, producing heat and carbon monoxide. This is a severe hazard to the skin and the eyes. LD_{50} 2.14 g/kg.

Incompatibilities: Incompatibility Group 6-B. Water will boil upon the addition of concentrated sulfuric acid. Sulfuric acid is incompatible with organic compounds, organic acids (producing carbon monoxide), sodium hydroxide, ammonium hydroxide, nitric acid (inorganic acids due to water content), sodium, potassium, and lithium metals, ethylenediamine, and methylamine.

Used in

Amphetamines:
Synthesis 0.7 ,Variations 1 and 2
Synthesis 6
Synthesis 16
Synthesis 18, Variation 1
Synthesis 18, Supplementary 5
Synthesis 20, Supplementaries 3 and 5
Synthesis 21, Variations 1 and 2

Sulfuric acid is used primarily as a gas generator with sodium chloride to generate hydrogen chloride gas for salting-out methamphetamine base. The generation of hydrogen chloride gas is done in modified 5-gallon propane tanks or in 5-gallon plastic gasoline containers, where it is mixed with sodium chloride to generate hydrogen chloride gas. Sulfuric acid could be used to enhance the production of methamphetamine from P-2-P and methylamine as the second step (Synthesis 6). Sulfuric acid is used to enhance the second step in Synthesis 16 to make P-2-P from the intermediate α-phenylacetonitrile. Sulfuric acid enhances the production of phenylacetic acid from benzyl cyanide (Synthesis 18) and enhances the production of methamphetamine from methylaminoethane and Grignard reagent (Synthesis 21). Synthesis 0.7, Variations 1 and 2; Synthesis 18, Supplementary 2.

Chemical Weapons:
Facility 8, Variation 4
Facility 10, Variations 1 and 2

Date Rape:
Chloral hydrate

Explosives:
Process 2, Variations 1 and 2
Process 3
Process 5
Process 6

Acetone peroxide, DATB, DDNP, McGyver pop bottle, MNA, nitroglycerin, nitrated milk powder, picric acid, PNT, RDX, TNT

98% Sulfuric acid:
A-NPNT, acetone peroxide, ADN, CNTA, DATB, DATBA, DMMD, DNR, DNPU, EDGN, ETN, hexanitrate, HNTCAB, isoitol nitrate, KDN, MON, MX, NDTT, NENA, nitrocellulose, nitroglycerin, Nitro-PCB, Nitroquanidine, nitrostarch, NMHAN, NU, PEN, PETN, PGDN, picric acid, PNT, quebrachitol nitrate, RDX, styphnic acid, TAEN, TCTNB, tetraniline, tetryl, TEX, TPG, TNEN, TNN

20% Fuming:
TNM

30% Fuming:
A-NPNT, DATB, F-TNB, HNTCAB, MNA, NDTT, picryl chloride, PNT, RDX, TCTNB, TNT, UNTNB

Incendiary Devices:
Situation 11

Illegal Drugs:
Cook 1, Variations 1 and 2
Cook 1, Supplementary 1
Cook 4, Supplementaries 1 and 2
Cook 5, Variation 1
Cook 6, Variations 1 and 2
Cook 9

Recreational Drugs:
Rush

Regulations: RQ 454 kg; Waste Code D002; Hazard Class 8; PG II (>51% 8, UN 1830) (<51% 8, UN 2796) (Spent 8, UN 1832); ERG 137.

Confirmed by: Sulfate Test.

SULFURYL CHLORIDE
SO_2Cl_3
7791-25-5
UN 1834

Appearance: Colorless, very pungent liquid that is decomposed by water to sulfuric acid and HCl.

Odor: Pungent.

Hazards: Reacts with water; corrosive. May be fatal if inhaled.

Incompatibilities: Violently water reactive. Acids, alcohols, bases, metals, amines.

Used in

Explosives:
Benzoyl Peroxide, Supplementary 1

Regulations: Hazard Classes 4.3 and 8; Corrosive label; ERG 137.

T

TABLE SALT
See Sodium chloride.

TALC
$MgSiO_3 \cdot nH_2O$

Appearance: Finely powdered hydrous magnesium silicate. White-to-grayish-white very fine crystalline powder that adheres readily to the skin. Insoluble in water. Lumps are called soapstone. Sold as baby powder.

Odor: None.

Hazards: Inert.

Used in

Ammunition:
Load 3

Regulations: OSHA considers some forms of talc to be an asbestos form, which generally is not heavily regulated. However, this can be a concern when buying baby powder.

TANNIC ACID
$C_{76}H_{52}O_{46}$
1401-55-4

Appearance: Yellowish-white to light brown, amorphous, bulky powder or flakes or spongy masses. Derived from tree bark.

Odor: Faint and characteristic.

Hazards: Tannic acid is considered an antidote for many poisons.

Incompatibilities: Strong oxidizers.

Used in

Hobbies:
Photography

TARTARIC ACID
$C_4H_6O_6$
87-69-4

Appearance: Colorless, transparent monoclinic crystals, white granules, or powder. Slowly soluble in water. This is used for baking and brewing.

Odor: None.

Hazards: Tartaric acid is used in food preparation.

Incompatibilities: Oxidizers, bases, and strong reducers.

Used in

Hobbies:
Beer brewing

Illegal Drugs:
Cook 2, Variations 1, 2, and 3
Cook 2, Supplementaries 1, 2, and 3

Regulations: Nonhazardous for air, sea, and road freight.

tert-BUTYL AMINE
See Butylamine.

TETRAAMINOMETHANE (w)
$C(NH_2)_4$

We could find no distributor or manufacturer of this compound. It would be easy to make in house. It is a logical intermediate for nitroform.

Used in

Explosives:
Nitroform

TETRA ANILINE (w)

Used in

Explosives:
ADNBF

TETRAETHYL LEAD
Industrial chemical weapon
$Pb(C_2H_5)_4$
78-00-2
UN 1649

Appearance: Yellow liquid, sinks in water.

Odor: Not found.

Hazards: Fatal by inhalation, ingestion, and skin contact. LD_{50}(rat) 12.3 mg/kg. Considered highly flammable despite a flash point of 93°C and vapor pressure of 1 mmHg @ 38°C. Possible carcinogen.

Incompatibilities: Strong oxidizing agents and concentrated acids. Detonates if confined at temperatures above 110°C.

Used in

Hobbies:
Racing

Tetraethyllead is used illegally to enhance motor performance for both boats and cars.

Regulations: Hazard Class 6.1; PG I; ERG 131. One of the EPA's TLCP chemicals.

Confirmed by: Tetraethyllead Test.

TETRAHYDRO FURAN
$OCH_2CH_2CH_2CH_2$
109-99-9
UN 2056

Appearance: Colorless viscosity 1 liquid that is soluble in water. The solvent used in many PVC plastic glues. Most likely to come from laboratory supply house.

Odor: Ether-like, odor threshold 2 to 50 ppm.

Hazards: Flammable, flash point 5°F. Irritating to the eyes and mucous membranes. Narcotic in high concentrations. PEL 200 ppm, Vapor pressure 129 mmHg @ 20°C, LD_{50}(rat) 1650 mg/kg. Can form peroxides; do not allow to evaporate until dry until after testing for peroxides (see Peroxide Test). NFPA Health 2, Fire 3, Reactivity 1.

Incompatibilities: Reacts violently with air upon standing. Incompatible with lithium aluminum hydride, oxidizers, sodium aluminum hydride, and potassium hydroxide.

Used in

Amphetamines:
Synthesis 2

Recommended as the extraction solvent.

Chemical Weapons:
Facility 6, Variation 3

Date Rape:
GHB, Variation 2

Designer Drugs:
Group 2, Variation 2

Explosives:
Hexaditon, TNA, TNBCl

Regulations: Hazard Class 3; PG II; RQ 400 lb; Flammable label; ERG 127.

TETRAMETHYLLEAD
Industrial chemical weapon
$Pb(CH_3)_4$
75-74-1
UN 1649

Appearance: Clear liquid that sinks in water.

Odor: Not found.

Hazards: Lethal by inhalation, ingestion, and skin contact. Flammable, flash point 38°C.

Incompatibilities: Strong oxidizing agents.

Used in

Hobbies:
Racing

Tetramethyllead is used illegally to enhance motor performance.

Regulations: Hazard Class 6.1; PG I; ERG 131. Motor antiknock mixture. One of EPA's TLCP chemicals.

Confirmed by: Tetra ethyl lead test.

TETRANITROMETHANE
Proposed chemical war gas
$C(NO_2)_4$
509-14-8
UN 1510

Appearance: Colorless liquid insoluble in water. Difficult to purchase but can be made; *see* Explosives TNM, Variations 1 and 2.

Odor: Pungent.

Hazards: Toxic by ingestion, inhalation, and skin absorption. Dangerous fire and explosion risk. PEL 1 ppm in air, IDLH 4 ppm. Oxidizer.

Incompatibilities: Very dangerous with any of the compounds listed in this document (such as sugar, flour, cellulose, etc.) as incompatible with strong oxidizers. Almost any of them, such as paper, cotton, or cloth, when wet with TNM, become explosives. Also incompatible with alkalis, aluminum hydrocarbons metals, other oxidizers, toluene.

Used in

Explosives:
Nitroform, KNF

Regulations: Hazard Classes 5.1 and 6.1; PG I; ERG 143; forbidden in passenger planes. Rail: Oxidizer label.

THAJONE
Active ingredient in absinthe.

THALLIUM
Tl
744-02-80
UN 1707

Appearance: Bluish-white, very soft, easily fusible heavy metal, leaves a streak on paper.

Odor: None.

Hazards: Poisonous!!!! 14 mg/kg may be fatal, by ingestion, inhalation, or skin absorption.

Incompatibilities: Inert.

Used in

Toxins:
Chemical poisons
Difficult to diagnose

Regulations: Hazard Class 6.1; Shipped with Poison label. ERG 151.

Confirmed by: Flame Test.

THALLIUM NITRATE
$TlNO_3$
10102-45-1
UN 2727

Appearance: White hygroscopic cubic crystals that turn brown in water.

Odor: Should be none; a strong garlic odor should be a serious warning.

Hazards: Poisonous!!! Strong oxidizing agent. Considered fire risk.

Incompatibilities: Strong reducing agents, organic materials, and combustible materials.

Used in

Amphetamines:
Synthesis 18A

Fireworks:
Green flame

Toxins:
Chemical poisons
Difficult to detect

Regulations: Hazard Classes 6.1 and 5.1; PG II; ERG 141; Shipped with Poison label.

Confirmed by: Flame Test; crystals will turn brown upon addition of water.

THALLIUM SULFATE
Tl_2SO_4
7446-18-6

Appearance: Colorless crystals that are soluble in water. Now unavailable, but in the past was available in rat poisons and in roach poisons.

Odor: Not found.

Hazards: Poisonous!!! Ingestion, inhalation, or skin contact. A dose greater than 500 mg is considered fatal.

Used in

Toxins:
Chemical poisons
Difficult to detect

Regulations: Hazard Class 6.1; shipped with Poison label. Considered Dangerous to the environment.

THF
See Tetrahydrofuran.

THIODIETHANOL
See Thiodiglycol.

THIODIGLYCOL
$S(C_2H_4OH)_2$
111-48-8
UN 3334

Appearance: Viscosity 2 clear liquid will be in brown glass bottles or jugs. Used in printers' ink and is the solvent in the ink in ballpoint pens.

Odor: Sulfurous, foul.

Hazards: Eye irritant.

Incompatibilities: Strong oxidizers. Reacts with a wide variety of compounds, including thionyl chloride.

Used in

Chemical Weapons:
Facility 4, Variation 4

Regulations: Hazard Class 9 (IATA). Nonhazardous for road and sea freight. Australian Group Schedule 2B.

THIONYL CHLORIDE
$SOCl_2$
7719-09-7
UN 1836

Appearance: Fuming clear-to-yellow liquid. On a humid day this material reacts so quickly with water that visible clouds of hydrogen chloride will whistle as it comes out of the bottle. Usually seen in clear glass chemical reagent bottles or jugs that probably will be etched at the top of the bottle.

Odor: Very irritating.

Hazards: Water-reactive liquid. Primary hazard is a copious amount of corrosive vapors which on a very humid day can come screaming out of a newly opened container. Very concentrated hydrogen chloride gas and sulfur dioxide gases are formed in the air upon contact. Thionyl chloride boils water, creating a wet Roman candle effect.

WARNING: This compound is far more dangerous than it appears to be.

Incompatibilities: Incompatibility Group 3-B. According to 49 CFR, this compound is not to be shipped or stored with any other compound. If placed in containment, it must be marked specifically . There are also evacuation zones established for this compound in the DOT regulations.

Used in

Amphetamines:
Synthesis 3
Synthesis 18, Variation 6

Chemical Weapons:
Facility1
Facility 4, Variation 4
Facility 4, N1-1
Facility 4, N2-1
Facility 4, N3

Illegal Drugs:
Cook 4, Supplementary 2

Regulations: Waste Code D002; Hazard Class 8; PG I; ERG 137; forbidden on passenger planes; cannot be sent by U.S. mail. Australian Group Schedule 3B.

Confirmed by: GasTec Polytec IV tube stripe 2 turns yellow to red; simultaneously, stripe 4 turns blue to yellow.

THIOPHOSPHORUS TRICHLORIDE (w)
$PSCl_3$

Appearance: Colorless liquid that is decomposed by water.

Odor: Penetrating.

Hazards: Highly toxic, corrosive, strong irritant to the skin and all tissues.

Incompatibilities: Water.

Used in

Explosives:
NPF

Regulations: Hazard Classes 4.3 and 8.

THIOUREA
$(NH_2)_2CS$
62-56-6
UN 2811

Appearance: White lustrous crystals or powder, soluble in cold water. Found in bottles and in paper bags.

Odor: Not found.

Hazards: A known carcinogen. Skin irritant (allergenic), may cause skin ulcers. May be fatal if swallowed. Causes liver damage. LD_{50}(rat) 125 mg/kg.

Incompatibilities: Strong acids, strong bases, strong oxidizing agents, metallic salts, proteins, hydrocarbons. Reacts violently with acrolein.

Used in

Hobbies:
Photography

Regulations: Hazard Class 6.1; PG III; Toxic Solid n.o.s. ERG 154. OSHA Carcinogen, FDA does not a allow uee in food products.

THORIUM NITRATE
Radioactive material
$Th(NO_3)_4, 4H_2O$
13823-29-5
UN 2976

Appearance: White slightly deliquescent crystals.

Odor: None.

Hazards: Weakly radioactive (radioactivity 3300 Bq/g). Toxic by inhalation, ingestion, and skin contact. Possible carcinogen.

Incompatibilities: Reducing agents.

Used in

Amphetamines:
Facility 14

Used as a catalyst. This is one of only two radioactive compounds associated with the production of methamphetamine.

Regulations: Not specifically regulated.

THORIUM OXIDE
Radioactive material
ThO_2
1314-20-1

Appearance: White, heavy, infusible crystals and powder that is insoluble in water. Was used in gas mantles for Coleman® lanterns and as a stabilizer for TIG welding. Both are being phased out due to radioactivity.

Odor: None.

Hazards: Radioactive. Carcinogen. Gastrointestinal and liver toxin.

Incompatibilities: Inert.

Used in

Amphetamines:
Used as a catalyst. This is one of only two radioactive compounds associated with the production of methamphetamine Syn 14.

THYMOL CRYSTALS
$(CH_3)_2CHC_6H_3(CH_3)OH$
89-83-8

Appearance: White crystals very soluble in water.

Odor: Aromatic, camphor, pungent.

Hazards: Combustible, flash point 107°C. Harmful if ingested, inhaled, or if absorbed through the skin. LD_{50}(rat) 980 mg/kg.

Incompatibilities: Strong oxidizers, strong bases, and organic materials.

Used in

Hobbies:
Photography, beekeeping

Regulations: Hazard Class 6.1; PG III; harmful to the environment.

TIN (Powdered)
Sn
7440-31-5

Appearance: Almost silver-white, lustrous, soft, very malleable, and ductile metal. Probably will be found as a gray amorphous powder.

Odor: Metallic. Not everyone can smell tin.

Hazards: All very finely ground metals can burn. Harmful by ingestion and inhalation, irritating to mucous membranes, PEL 2 mg/m^3, and skin absorption may cause irritation. NFPA Health 1, Fire 2.

Incompatibilities: Strong oxidizing agents, sulfur, strong bases, and halogens.

Used in

Ammunition:
Load 9
Load 10

Chemical Weapons:
Facility 7, Variation 3

Regulations: Not specifically regulated.

Confirmed by: Tin Test.

TIN CHLORIDE
See Stannic chloride.

TIN DIOXIDE
SnO_2
242-159-0

Appearance: White powder insoluble in water.

Odor: None.

Hazards: Harmful if swallowed LD_{50}(rat) 20,000 mg/kg. Eye and skin irritant.

Incompatibilities: Inert.

Used in

Ammunition:
Load 3

Regulations: Nonhazardous for air, sea, and road transport.

Confirmed by: Tin Test.

TINCTURE OF IODINE
UN 1293

Appearance: Yellow viscosity 1 liquid that stains brown anything it touches. In brown or black plastic bottles obtained from a pharmacy.

Odor: Ethanol.

Hazards: This is a topical medicine that is toxic by ingestion and is flammable.

Ingredients: Alcohol and iodine.

Used in

Amphetamines:
Synthesis 1, Supplementary 4

The iodine is released from the solution with the addition of hydrogen peroxide. Tincture of iodine is a source of iodine (Synthesis 1). *See* Synthesis 0.5.

Regulations: Flammable label; ERG 127.

Confirmed by: Iodine Test.

TITANIUM
Ti
7740-32-6
UN 2546

Appearance: Metallic element; silvery solid or dark gray amorphous powder. Will be stored as a wet powder.

Odor: None.

Hazards: Dry powder is flammable and a serious fire risk. Dust is an eye irritant. The metal is inert (what the Star Ship *Enterprise* is made of).

Incompatibilities: Mineral acids, halogens, carbon dioxide, strong oxidizing agents.

Used in

Hobbies:
Rocket fuels, Variation 7

Fireworks:
Rockets, dark and flash effects.

Regulations: Hazard Class 4.2; ERG 135; Shipped as flammable solid, wet or dry. Forbidden in passenger transportation.

Confirmed by: Titanium Test.

TITANIUM DIOXIDE
TiO_2
13463-67-7

Appearance: Fine very white powder. Used as a paint pigment. Insoluble in water, will form a perfect suspension in water. Titanium dioxide is the forty-fourth most produced compound in the United States.

Odor: None.

Hazards: One of the most stable compounds known. Nuisance dust only.

Incompatibilities: Inert.

Used in

Ammunition:
Load 4

Regulations: No specific regulations.

Confirmed by: Titanium Test.

TITANIUM TETRACHLORIDE
Industrial chemical weapon (corrosive smoke)
$TiCl_4$
7550-45-0
UN 1838

Appearance: Colorless liquid; fumes strongly when exposed to moist air, forming a dense and persistent white cloud. Sold in glass bottles.

Odor: Pungent.

Hazards: Decomposes in air and water with the generation of heat. Solutions are corrosive. Toxic by inhalation, ingestion, and a strong irritant to the skin and tissue. Absorbed through the skin.

Incompatibilities: Reacts violently with water. Incompatible with ammonia, amines, alcohols, potassium, and other chemically active metals.

Regulations: Hazard Classes 4.3 and 8; ERG 137.

TOLUENE
$C_6H_5CH_3$
108-88-3
UN 1294

Appearance: Colorless viscosity 1 liquid. Toluene will be in 1-pint, 1-quart, or 1-gallon metal cans from a hardware store, or in 1-pint, 1-quart, or 1-gallon brown glass bottles from a laboratory supply house.

Odor: Airplane glue.

Hazards: Flammable liquid. Inhalation intoxication. Flash point 40°F, vapor pressure 36.7 mmHg @ 30°C. LEL 1.27%. This is what kills glue sniffers.

Incompatibilities: Incompatibility Group 4-A. Chromium trioxide.

Used in

Amphetamines:
Synthesis 18, Supplementary 3

Ammunition:
Load 3

Chemical Weapons:
Facility 1
Facility 2
Facility 4, Variation 4

Designer Drugs:
Group 1, Variation 1

Explosives:
Sulfur nitride, TATB, TNA, TND, TNT

Hobbies:
Rocket fuel, Variation 3

Illegal Drugs:
Cook 3, Variation 1
PCP

Regulations: RQ 454 kg; Waste Codes D001 (0.5 mg/L), U019, D018, and F005; Hazard Class 3; PG II; Labels 3; ERG 130.

Confirmed by: Determination by iodine.

TOLUENE-2,4-DIISOCYANATE
Industrial chemical weapon
$(CNO)_2C_6H_3CH_3$
784-84-9
UN 2078

Appearance: Watery clear-to-pale yellow liquid; reacts with water, producing CO_2. Found also in urethane plastics.

Odor: The PEL of this material is so far below the odor threshold that the odor provides no warning.

Hazards: Highly toxic by ingestion and inhalation. Considered to be a sensitizer; upon repeated exposure, the reaction in the lungs is allergic and can occur even below the PEL of 0.005 ppm.

Incompatibilities: Strong oxidizers.

Used in

Hobbies:
Art, urethane plastics

Regulations: Hazard Class 6.1; ERG 156.

p-TOLUENESULFONIC ACID (w)
$C_6H_4(SO_3)(CH_3)$
104-15-4
UN 2583, UN 2584, UN 2585, UN 2586

Appearance: Colorless leaflets that are soluble in water. Generally sold as a liquid or solid containing 5% or more sulfuric acid.

Odor: Not found.

Hazards: Moderately toxic; skin irritant; combustible.

Incompatibilities: Bases.

Used in

Ammunition:
Load 4

Illegal Drugs:
Cook, Variations 1 and 2

Regulations: Hazard Class 6.1; ERG 153.

TRANS-8-METHYL-*N*-VANILYL-6-NONENAMIDE

Tear Agents:
Pepper spray

TREE STUMP REMOVER
See Sodium nitrate.

TRIAZOETHANOL (w)
Solid used in experimental rocket fuels.

Used in

Explosives:
TAEN

Hobbies:
Rocket fuels

TRIBUTANOL AMINE (w)

Used in

Chemical Weapons:
HN_4

TRICALCIUM PHOSPHATE
$Ca_3(PO_4)_2$
12167-74-4

Appearance: White crystalline powder that is insoluble in water.

Odor: None.

Hazards: Nontoxic, nonflammable.

Incompatibilities: Inert.

Used in

Fireworks:
Anticaking

Regulations: No specific regulations.

2,2'2''-TRICHLOROETHYL AMINE (w)
Chemical weapon (*N*-mustard)

Used in

Explosives:
Azidoethyl

1,3,5-TRICHLORO, 2,4, 6-TRINITROBENZENE (w)

Used in

Explosives:
TATB, TNTPB

TRIETHANOL AMINE
$C_3H_3NH_2(OH)_3$
102-71-6

Appearance: Viscosity 5 or 6 clear liquid soluble in water, Comes in brown bottles.

Odor: Slight "dying" animal odor.

Hazards: Alkaline, pH near 11, used in soaps, cosmetics, and shampoos. LD_{50}(rat) 4920 mg/kg.

Incompatibilities: Strong oxidizers and acids.

Used in

Chemical Weapons:
Facility 4, NI-3

Hobbies:
Rocket fuel, Variation 2; photography

Regulations: Nonhazardous for air, sea, and road freight.

TRIETHANOL AMINE HYDROCHLORIDE (w)
$C_3H_3NH_2(OH)_3 \cdot HCl$
The salt of triethanol amine.

Regulations: Requires export permits from the DTCC.

TRIETHYL AMINE
$(C_2H_5)_3N$
121-44-8
UN 1296

Appearance: Colorless liquid that is soluble in water. Sold in 1- to 5-gallon metal containers.

Odor: Strong ammoniacal odor; lachrymator.

Hazards: Flammable, flash point −8°C; toxic by ingestion and inhalation, LD_{50}(rat) 460 mg/kg; strong irritant to tissue, causes burns.

Incompatibilities: Readily forms explosive mixtures with air. Incompatible with oxidizing agents, strong acids, ketones, aldehydes, and halogenated hydrocarbons.

Used in

Explosives:
Binary explosives, Variation 3

Regulations: Hazard Class 3; PG II; ERG 132.

TRIETHYL BENZYL AMMONIUM CHLORIDE (w)

Used in

Explosives:
HNS

TRIETHYL PHOSPHITE (w)
$C_6H_{18}O_3P$
122-52-1

Used in

Chemical Weapons:
Facility 1, Variation 2

Regulations: Australian Group Schedule 3B.

TRIFLUOROACETIC ACID (Anhydrous)
CF_3COOH
407-25-0
UN 1760

Appearance: Colorless, fuming liquid, (hygroscopic) that reacts with water.

Odor: Strong, pungent.

Hazards: Strong nonoxidizing acid, irritant to skin, toxic LD_{50}(rat) 100 mg/kg.

Incompatibilities: Reacts with water. Incompatible with strong oxidizing agents.

Used in

Explosives:
DATBA, DNFA-P, NDTT, TNTC

Illegal Drugs:
Cook 2, Variations 4 and 7

Regulations: Hazard Class 8; ERG 154; Corrosive Liquid n.o.s. CDTA drug precursor.

1,3,5-TRIFLUOROBENZENE
$C_6H_3Fl_3$
367-23-71

Appearance: Liquid.

Hazards: Flammable liquid, irritant.

Used in

Explosives:
NDTT, F-TNB, TCTNB, UDTNB

3,4,5-TRIHYDROXYBENZOIC ACID
$C_6H_2(OH)_3CO_2 \cdot H_2O$
149-91-7 (anhydrous)
5995-86-8 (monohydrate)

Appearance: Colorless or slightly yellow crystalline needles or prisms, sparingly soluble in water. Gallic acid.

Odor: None.

Hazards: Harmful by ingestion, LD_{50}(rat) (anhydrous) 5mg/kg. NFPA Health 1.

Incompatibilities: Strong oxidizers, ferric salts, ammonia, alkalis, nitrous ether, lead acetate, silver salts, chlorates, permanganate.

Used in

Fireworks:
Whistle formula

Hobbies:
Photography

Illegal Drugs:
Cook 6, Variation 1

Regulations: None specific.

TRIMETHOXYBENZALDEHYDE

$(CH_3O)_3C_6H_2CHO$

830-79-5 and/or 86-81-7

Appearance: Neither formula quoted here specified 2,4,6- or 3,4,5- trimethoxybenzaldehyde, although 3,4,5- is the most likely for use to make mescaline.

Used in

Fireworks

Illegal Drugs:
Cook 6

3,4,5-TRIMETHOXYBENZALDEHYDE (w)

$(C_3O) C_6H_2CHO$

86-81-7

Used in

Amphetamines:
Synthesis 18

TRIMETHYLBENZENE
See Mesitylene.

TRIMETHYL PHOSPHITE

$PO_3(CH_3)_3$

121-45-9

Appearance: Liquid that is moisture and air sensitive.

Odor: Very unpleasant.

Hazards: Carcinogen. Harmful if swallowed, inhaled, or absorbed through the skin, LD_{50}(rat) 1600 mg/kg. Flammable, vapor pressure 17 mmHg @ 20°C.

Incompatibilities: Strong oxidizers and strong bases.

Used in

Chemical Weapons:
Facility 1, Supplementary 1

Regulations: Hazard class 3; PG III.

TRINITROBENZYL CHLORIDE (w)

$(NO_2)_3C_6H_2CH_3Cl$

Generally, we could not find trinitro compounds being either manufactured or sold. We expect that this is because most will be explosives. We cannot guess if they will be primary or secondary explosives. Most will probably have to be made on site, probably using concentrated nitric acid and 98% sulfuric.

Used in

Explosives:
HNS

4,4,4-TRINITROBUTYLALDEHYDE (w)

Generally, we could not find trinitro compounds either being manufactured or sold. We expect that this is because most will be explosives. We can not guess if they will be primary or secondary explosives. Most will probably have to be made on site, probably using concentrated nitric acid, and 98% sulfuric.

Used in

Explosives:
TNB

TRINITROTOLUENE (TNT)

$C_6HCH_3(NO)_3$

UN 1356

Appearance: Yellow monoclinic needles that are insoluble in water.

Hazards: Toxic by ingestion, inhalation, and skin absorption. Flammable, will detonate if heated to 450°F.

Used in

Ammunition:
Load 5
Rim-fire Primer mix

Regulations: Shipping (rail) (dry) Explosive A label. Forbidden in passenger transportation. (wet with not less than 10% water) Flammable Solid label. ERG 113.

TRIOXANE (as Formaldehyde)
$C_3O_3H_6$

Appearance: Crystalline solid very soluble in water; sublimes. Will have to be made; *see* Explosives process 5. This is the cyclic trimer or formaldehyde.

Odor: Chloroform.

Hazards: *See* Formaldehyde.

Used in

Explosives:
BDPF, RDX, TNEN

TRIPROPIONYLHEXANE HYDROTRIOZINE (w)

Used in

Explosives:
RDX

TUBES
See Pipes.

TUNGSTEN
W
7740-33-7

Appearance: Metallic element; hard, brittle, gray solid. Most likely to be found in cans, barrels, or drums (as a gray powder). Tungsten is very heavy.

Odor: None.

Hazards: Finely divided tungsten is very flammable and may ignite spontaneously. Tolerance 5 mg per cubic meter of air; dust may act as a respiratory irritant.

Incompatibilities: Inert. Keep away from flame.

Used in

Fireworks:
Fuse

Regulations: Not specifically regulated; however, as a flammable metal, is hazardous and requires a label.

TUNGSTEN HEXAFLUORIDE
Industrial chemical weapon
WF_6
7783-82-6
UN 2196

Appearance: Colorless gas, or yellow liquid (boiling point 63°F) that reacts violently with water.

Odor: Odorless gas.

Hazards: Oxidizer. Toxic by ingestion and inhalation (poison gas), forms methemoblobin in the blood, with resulting cyanosis, pulmonary edema. Ingestion, produces burns in the mouth, throat esophagus, and stomach; skin contact, burns; may cause permanent damage to eyes. PEL 2.5 mg/m³ (as fluoride). Strong irritant to tissue.

Incompatibilities: Water. Oils, grease, and combustibles (accelerates combustion).

Regulations: Hazard Class 2.3. (Poison gas, corrosive) ERG 125. Empty containers must be returned to the manufacturer.

TURPENTINE
$C_{10}H_{16}$
8006-64-2
UN 1299

Appearance: Yellowish, opaque, sticky masses that are insoluble and float on water. Generally a mixture of pinene and diterpene.

Odor: Characteristic odor that is usually in liquid form.

Hazards: Moderately flammable, flash point 90 to 115°F. Generally, constants are variable for this product, as there is some variation. It can be very reactive with oxidizers, especially halogens. Toxic by ingestion only.

Used in

Incendiary Devices:
Situation 3

Hobbies:
Art

Regulations: Hazard Class 3; Flammable label. ERG 126.

U

URANIUM-238 (depleted uranium)
Radioactive material
U

Appearance: A very heavy strong brittle metal. It is the most common isotope of uranium found in nature. Around 99.28% of natural uranium is uranium-238. Only one source suggested this as a component of a dirty bomb, but the large availability and easy accessibility to this element precluded our ignoring this possibility.

Emits: Alpha rays.

Half-life: 4.46 billion years.

Applications: Used in armor-piercing ammunition. Has found use in many civilian activities, mostly due to its very high mass (it weighs a lot). It has been used to make the keels of racing sailboats, to weight one side of a racing car, and as a radiation shield since the alpha radiation can be easily shielded. The mass of the uranium is a good shield (about five times as effective as lead) for radioactive materials.

Used in

Dirty Bombs

Regulations: Radioactive elements are highly regulated and have specific requirements for both transportation and disposal.

UREA
$CO(NH_2)_2$
77-13-6

Appearance: White or colorless crystals, or waxy prills that are slowly soluble in water. Most often seen as an aqueous solution.

Odor: odorless

Hazards: Skin irritant. Used as a fertilizer.

Used in

Explosives:
ADN, DNR, MON, NDTT, NU, PEN, TEX, UDTNB, Urea nitrate

Regulations: Not specifically regulated.

URINE

Appearance: Yellow-to-colorless viscosity 1 liquid. May be found as containers of yellow liquid that may or may not contain sodium hydroxide pellets. May be seen as a yellow liquid with a clear layer above or below it in a separation funnel.

Odor: Urine!!

Hazards: Infectious viruses.

Incompatibilities: Chlorine bleach produces chlorine gas.

Used in

Amphetamines:
Synthesis 0.2

Method of production: About 80% of methamphetamine is excreted in the urine. Tweekers will collect their urine in order to recycle the meth.

Confirmed by: Urine Test.

URINAL CAKES
See PARA dichlorobenzene.

V

VANADIUM PENTOXIDE
V_2O_5
1314-62-1
UN 2862

Appearance: Yellow-to-red crystalline powder that is slightly soluble in water. Comes in drums and paper sacks.

Odor: None.

Hazards: Extremely toxic by inhalation, ingestion, or if absorbed through the skin. PEL 0.05 mg per cubic meter of air, LD_{50}(rat) 10 mg/kg.

Incompatibilities: Chlorine, chlorates, acids, and alkali metals.

Used in

Explosives:
Process 3, Variation 1

Regulations: Hazard Class 6.1; PG III; ERG 151. Hazardous to the environment.

VASOLINE®

Used in

Explosives:
McGyver plastic explosive

VICKS® SINEX and/or AFRIN

Used in

Toxins:
Toxic household chemicals

VICKS® VAPOR INHALER
$C_6H_5C_3H_6NCH_3$
l-Methamphetamine is the mirror image of d-methamphetamine. l-Methamphetamine is of no value in producing the tweeker's desired effects. Since it has no narcotic qualities, it can be sold over the counter. It can be used in conjunction with Ecstasy for a rush.

Used in

Amphetamines:
Synthesis 0.1

A 50–50 ricimic mixture of l- and d-methamphetamine can be made by ricimizing the l-mehtamphetamine.

VINEGAR

Appearance: Vinegar is a 5% solution of acetic acid. *Also see* Acetic acid.

Odor: Vinegar.

Hazards: Although acetic acid is hazardous, vinegar is consumed as food. Irritating in the eyes.

Incompatibilities: Vinegar is sufficiently acidic that it will effervesce most carbonates, and contains sufficient water to react with water reactive compounds.

Used in

Amphetamines:
Synthesis 11

Can be mixed with sodium hydroxide and used as a substitute for sodium acetate in Synthesis 11.

Explosives:
Nitrated milk powder

Regulations: Should be stored with foodstuffs.

Confirmed by: Organic Acid Test.

VINYL CHLORIDE (gas)
75-01-4
UN 1086

Appearance: Colorless gas that would most likely be in a lab bottle from a laboratory or gas supply house.

Odor: Odorless.

Hazards: Strong carcinogen; highly flammable gas. It is reported that large fires of vinyl chloride are practically inextinguishable. Will cause intoxication by inhalation PEL 1 ppm, LD_{50}(rat) 500 mg/kg. May undergo auto polymerization.

Incompatibilities: Strong oxidizers, chemically active metals,

Used in

Chemical Weapons:
Vinylarsine

Regulations: Hazard Classes 2 and 3. Forbidden on passenger planes. OSHA

Carcinogen Standard Carcinogen On the TCLP list

Monitored by: Vinyl chloride indicator tubes. PID, nonspecific.

VISINE®

Used in

Toxins

VITAMIN B

Appearance: White crystals or crystalline powder.

Odor: Slight yeast-like or nutty odor.

Hazards: This is food.

Used in

Amphetamines:
Chemicals used to cut methamphetamine

W

WATER
H_2O
7732-18-5

Appearance: Colorless viscosity 1 liquid. Water is obviously going to be found in ALL manufacturing locations: We have listed water only when it is a crucial part of the formula.

Odor: None.

Hazards: Primary hazard: drowning. May cause pronounced reversible reddening and wrinkling of the skin upon prolonged contact. Very high levels of water in the air have been known to cause inability to see well.

Incompatibilities: Alkali metals, sodium borohydride, and hot bacon grease in a frying pan.

Used in

Amphetamines:
A polar extraction solvent. Clear, pink, or slightly blue liquid over a powder or being

heated to form a paste. Color is from tablets. Since the use of a solvent is to evaporate as quickly as possible, one is more likely to find alcohol than water.

Ammunition:
Load 4

Chemical Weapons:
Facility 14

Explosives:
Benzoyl peroxide, Supplementary 2
Plastic explosives, nitrogen trichloride, styphnic acid

Illegal Drugs:
Cook 8, Variations 3 and 4

Recreational Drugs:
Nitrous oxide 2

Regulations: Regulated more as a commodity than as a waste.

Confirmed by: Water Test.

WAX

Used in

Ammunition:
Load 3
Exploding bullets

Explosives:
Tetraminecopper(II) chlorate, plastic explosives

Hobbies:
Photography

Incendiary Devices:
Situation 4
Car fire; electric wires in wax on a warm surface

WEDDING BELLS (Morning Glory)
Seeds

Used in

Illegal Drugs:
Cook 2

WEED KILLER
See Sodium chlorate.

WHITE GAS
See Nonpolar liquids that float.

WIKINITE
See Bentonite.

WINTERGREEN LEAVES

Used in

Hobbies:
Brewing root beer

WOOD ALCOHOL
See Methanol.

WOOD BLEACH
See Hydrogen peroxide.

WOOD PULP

Used in

Explosives:
Nitrocellulose

WORMWOOD
See Artemisia pontica.

X

XYLENE
$C_6H_4(CH_3)_2$
108-38-3 (*meta*)
95-47-6 (*ortho*)
106-42-3 (*para*)
UN 1307

Appearance: Colorless, mobile, flammable liquid that floats on water with a sharp curved line. *Also see* Nonpolar liquids that float. Xylene is actually a mixture of *p*-, *m*- and *o*- xylene and some ethyl benzene (100-41-4). As such, it has its own CAS number: 1330-20-7.

Odor: Aromatic, characteristic odor.

Hazards: Flammable liquid, flash point 85 to 115°F, depending on the mixture. Vapor pressure 5.1 mmHg @ 20°C. May act as a narcotic if inhaled, PEL 100 ppm. Harmful if ingested, LD_{50}(rat) 4300 mg/kg.

Used in

Explosives:
TABN

Illegal Drugs:
Cook 5, Variation 1

Regulations: Hazard Class 3; Flammable label. ERG 130. BTEXs are only allowed in certain incinerators.

Confirmed by: Determination by iodine.

Y

YEAST EXTRACT

Used in

Hobbies:
Brewing beer

Illegal Drugs:
Cook 5, Supplementaries 1 and 2
Psilocilin

YEW
Poisonous plant

Cones and berries are rarely seen

Description: Yew is an evergreen tree, up to 32 m high. Its treetop is thick and oval;

its leaves are dark green and shiny above, matte and milky green beneath. The plant produces bright red berries.

Where found: Throughout the northern hemisphere. The European yew is not a common plant in the United States. However, all species of yew are closely related and therefore very poisonous. The golden yew is a very common garden variety in the United States; the golden yew can be any one of several subspecies, including the Japanese yew, Mexican yew, Sumatran yew, or others. These tend to have yellow leaves.

Deadly part: All parts, except the red fruit. The seeds in the fruit are poisonous.

Symptoms: Symptoms start within one hour. The poison is taxine. Nausea and vomiting, increased salivation, stomach ache, diarrhea, sleepiness, and shortage of breath. Trembling, spasms, malfunction of the cardiovascular system which leads to collapse and death.

Antidotes: Survival after poisoning is rare.

Field test: Turns red in concentrated sulfuric acid.

Used in

Toxins:
Taxus baccata

Z

ZALEPTRON
Date rape drug

This is a prescription sleeping pill that has been implicated in date rape. The drug is sold as Sonata®.

ZINC (Zinc metal, powder)
Zn
7440-66-6
UN 1436

Appearance: Dull gray metal; usually found as a granular or a powder.

Odor: None.

Hazards: Zinc is generally a nuisance dust. Reacts with weak acids to produce hydrogen gas. Zinc fumes will cause a metal fume fever that is reversible.

Incompatibilities: Effervesces off flammable hydrogen gas in weak acids, and more vigorously with strong acids. Incompatible with amines, cadmium, sulfur, chlorinated solvesnts, strong acids, and bases.

Used in

Amphetamines:
Synthesis 1, Supplementary 5
Synthesis 3

Can be used as a hydrogenator in place of lithium aluminum hydride. This gives a very low yield.

Chemical Weapons:
Facility 8, Variation 5
Facility 14
Facility 15

Illegal Drugs:
Cook 4, Supplementary 1

Regulations: Zinc dust or powder is Hazard Class 4.3. Water-reactive, releasing flammable gas. ERG 138. Zinc metal is not specifically regulated.

ZINC CHLORIDE
$ZnCl_2$
7646-85-0
UN 2331

Appearance: White powder crystals or graduals that are soluble in water.
Odor: Not found.

Hazards: Poison by ingestion, lung hazard. Corrosive causes burns. LD50(rat) 350 mg/kg.

Incompatibilities: Potassium.

Used in

Explosives:
Benzoyl peroxide; Supplementary 2

Regulations: Hazard Class 8; PG III; ERG 154 (anhydrous). Very harmful in the environment; very toxic to aquatic organisms.

Confirmed by: Zinc Test.

ZINC SULFATE (Hydrated)
$ZnSO_4 \cdot 7H_2O$
7446-793-3
UN 3077

Appearance: Colorless crystals, small needles, or granular crystalline powder that are soluble in water.

Odor: None.

Hazards: Low toxicity, severe eye irritant, may cause damage. LD_{50}(rat) 1260 mg/kg.

Incompatibilities: Strong oxidizers.

Used in

Illegal Drugs:
Cook 2, Supplementary 3

Regulations: Hazard Class 8; PG III; ERG 171, Environmentally Hazardous Substance, Solid n.o.s.

Confirmed by: Zinc Test.

ZIPPO® LIGHTER FLUID
Mixed hydrocarbons with an average around C_5
See Nonpolar liquids that float.

ZIRCALIUM
ZrAl

Appearance: A metal alloy specifically made for fireworks.

Used in

Fireworks

ZIRCONIUM
Zr
7440-67-7
UN 1358, UN 2008, UN 2007

Appearance: Hard lustrous metal; grayish crystalline scales, or gray powder that is insoluble in water.

Odor: None.

Hazards: Suspected carcinogen. Dry powder is flammable or explosive as a dust the air. Ingestion, irritation of the digestive tract, low hazard. Inhalation, respiratory tract irritation, low hazard.

Incompatibilities: Air, oxygen.

Used in

Fireworks:
Burn-rate regulator, sparks

Regulations: Powder shipped wet, ERG 170; powder shipped dry, ERG 135; metal stripes/sheets, ERG 170. FDA does not permit in cosmetics.

3

MIND-ALTERING DRUGS

AMPHETAMINES[†]

Includes amphetamines, Ecstasy, ephedrine, MDA, methamphetamine, methcatanone, pseudoephedrine, speed.

Miscellaneous Amphetamine-Related Methods

Synthesis 01. Ricimizing *l*-Methamphetamine to *l*,d- Methamphetamine

Vicks Vapor Inhaler® consists of *l*-methamphetamine, which does not have narcotic qualities. The product is extracted and then ricimized in dilute hydrochloric acid. The ricimized methamphetamine is salted out.

Chemicals required

Vicks Vapor Inhaler *
Hydrochloric acid
Sodium hydroxide
Ether
Water

Synthesis 0.2. Piss Lab

Methamphetamine passes through the body and as much as 80% is excreted in the urine. The urine can be collected and made into

†Portions of this section are included with permission of HazTech Systems, Inc.

*The asterisks indicate least/most important clue in the chemical formula.

methamphetamine base by adding sodium hydroxide.

Chemicals required
Urine*
Sodium hydroxide
Hydrochloric acid

Synthesis 0.3. Dirt Lab

The soil at a mass production lab (*See* Synthesis 1, Variation 3) can be very contaminated with methamphetamine. The methamphetamine can be extracted with nonpolar solvents, then made into methamphetamine base by adding sodium hydroxide.

Chemicals required

Soil from mass production lab*
Nonpolar organic liquid
Sodium hydroxide
Hydrochloric acid

Synthesis 0.4. Extracting Amphetamines from Pills

Preparing the ephedrine or pseudoephedrine

Ephedrine labs allow a person to make money from meth without having to make meth. Selling the ephedrine to other cooks can be very profitable. Being caught in an ephedrine lab grinding and pulverizing commercial decongestant pills and tablets to free up ephedrine using mixers, coffee grinders, meat grinders, and food processors will probably not lead to prosecution.

Chemicals Used for Illegal Purposes: A Guide for First Responders to Identify Explosives, Recreational Drugs, and Poisons, By Robert Turkington
Copyright © 2010 John Wiley & Sons, Inc.

Ephedrine and pseudoephedrine can be extracted from tablets more quickly if they are ground up. Meth cooks are likely to be on methamphetamine and therefore impatient. Red, white, or blue powders found in blenders or caked on filters (called *cookie dough*) are more than likely pulverized ephedrine or pseudoephedrine precursor. Besides the precursor, the powder may contain, other medications acting as active ingredients and inert compounds, referred to as *binders* (e.g., cellulose, cornstarch, carnauba wax, lactose). This would be the result of the first step in preparing to extract ephedrine from tablets. Once extraction has been completed, the ephedrine powder may once again be ground. This is not necessary.

Extraction of ephedrine or pseudoephedrine.

i. The base is extracted in a polar solvent such as methanol, ethanol, isopropyl alcohol, or water.
ii. A polar solvent is added to the pulverized pills. Unpulverized pills can be extracted using solvents; it just takes longer. Water can be used, but it takes much longer to evaporate after the extraction. Everclear®, vodka, proprietary solvent, isopropyl alcohol, and methanol are readily available and can all be used. The polar solvent can be added directly to the grinding device, or the powder can be poured into any cupped container to extract the ephedrine. Once again the light red (pink) and blue colors of either a liquid over a residual powder or paste or just the powder and paste are indicators of this preparation. The cookie dough–like material remaining after this stage will have sufficient ephedrine to be positive in a NIC Test.
iii. There is a concerted effort by the DEA and manufacturers of the pills to make it difficult to extract the ephedrine. Besides the filler that is often a carbonate, there is carnauba wax, water-soluble fiber, and a compound called *guarfenesin*, which mimics the reactions of ephedrine and

becomes a real problem for the cook when trying to separate the methamphetamine in the final stages. To overcome these other compounds, the amphetamine that is usually in the form of the hydrochloride can be changed to the base, extracted with a nonpolar liquid, then extracted again with a polar solvent or in the presence of bicarbonate of soda. The nonpolar solvent could be mineral spirits or toluene. The base is made using sodium hydroxide. Hydrochloric acid is necessary to return to the ephedrine hydrochloride form in an operation that is the same as salting.

Filtering of ephedrine liquid to remove other tablet materials

i. Funnels, cloth, coffee filters, and Whatman filters have all been used as filters at this point and in other parts of the cook.
ii. The polar solvent containing the ephedrine or pseudoephedrine must be separated from the fillers in the pulverized pills. There are three types of funnels: restricting, addition, and separatory. The standard household restricting funnel with Whatman filter paper is probably the most effective for this operation. However, the funnel used may be a coffee filter paper holder with coffee filter, a handkerchief held in place with a rubber band over a plastic pop bottle that has been cut in half, a household funnel, or any hoop. Once again, look for the red (pink) or blue coloring on the filtering material. A white filtrate could be anything. A dark red filtrate is probably phosphorus from another step.

Reclamation of ephedrine and pseudoephedrine

i. Flat casserole dishes, hot plates, or some other form of heat, and an impervious surface. Teflon pans are becoming increasingly popular for this activity.
ii. The flat casserole dish is the preferred method of recovering the ephedrine by evaporation of the polar solvent. This utensil has many other uses in the

methamphetamine kitchen. Any residual powder on a casserole dish should be tested for ephedrine as well as methamphetamine. If water was used as the solvent, the residual powder will often be slightly blackened, as the cook will overheat the material to speed up the evaporation process.

iii. Hot plates are most commonly associated with the evaporation of a polar solvent from ephedrine. Generally, hot plates are not used for temperature control in other reactions. They are associated with the evaporation of solvents from either the ephedrine extraction or methamphetamine extraction.

iv. Other heat sources include Sterno® candles, cook stoves, and oven ranges. Oven ranges are the most dangerous method and have been associated with many "kitchen of death" explosions and fires.

Chemicals required

Ephedrine or pseudoephedrine*
Polar solvent (water, ethanol, methanol)
Nonpolar liquid that floats
Bicarbonate of soda
Sodium hydroxide
Hydrochloric acid

Synthesis 0.5. Making Speed

Place a piece of soft coal in a bowl. In another bowl, add table salt, Blue boy® or Rosebud bluing®, water, and concentrated ammonia. Pour the solution in bowl 2 over the coal; crystals of methamphetamine will form.

Chemicals required

Soft coal
Blue boy or Rosebud bluing*
Water
Table salt
Ammonium hydroxide

Synthesis 0.6. Cleaning Up Low-Grade Meth

Tweeker-made methamphetamine can be yellow, sticky, and tarry. A person choosing to try to clean it can extract the material with a nonpolar liquid and then treat it with bicarbonate of soda, sodium thiosulfate, hydrochloric acid, and sodium hydroxide. There is not much apparatus associated with this operation.

Chemicals required

Nonpolar liquid that floats
Bicarbonate of soda
Hydrochloric acid
Sodium thiosulfate
Sodium hydroxide

Synthesis 0.7. Methcatinone

If ephedrine is oxidized instead of reduced, the end product is methcathinone (Cat). Oxidation requires chromium +6 and sulfuric acid. The product can be purified in the same way as methamphetamine. Indicators of this synthesis are a green waste. People using methcathinone have a characteristic unpleasant body odor.

Chemicals required

Variation 1

Ephedrine
98% Sulfuric acid
Sodium dichromate* or
Potassium dichromate*
Sodium hydroxide
Ether
Magnesium sulfite
Hydrogen chloride gas
Methylene chloride

Variation 2

Ephedrine
Ether
Hydrogen chloride gas
Glacial acetic acid
Potassium permanganate*
Sodium bisulfite
Sodium hydroxide
5% Sulfuric acid
Magnesium sulfate
Hexane

List of materials used to cut methamphetamine (these will not be used all together)

Mamitol (baby laxative)
Caffeine
Lactose
Dimethyl sulfone
Bicarbonate of soda
Vitamin B
Procaine
Lidocane
Methylsulfonylmethane (MSM)

METHOD 1. Reduction: Ephedrine or Pseudoephedrine Hydration

There are three principal methods of making methamphetamine. Each method has several syntheses, and many syntheses have several variations and supplementary reactions. Determining the dividing line between a synthesis and a variation is difficult in many cases. For ease of understanding we have grouped the various synthesis as best as science and art will let us. It is likely that other methods, synthesis, or variations of these methods will appear as essential ingredients become more difficult to obtain, or simpler modifications of existing syntheses are found.

The quickest, easiest, and therefore most popular syntheses are the first two syntheses presented. These syntheses involve the hydrogenation (reduction) of ephedrine or pseudoephedrine. Ephedrine and pseudoephedrine are commercially and legally available and can be used to produce methamphetamine.

The only structural difference between ephedrine or pseudoephedrine and methamphetamine is the hydroxyl group (OH) on the first carbon of ephedrine or pseudoephedrine. The removal of a functional group and its replacement by hydrogen in organic chemistry is called *reduction* or *hydrogenation*. There are many ways in which this can be accomplished. Benzyl alcohols are the easiest of all alcohols to reduce to their corresponding hydrocarbon.

Five-hydrogenation syntheses for the reduction of ephedrine are presented here.

Pseudoephedrine does not reduce as efficiently as ephedrine.

Synthesis 1. Red P Hydriodic Acid

Currently, the most commonly encountered clandestine laboratory method of production is the hydriodic acid and red phosphorus method. There are a number of variations on this basic theme, but the principles remain the same. Hydriodic acid dissociates at high temperatures. The hydrogen causes the reduction; the iodine is pulled away by the formation of PI_3, which reacts further with water to form phosphorous acid. It is interesting to see how these components appear in the variations. Ephedrine and pseudoephedrine are reduced with hydriodic acid in the presence of red phosphorus to produce methamphetamine. It appears that the presence of either iodine or phosphorus in another compound make that compound a good candidate to use for this synthesis. In the cold cook method, simply placing the two elements together with ephedrine is sufficient to produce methamphetamine. Variations result from improvisations in the production of hydriodic acid, substitute sources of phosphorous, and radical differences in equipment and apparatus.

Because it is so common, this method has a wide range of users. Pros, amateurs, commercial enterprises, and dabblers each perform this synthesis with their own characteristic level of ingenuity and skill. Both the high-tech operation (California method) and the low-tech operation (Bevis and Butthead) are presented to show the wide range in materials and apparatus. Usually, an operation will be somewhere between these two extremes.

The idealized reaction consists of pure imported ephedrine refluxed in a round-bottomed reaction vessel.

Variation 1. High End

California lab

The high-end California labs are well-stocked facilities with the intention of obtaining maximum high-quality product yield for cut and distribution. Apparatus setups consist of a round-bottomed double-necked reaction vessel. A condenser column in the reflux configuration is located in one of the necks. This allows for maximum vapor recovery. The second neck will hold an addition funnel for the hydriodic acid. This allows the cook to add the hydriodic acid slowly. The contents of the funnel will be clear to yellow if the operator has high-quality hydriodic acid, but will be the color of grape juice if the hydriodic acid is homemade. The reaction vessel will be nested in a heating mantle. The contents will be a mushy red from the red phosphorus. Generally, glassware and many of the reagents will be of laboratory quality. The product is separated in a reparatory funnel and filtered. Salting is done with hydrogen chloride gas.

Chemicals required

Ephedrine or pseudoephedrine
Hydroiodic acid*
Red phosphorus*
Sodium hydroxide
Nonpolar liquids that float
Hydrogen chloride gas

Variation 2. Low End

Mom and Pop, Tweaker, or Bevis and Butthead lab

Usually, these labs make their own hydriodic acid. Sudafed® pills are the source of pseudoephedrine, which is extracted with vodka. They may have pop bottles for hydrogen chloride generators, and many matchbooks as a source of red phosphorus. Methamphetamine is extracted with ether, usually from containers of engine starter fluid. A metal or glass mixing bowl filled with cooking oil may serve as the heating mantle. Often, the reaction vessel will be vented through a container of kitty litter to keep odors from attracting neighbors.

These labs are often found in the kitchens of homes. The stove and refrigerators are often part of the apparatus. Hobbies, automotive repair, and general kitchen paraphernalia may be mixed in with the debris of the meth lab. The lab will generally be colored red from the red phosphorus. Look for a container of urine for recycled meth. There may be a hydrogen sulfide odor, as match strike plates also contain antimony sulfide that can release some hydrogen sulfide during the cook.

Chemicals required

Ephedrine or pseudoepherdrine
Matches*
Tinture of iodide*
Sodium hydroxide
Nonpolar liquids that float
Table salt
Sulfuric acid

Variation 3. Mass Production Lab

Mexican national lab

The variations of this synthesis may look remarkably different, but they are essentially the same. A mass production lab is a seriously commercial operation and is massive in size. There will be several reaction vessels. There is no need to reflux the product, as the small amount of methamphetamine escaping through the exhaust hoses is small compared to the total operation. Most likely the responder will find single-necked reaction vessels hooked up to vacuum cleaner hoses or other 2-inch hoses, which are vented to kitty litter or charcoal in a pail to keep the odors of the laboratory to a minimum. Filtering removes the majority of hydrogen iodide gas and can prevent cooks from getting a free high. A minimum is still more than you would like to be in, as hydriodic acid and methamphetamine vapors do escape, and people in these labs suffer from exposure.

Chemicals approaching methamphetamine in reaction vessels tend to take on the smell of P-2-P and phenylacetic acid, a sweet honey-to-cat urine smell that can attract the attention of neighbors. Ephedrine will usually be in 55-gallon fiberboard drums imported from China through

Mexico. Mass production labs are most likely to be in country areas where the operation is less likely to be noticed. They have been located in storefront buildings in urban areas. Recently, there is a growing tendency for these labs to remain in Mexico, transporting the finished product across the border.

Chemicals required

Ephedrine or pseudoephedrine
Hydroiodic acid
Red phosphorus
Sodium hydroxide
Nonpolar liquids that float
Hydrogen chloride gas

Variation 4. Cold Cook

The cold cook shows how effective iodine is at allowing hydrogen to replace the hydroxyl group in ephedrine or pseudoephedrine. It has been possible for prisoners to assemble the necessary materials while in jail. There is a report of a prisoner making methamphetamine in the window of his prison cell using this variation. This method uses only a few simple vessels and includes extraction and reactions with a minimum of purification and manipulation. Sports cups, Erlenmeyer flasks, or similar vessels can be used for this method, but generally all the vessel needs to have is a restricted upper end. Ephedrine, red phosphorus, and iodine are added to a closed or restricted vessel in the presence of water. The reaction can generate its own heat; however, heat may be applied to encourage the reaction. The reaction vessel can be two sports cups connected at the top by a common tube, with a second connection using a tube somewhere near the middle of the cups. Any sealed container could be used. This is a cold cook reaction vessel, a low-yield but effective mixture. One cup will contain the ephedrine, iodine, and red phosphorus. The other cup will contain water. After manipulation and a series of cleansing and salting operations, the water solution in the second cup will produce methamphetamine.

Chemicals required

Ephedrine or pseudoephedrine
Iodine
Red phosphorus
Water
Sodium hydroxide
Hydrochloric acid
Nonpolar liquid that floats

Variation 5. Bottle Pop

The bottle pop method uses the same ingredients and has the same results as the traditional cold cook, but the cook time is greatly reduced and the yield is higher because it is done hot. Ephedrine, red phosphorus, and iodine are placed in beer bottles. The bottles are taped with pressure caps lined with inner-tube rubber. The bottles are then placed in stainless-steel soda pop dispensers (the type that can be pressurized) under just enough water to cover them by ½ inch or so. The container is sealed and the water is heated, pressurizing the container. The reaction is over when the pressure seals on the bottles "pop" off. The container is left to cool and the methamphetamine is formed.

Things to look for at bottle pop labs include all the same items as the cold cook lab with three additional items: beer bottles, pressure caps for the beer bottles, and a pressure vessel into which the beer bottles can be placed. Containers that have been encountered include pressurized soft drink dispensers, fire extinguishers, and pressure cookers.

Chemicals required

Ephedrine or pseudoephedrine
Iodine
Red phosphorus
Water
Sodium hydroxide
Hydrochloric acid
Nonpolar liquid that floats

Variation 6. Hypophosphorous Acid

Hypophosphorous acid is the source of phosphorus in this synthesis. The advantage of this method over Variation 1 is that there is almost no need for preparation. The ingredients (ephedrine, iodine, and hypophosphorous acid) are placed into a reaction vessel and the reaction occurs. The disadvantage of this synthesis is that it produces more poisonous gas (phosphine) than do other syntheses.

Chemicals required

Ephedrine or pseudoephedrine
Iodine
Hypophosphorus acid*
Sodium hydroxide
Nonpolar liquids that float
Hydrogen chloride gas

Supplementary 1. Preparing Hydriodic Acid

i. Homemade hydriodic acid. Hydriodic acid can be manufactured from iodine in the presence of red phosphorus. Very poor quality can be obtained simply by adding red phosphorus and iodine to water or hydrochloric acid. This product is often red or the color of grape juice, and may be oxidizing due to the presence of excess iodine.

ii. Much higher quality can be obtained by refluxing red phosphorus and iodine in water, together under a hydrogen or acetylene blanket. Look for a balloon on the top of the reflux column to hold in the gas. Distillation of any of these mixtures provides a high-quality product.

iii. Hydriodic acid can be manufactured by combining iodine and hypophosphorous acid and then distilling.

iv. Bubbling hydrogen sulfide gas through a water solution of suspended iodine. Iron sulfide and sulfuric acid are the products most likely to be used to generate the hydrogen sulfide gas.

Chemicals required

i.
Iodine
Red phosphorus
Hydrochloric acid

ii.
Water
Iodine
Phosphorus
Hydrogen gas or acetylene gas*

iii.
Hypophosphorous acid
Iron sulfide
Iodine
Sulfuric acid

Supplementary 2. Obtaining Hydrogen Chloride Gas

Homemade hydrogen chloride. Hydrogen chloride gas can be generated by putting table salt in sulfuric acid. Entire books have been written on this subject. This is a dangerous practice; the reaction will stop when the salt crusts over. If the container is disturbed, the reaction regenerates, suddenly producing considerable amounts of hydrogen chloride gas.

Chemicals required

Table sale
Sulfuric acid

Supplementary 3. Obtaining Hydrogen Gas

i. Aluminum in hydrochloric acid. Hydrogen gas can be generated with aluminum and any acid, except nitric, in a plastic pop bottle.

ii. Lithium aluminum hydride. Hydrogen gas can be generated by the introduction of lithium aluminum hydride to the chemical process. This is the best method if chloroephedrine is involved.

Chemicals required

i.
Hydrochloric acid
Aluminum foil
ii.
Lithium aluminum hydride*

Supplementary 4. Preparing Iodine

Iodine can be separated from tincture of iodine by adding 30% hydrogen peroxide.

Tincture of iodine can be purchased at a pharmacy.

Chemicals required

30% Hydrogen peroxide
Tincture of iodine

Supplementary 5. Making Ephedrine from Scratch

Although not seen in the field, this seems a fairly logical next step in the battles between the DEA and the cooks. This method has been published and is available. This is a straightforward method using flasks and considerable filtration and evaporation. The end product is a mixture of *l*- and *d*-ephedrine.

Chemicals required

Benzaldehyde
Nitroethane
Sodium hydroxide
Diethyl ether
Sodium bisulfite
Magnesius sulfate
Formaldehyde
Zinc (finely divided)
Glacial acetic acid
Hydrogen sulfide gas
Hydrochloric acid
Sodium carbonate
95% Ethyl alcohol

Synthesis 2. Alkali Metal and Ammonia

Variation 1

Nazi method

This synthesis is easy and quickly produces 100% yield (according to street literature) of *d*-methamphetamine. Unlike the previous synthesis and the one following, this synthesis produces fewer problematic by-products. All the necessary materials are easily accessible and can be obtained with a minimum of attention. The main drawback is the strong odor of ammonia. The odors, along with easier access to ammonia in rural areas, are the two main reasons that these operations are usually seen in the countryside.

The toxicity of anhydrous ammonia and reactivity of lithium create conditions that require safety precautions and special handling. However, the reaction phase is of relatively short duration and generates minimal hazardous waste.

An alkali metal and anhydrous ammonia are combined in a Pyrex® container or sometimes a crockpot. The resulting solution becomes dark blue. It is unlikely that law enforcement will enter during this short period unless this is a production operation. Ephedrine or pseudoephedrine, either the refined product or the alcohol solution from the extraction, is added to the blue solution formed when the ammonia is added to the alkali metal. A short hot reaction produces methamphetamine. The product remaining is a white crystal that becomes methamphetamine base as the water is added. This waste would look like cookie dough and could be identified by the presence of lithium. The ammonia is allowed to evaporate and the remaining substance is quenched with water to react off any remaining alkali metal and enhance the production of meth base. Lithium and sodium are easily made or obtained. Liquid ammonia can be collected in propane cylinders for transport. More recently, thermos bottles have been used to transport the refrigerated liquid. PVC pipe has also been used. The last two methods can be very dangerous if either container is left in the direct sun. The propane bottles may fail in a relatively short period of time, due to the corrosive nature of ammonia.

Flat Pyrex® dishes are filled with liquid ammonia sprayed using a hose or wand or poured from the transport container. The alkali metal and ephedrine are added. The order in which the ingredients are added is actually unimportant. Dry ice is sometimes used to keep the ammonia in the liquid phase and air from the alkali metal during the reaction.

Chemicals required

Ephedrine or pseudoephedrine
Alkali metal*
Ammonia gas*
Nonpolar liquids that float
Hydrogen chloride gas

Variation 2. Alkali Metal and Ammonia Cold Cook

Neo-Nazi method

This is the equivalent of the cold cook for the alkali metal and ammonia method. All that is required is a pressurized plastic pop bottle and other easily available ingredients. To the bottle add ephedrine or pseudoephedrine tablets, water (which puts the ephedrine into solution), any ammonium salt (ammonium nitrate fertilizer is usually used), and sodium hydroxide (drain cleaner), which will release ammonia gas. Add ether (from an engine starter), then add lithium foil (from lithium batteries). Cap the bottle and allow to stand for a couple days. There are reports that this reaction can be very fast.

Chemicals required

Plastic pressure pop bottle
Ephedrine or pseudoephedrine
Lithium batteries
Engine starter fluid (ether)
Drain cleaner (sodium hydroxide)
Water

Supplementary 1. Down's Cell: Making Sodium from Sodium Hydroxide

Sodium metal can be made from sodium hydroxide (a popular source of pure sodium hydroxide is Red Devil® drain cleaner). There are many sources for sodium hydroxide, including industrial supply houses, where it is available in multilayered paper bags. The sodium hydroxide is made molten in a frying pan by heating it over a stove. While molten, the frying pan is attached to two electrodes on a car battery or a battery charger (the electrodes are usually jumper cables with one attached to the metal of the frying pan and the other to a graphite rod submerged in the molten sodium hydroxide). This is called a Down's cell. The result is patches of shiny sodium metal appearing in the salted pan. The metal is pried out of the cooled salt for use in the alkali metal ammonia reaction. The sodium metal will be stored under a nonpolar, nonchlorinated solvent. Kerosene is used in the universities. There obviously is no specified standard for meth labs.

Chemicals required

Sodium hydroxide
Kerosene*

Synthesis 3. Shake and Bake

This clandestine methamphetamine laboratory method of production is a two-step process starting with ephedrine (pseudoephedrine), thionyl chloride, chloroform, alcohol, hydrogen, and a metal catalyst. This method is generally referred to as the *hydrogenation method*, although there are many hydrogenation methods. The slang "shake and bake" is commonly for this method. This synthesis was common a few years ago but is seldom seen now. The name comes from the use of paint shakers as the hydrogenators used in step 2. There are a number of variations and chemical substitutions used in this synthesis, especially in part 2.

BE AWARE! The first step requires thionyl chloride, an extremely hazardous water-reactive material hydrolyzer. Equally hazardous phosphorus pentachloride, phosphorus oxychloride, phosphorus trichloride, phosphorus tribromide, or phosphorus pentabromide may be substituted. Any of these chemicals that contain phosphorus may also produce phosphine gas during the reaction. The second step requires a metal catalyst. One of the metals used, Raney nickel, is pyrophoric and will be found stored under alcohol or water. The reaction is conducted under a hydrogen blanket. Some of the most hazardous of the methamphetamine chemicals are seen in this synthesis.

Step 1 involves the production of an intermediate called *chloroephedrine hydrochloride*. Ephedrine is combined with thionyl chloride and chloroform. Alcohol is added, and the reaction is allowed to progress while being stirred.

The reaction vessel of choice here is a 2000-mL Erlenmeyer flask, but beakers have been used. The vessel must be placed in an ice bath, as the reaction is very hot. With the addition of ether, crystalline chloroephedrine hydrochloride is formed. The product of this reaction is a white froth on the top of a liquid in the vessel. The froth is filtered, cleansed with alcohol, and thoroughly dried. To recover the most chloroephedrine, the crystals can be extracted in ether and stored in a freezer overnight. Look for a sealed 1-gallon glass jug of a crystalline slush in the freezer.

Step 1

Step 1 of this synthesis has a great deal of hazardous waste associated with it. The chloroform is an inhalation hazard as well as a carcinogen. Most hazmat teams do not have good air-monitoring equipment to detect chlorinated compounds. Unfortunately, many of the chlorinated compounds have a very high odor threshold, which means that exposure may occur without sensory warning. Reactive materials that may be stored under water or an organic liquid pose another serious problem.

There are two by-products associated with this method, which in order to make a good product will require some special cleanup. One by-product, 1,2-dimethyl-3-phenylaziridine, can be avoided if bicarbonate of soda is used to make the base in place of sodium hydroxide. Chloroephedrine, the other by-product, requires distillation to be removed. Sodium thiosulfate can also be used to neutralize chloroephedrine.

Step 2

Step 2 involves the conversion of the intermediate chloroephedrine hydrochloride to methamphetamine hydrochloride. This reaction can also be used to make Ecstasy from bromosafrole using the same materials and apparatus. This is accomplished by combining chloroephedrine with palladium black catalyst and alcohol in a container that can be sealed while under low pressure. Hydrogen is injected under pressure and agitated. As the agitation continues, the pressure within the vessel diminishes and more hydrogen is added. This activity continues until no more hydrogen will be accepted and the pressure begins to rise. The resulting liquid is filtered and heated. Methamphetamine hydrochloride crystals are formed as the liquid cools. Note that some pseudoephedrine is carried through these reactions.

There are several variations in both apparatus and in reagents. One of the more common homemade hydrogenators is a paint can shaker. Any metal or plastic container that can be under pressure and can be agitated is a potential hydrogenator. Agitation is accomplished by hanging containers on bungee cords and all sorts of Rube Goldberg machines to shake the containers. Hydrogenation can be done in a champagne bottle used as a pressure vessel. This method requires a magnetic stirrer to keep the contents agitated. A careful cook may wrap the bottle with fiberglass resin. Lithium aluminum hydride placed into the vessel as a hydrogen source will use up the chloroephedrine completely. Fine zinc metal can be used, but gives a poor yield.

The reaction is not complete until the contents begin to increase pressure. Hydrogen can be produced in any number of ways. The gas can be piped in from a generator, from aluminum foil in hydrochloric acid in a plastic drinking bottle, or it can come directly from a hydrogen cylinder. If a hydrogen cylinder is being used, leaking hydrogen should be a primary concern for the responder. Hydrogen is under pressure in a reaction vessel. If the cook is following directions, it will be three times atmospheric pressure. Be careful if this reaction is going on, as when it is finished the pressure of the hydrogen will go up.

Chemicals required

Ephedrine or pseudoephedrine
Chloroform
Thionyl chloride*
Ether
Acetone
Palladium chloride, lithium aluminum hydride, or Raney nickel*
Hydrogen gas
Sodium thiosulfate
Bicarbonate of soda

Synthesis 4. Perchloric Acid and Palladium on Barium Sulfate Under Hydrogen

This method was first reported in California in the mid-to-late 1980s. This reaction can be run using phosphoric or acetic acid. It is necessary to bubble hydrogen chloride gas through the acid before putting it into a container. The reaction container is placed in a commercial drink cooler to keep the reaction cool by convection within enclosed steam. The cooler will be sitting on the magnetic stirrer. The drink cooler must be made of plastic or Styrofoam® to allow a magnetic stirrer to agitate the chemicals in the bottle. Some potassium salt will be needed to remove the perchlorate, as potassium perchlorates are insoluble.

Chemicals required

Ephedrine or pseudoephedrine
Perchloric acid*
Palladium on barium sulfate*
Hydrogen
Any potassium salt

Synthesis 5. Hydrazine or Hydrazine Hydrate

This is a Wolff–Kishner reduction. There are two main concerns for the responder at a lab

using this reaction. Although not quite as reactive as hydrazine, hydrazine hydrate can be very hazardous, and it is without many warning properties. It is a strong reducing agent. Most people are not as familiar with reducing agents as they are with oxidizers. The boiling temperature is very high. The pan with the reactor contains a high concentration of sodium hydroxide in order to obtain a temperature of 245°C. (Water boils at 100°C.) A splash from the diethylene glycol bath in which the reaction vessel is located would be nasty. The product is first refluxed in a diethylene glycol bath. Then the condenser column is rearranged and the product is distilled. Hydrazine can be made; *see* Explosives, Astrolite.

Chemicals required

Ephedrine or pseudoephedrine
Diethylene glycol*
Hydrazine hydrate*
Potassium hydroxide

METHOD 2. Substitution: The Substitution of Methylamine for the Carbonyl Group on P-2-P

When looking at the structures of P-2-P, methylamine, and methamphetamine, it is easy to see that this is a simple substitution of methylamine and a hydrogen for a carbonyl group (=O). Since the oxygen is double-bonded to the carbon, there is no specific direction to a newly attached methylamine functional group. This means that there are two possible outcomes to this reaction. The *l*- and *d*-forms of methamphetamine are formed in equal amounts, making the final product only half as potent as the methamphetamine produced by hydrogenation of ephedrine or pseudoephedrine.

The relative length of Method 2 in this document to Method 1 is deceptive. Despite the fact that there are more methods listed here, most are far more difficult than could be attempted by an ambitious tweeker. One thing to consider when encountering one of the clandestine labs listed in this section is that it may be a production lab, as these methods are not restricted by the number of ephedrine® pills that are obtainable.

Synthesis 6. Biker Method

This clandestine methamphetamine laboratory method of production is generally referred to as a *P-2-P*, *biker*, *amalgam*, or *prop dope lab*. Some of the chemical ingredients are highly toxic, difficult to manage, and difficult to obtain. Therefore, this method of production is encountered less frequently. The simpler and more productive nonricimic methods used in Syntheses 1 and 2 are preferred.

The reaction vessel may be a beaker or a beer pitcher placed in cooking oil to heat the contents, although a round-bottomed reaction vessel placed in a heating mantel is most commonly used. Aluminum foil, phenyl-2-propanone, methanol, mercuric chloride, and methylamine are placed in a reaction vessel and heated. This mixture of ingredients is allowed to react until the aluminum becomes a gray sludge and the resulting bubbling diminishes. The aluminum amalgam, called *freshly turned aluminum*, provides hydrogen to the reaction. If it is not vented, hydrogen gas may be an explosion hazard. A biphasic liquid is formed which can be enhanced with the addition of salt water. Methamphetamine is in the top golden layer. The vessel is allowed to cool and the liquids to separate. The residue is a sludgy material that will be difficult to separate.

A vacuum pump is hooked to a vacuum flask and Buchner filter funnel assembly to separate the two layers. When separated, the top clear oily liquid is mixed with a nonpolar solvent to extract the methamphetamine base. The mixture is then shaken and separated using a reparatory funnel to collect the extraction solvent. There are many variations to this method. They seem so much alike that they are not broken down further here. The main differences are the

forms of methylamine, which can be the gas, the 40% aqueous solution, or methylamine hydrochloride, a crystal.

The contents of the reaction vessel will become separated with a red layer on top as the reaction progresses. The consistency of this material makes vacuum filtration necessary. Although a fine cook would distill out the P-2-P, you won't see that. The Buchner funnel and a suction pump are the main clues to recognize this synthesis.

Chemicals required

Phenyl-2-propanone
40% Methylamine or methylamine·HCl or methylamine gas
Aluminum foil
Mercuric chloride*
Ether (nonpolar solvent)
Acetone
Sodium chloride
Vacuum pump*

Synthesis 7. Formic Acid and Dilute Acid Reflux

This synthesis is not considered to be common, despite the fact that Variation 2 is the lead-off method in the underground "Uncle Fester" book, *Secrets of Methamphetamine Manufacture*. This is a two-step method. P-2-P, methylamine, and formic acid are placed into a round-bottomed flask and heated to produce an intermediate, *n*-formylmethamphetamine. Dilute acid is added and the mixture is refluxed to produce methamphetamine. The temperature is very important in this reaction, requiring the use of a thermometer, which is also used as a stirring rod.

The contents of the reaction vessel will become black and then a form a brown layer, which is a methamphetamine base on a sodium hydroxide solution.

N-Methylformamide is an industrial solvent readily accessible to many people. It would be difficult to purchase from a chemical supply house without creating suspicion.

Chemicals required

Variation 1

40% Methylamine or methylamine·HCl or methylamine gas
Phenyl-2-propanone
Formic acid
N-Formylmethylamine*

Variation 2

Formic acid
Dilute acid

Supplementary 1. To make N-Methylforamide

Gaseous methylamine is bubbled through formic acid to create the amide. The same apparatus can also be used to make gaseous methylamine from the 40% stock solution in other operations. In the first flask, 40% methylamine is heated. The condenser is filled with ice water or some other cold liquid. The gas is passed through formic acid in the second container. If the cook has methylamine in the gas form, the first flask can be eliminated. The first flask without the second can be used to make methylamine gas for other reactions. Anhydrous ammonia for Synthesis 2 can be made using dry ice and alcohol in the reflux column.

If the glass tubing from an Erlenmeyer flask is connected directly to one reaction vessel, it is being used to insert dry methylamine gas. If a second vessel is attached, it contains formic acid to make N-methylforamide.

Chemicals required

Methyl amine gas
Formic acid

Synthesis 8. Using Sodium Cyanohydroborate

P-2-P and methylamine are placed into a round-bottomed flask with sodium cyanohydroborate and hydrochloric acid. The mixture is heated to produce methamphetamine.

Chemicals required

Phenyl-2-propanone
Methylamine
Sodium cyanhydroborate*
Hydrochloric acid

Synthesis 9. Using Sodium Metal

Chemicals required

Phenyl-2-propanone
Methylamine
Ethanol
Sodium metal*

Synthesis 10. Using Platinum Oxide

This reaction has so many variations that it seemed better just to generalize them here. The most important aspect is that some sort of pressurized hydrogenator will be employed. Champagne bottles that can handle up to 30 lb of pressure are easily accessible. Some of the reactions are a little strong for the champagne bottle and it may develop stress cracks. A hydrogenator or welded stainless steel container (a homemade hydrogenator) may be used. The responder should check to see if the valve is off and where the hydrogen is being vented. Hydrogen is an extremely flammable gas that burns with no color or luminosity.

The main variations involve the catalyst, which ranges from platinum made with borohydride to Raney nickel. Other variations that seem to be independent of the catalyst include the form of the methylamine that can be used as a 40% solution, a gas, or as the solid methylamine hydrochloride.

Finally, one variation uses the aluminum mercuric chloride amalgam to form the hydrogen in a pressurized sort of biker method. If aluminum is being used, the primary concern is that the waste stream may be contaminated with mercuric chloride. Look for gray sludges. Although not mentioned specifically, it is likely that lithium aluminum hydride could also work.

Chemicals required

Phenyl-2-propanone
Methylamine gas, 40% solution or methylamine·HCl
Methanol
Platinum oxide*
Hydrogen
Aluminum
Mercuric chloride

Manufacture of P-2-P

Synthesis 11. Phenylacetic Acid, Acetic Anhydride

This synthesis has many subtle differences and may have three or more variations. The end product can be red oil (orange) or pee (yellow) P-2-P. Each of these is ready for use or sale as P-2-P. A red solid in the P-2-P consists of the unreacted or polymerized junk. It is called *red tar*. The presence of red tar indicates a ruined batch. This is a common problem with this reaction. The end product may be distilled if quality control is a consideration. The variations blur in the literature and are presented here as a generalized single method. Significant differences seem to exist, and the proponents of each are quick to point out that the other methods do not work.

One variation uses sodium acetate and can either distill the final product or neutralize it with pyridine. The three ingredients acetic anhydride, phenylacetic acid, and sodium acetate are put into a round-bottomed flask, heated, and refluxed. An electric mixer is used for a short time. If the resulting mixture is distilled, yellow P-2-P called P or pee is made. If pyridine is used to make the solution basic, a product called *red oil* is made. The reaction vessel contents may be cooled in a crockpot. This and Synthesis 12 are the most common methods for making P-2-P. These methods for making P-2-P are characteristic of biker labs.

The method above simply does not work, according to Uncle Fester. It is difficult to determine how methods spread through the underground network. It seems more likely that cooks will be more familiar with Uncle Fester than with the Russian literature. With P-2-P methods on the wane, these discussions may become academic. Using pyridine instead of sodium acetate to make P-2-P creates many problems, including removal of the residual acetic anhydride and the pyridine, which is saved due to its high cost. Pyridine smells like rotten soy sauce. It is most likely to be found in a refrigerator in a tightly sealed container. The second variation uses pyridine exclusively without the sodium acetate. This variation requires several distillations and purifications. These duplications, along with the likelihood that the recovered pyridine will be stored for further use, increase the paraphernalia and the chemicals needed. Many of the chemicals used in the purification of P-2-P are found in other methods but used in very different ways. Look for a round-bottomed flask with orange-red contents, a reflux column, and the odor of cat urine. The end product of refluxing is a combination of P-2-P, acetic anhydride, and pyridine. These are easy to separate because acetic anhydride and pyridine have boiling temperatures much lower than that of P-2-P. The anhydride collected and the pyridine may be re-distilled to collect the pyridine. Look for a combination of distillation and reflux columns which indicates this process. Sodium hydroxide is added to the P-2-P remaining in the reaction vessel to make the anhydride water-soluble. The product will be removed using a separatory funnel. The wastewater may contain a red tar. Hydrochloric acid is used to remove the pyridine, and then a nonpolar organic compound is used to purify the remaining P-2-P. This final product is re-distilled.

Associated with this type of lab is a refrigerated 1-gallon jug of acetic anhydride and pyridine mixed together. Sodium acetate may be substituted with a mixture of vinegar and sodium hydroxide.

This is a distillation. In this reaction, if pyridine is used, the liquid in the collection vessel will be orange or yellow. The reaction vessel will be red.

Chemicals required

Pyridine*
Sodium acetate
Vinegar
Acetic anhydride*
Phenylacetic acid
Toluene
Sodium hydroxide

Synthesis 12. Dry Distillation of Phenylacetic Acid to Make P-2-P

This is a standard dry distillation. In this synthesis the reaction vessel will be blue-gray and viscous.

Chemicals required

Phenylacetic acid
Lead acetate*

Synthesis 13. Benzene and Chloroacetone to Make P-2-P

Benzene and chloroacetone are reacted in the presence of aluminum trichloride.

Chemicals required

Benzene
Chloroacetone*
Aluminum trichloride*
Sodium bisulfite
Hydrochloric acid

Synthesis 14. Tube Furnace

Phenylacetic acid and acetic acid are heated in a tube furnace with thorium oxide or manganous oxide as a catalyst. The method also incorporates carbon dioxide or nitrogen gas to move the product through the tube. Benzene is also used in this synthesis. Thorium nitrate may be used to make the thorium oxide. This is a low-cost high-production method. When encountered, this synthesis causes excitement because of the radioactive thorium.

A commercial tube furnace is about one-half the length needed for optimum production. More likely, a homemade furnace will be found.

A glass tube with thorium oxide catalyst packed inside is inserted into a metal tube. The metal tube is also packed on the outside to keep the glass tube tight. The metal tube is wrapped for insulation and wired for heat. Nitrogen is bubbled through H_2SO_4 to dry it. Cooling water goes through a condenser. The collected P-2-P will have residual benzene solvent. It should be distilled prior to use. A tube furnace should be cause for an immediate radioactivity survey.

Chemicals required

Acetic acid
Thorium oxide or thorium acetate
Sulfuric acid
Benzene
Sodium carbonate
Carbon

Synthesis 15. Benzaldehyde and Nitroethane to Make P-2-P

Two variations of this reaction have been encountered. Although all of the beginning and intermediate products are the same in these two variations, the chemicals used to manipulate the reactions in step 2 are different. The differences are subtle but significant. The second variation, described in Uncle Fester's *Secrets of Methamphetamine Manufacture*, is called the Knoevenagel reaction.

The intermediate 1-phenyl-2-nitro-propene crystals are not very stable and will start to become a tarry black if left out for any length of time. They are very irritating, having a tear gas effect on the eyes. The original mixture may be found in a dark place in a reaction vessel. It is supposed to remain in the dark for a substantial period of time.

Variation 1

1. Heat the two components in the presence of butylamine, reflux, and let stand for

several days to produce 1-phenyl-2-nitropropene. Reflux the 1-phenyl-2-nitropropene in a round-bottomed flask with ferric chloride, iron, and hydrochloric acid.
2. Heat the two components in the presence of ammonium acetate, reflux 2 hours, and wash the crystals with acetic acid to produce 1-phenyl-2-nitropropene.

Reflux the 1-phenyl 2-nitropropene in a round-bottomed flask with ferric chloride, iron, and hydrochloric acid to produce P-2-P.

Variation 2

1. Heat the two components in the presence of butylamine and absolute alcohol, reflux, and let stand for several days to produce 1-phenyl-2-nitropropene. Reflux the 1-phenyl-2-nitropropene in a round-bottomed flask with ferric chloride, iron, and hydrochloric acid.
2. Recover and wash the crystals with ether.
3. Reflux the 1-phenyl 2-nitropropene in a round-bottomed flask with ferric chloride, iron, and hydrochloric acid to produce P-2-P. Here it is suggested that the iron is reacting with the acid to form hydrogen gas to push the reaction.

Synthesis 15 requires a great deal of heat in the first reaction. Motor oil is used in a bath to maintain the high temperature. An extension on the bottom of the reflux column is called a *Dean Stark trap*. This device prevents water from returning to the reaction vessel. This reaction is very sensitive to water.

Variations 1 and 2

Chemicals required

n-Butylamine
Benzaldehyde
Nitroethane
Ferric chloride
Iron filings
Hydrochloric acid
Acetic acid
Ammonium acetate
1-Phenyl,2-nitropropane crystals
Ethyl ether

Absolute alcohol
Toluene/benzene

Synthesis 16. Sodium Ethoxide to Make P-2-P

The glass bowl contains ice water, and the separatory funnel contains absolute alcohol. The reaction vessel may be covered with wet towels and contains sodium metal to which the alcohol is being added. This reaction is very hot. If any water were to enter the reaction, an explosion would occur, throwing sodium metal and warm alcohol everywhere. This lab may also have sodium cyanide present. The general rule with clandestine labs is to allow the cook to continue to completion. This author does not agree. However, with this configuration, it is important not to interfere with the process.

Benzyl cyanide, ethyl acetate, and sodium ethoxide are combined to make α-phenylacetoacetonitrile sodium salt. This mixed with acetic acid produces α-phenylacetoacetonitrile. This compound mixed with sulfuric acid produces P-2-P. This is not a common method of producing P-2-P because of the tricky process with hazardous materials. Two hazardous substances associated with this reaction are sodium cyanide and sodium metal. Benzyl cyanide can be purchased, but sodium ethoxide must be made. Sodium cyanide could be used to replace benzyl cyanide in the reaction, but this shortcut is dangerous, as it produces hydrogen cyanide gas during the reaction. Benzyl chloride and sodium cyanide in the presence of ethanol are refluxed to produce the benzyl cyanide. Making sodium ethoxide from absolute alcohol and sodium metal is very hazardous. Any mistakes during the formation of sodium ethoxide could cause an explosion.

Chemicals required

Benzyl cyanide*
Sodium ethoxide
Ethyl acetate
Bromobenzene

Sodium metal*
Sodium hydroxide
Absolute alcohol
Benzyl chloride*
Acetic acid
Sulfuric acid

Synthesis 17. Benzene and Acetone to Make P-2-P

Chemicals required

Benzene
Acetone
Manganese(III) acetate

Synthesis 18. Ritter Reactions P-2-P from Methyl or Ethyl Nitrite and Allyl Benzene

Allyl benzene provides an easy starting chemical for the direct manufacture of amphetamine. This is called a Ritter reaction. More important, the Ritter reactions work the same for safrole, a major component of Sassafras oil, which is readily available. When reacted in the P-2-P synthesis, 3,4,methylenedioxy-1-phenyl 2-propanone, the end product from safrole in the Ritter reaction is ETC (Ecstasy). These reactions therefore may be found in the manufacture of amphetamine, MDA, Ecstasy, Rush (discussed below), and finally, methamphetamine.

Allyl benzene is commercially available, or it can be easily made using chemicals that do not have any association with clandestine labs. It can be made by adding aryl copper and allyl bromide, and benzene, or by degrading safrole. The Ritter reaction is the addition of acetonitrile to allyl benzene using a reaction vessel that may be as simple as a beer pitcher.

Ethyl or methyl nitrite is required to make allyl benzene into P-2-P. Nitrites are available in some states; however, they are controlled in many states due to their use as a "head rush" drug. It is most likely that the nitrites will have to be made for this reaction as described below. If this reaction is seen, the end product may only be Rush.

A reaction vessel contains sodium nitrite and ethanol, an addition funnel contains sulfuric acid, and the gas bubbling through the allyl benzene is methyl nitrite. The receiving vessel is on a magnetic stirrer. The resulting solution in the Erlenmeyer flask becomes dark and fizzes, looking very much like a freshly poured Coca-Cola. This same apparatus could be used to make recreational drugs such as Rush or Jock-a-Rama.

Synthesis 18 contains several interesting side dishes and has many variations. The dream that allyl benzene would be the new way to get around the DEA was lost when the Syntheses 1 and 2 reactions were so easy and made P-2-P methods a thing of the past. Uncle Fester recommends that this reaction be carried out in a beer pitcher.

Variation 1. Making Amphetamine

Chemicals required

Allyl benzene*
Acetonitrile*
Sulfuric acid

Variation 2. Making MDA

Chemicals required

Safrol*
Acetonitrile*
Sulfuric acid

Variation 3. Making P-2-P

Allyl benzene*
Palladium chloride*
Ethyl nitrite*

Variation 4. Making 3,4-Methylenedioxy-phenyl-2 -propanone Ecstasy

Chemicals required

Safrole*
Palladium chloride*
Ethyl nitrite*

Variation 5. Making MDA

These last two methods for making MDA are not related to allyl chloride. These come from Chewbacca Darth. These are entirely different approaches to making MDA. The methods above are much easier, but people will always work with what they have.

Chemicals required

Step 1
Piperonal*
Methanol
Nitroethane*
Sodium hydroxide
Concentrated hydrochloric acid

Step 2
Lithium aluminum hydride*
Ether
1.5 N sulfuric acid
Potassium sodium tartrate
Sodium hydroxide
Methylene chloride
Ether
Hydrogen chloride gas
Isopropyl alcohol

Variation 6. Making MDA

Piperonyl alcohol*
Ether
Pyridine*
Thionyl chloride*
Calcium chloride
Magnesium turnings
Iodine
Methylamine gas
Ether
HCl gas

Supplementary 1. Steam Distillation of Safrole

Grind the root bark into pieces. The oil and water collect in the receiver flask, whereupon the oil can be seen as droplets. The oil is denser than the water, so it will form droplets below the water. Place the oil/water mixture in a seperatory funnel, and recover the lower oil layer. Dry the oil by mixing with calcium chloride. Place the oil in a vacuum distillation apparatus. Note that the oil can also be separated by freezing.

Chemicals required

Sassafras oil
Root bark from a *Sassafras* tree
Calcium chloride

Supplementary 2. Synthesis of Safrole from Catechol

1. Making 4-allyl catechol. Prepared by reacting catechol with allyl bromide, extraction, and distillation.
2. Conversion of 4-allyl catechol to safrole. Reflux the 4-allyl catechol with methylene chloride and dimethyl sulfoxide in sodium hydroxide. Distill and steam distill to collect the oily safrole product.

Chemicals required

Catechol*
Allyl bromide*
Acetone
Potassium carbonate
10% Sulfuric acid
Ether
Sodium hydroxide
Methylene chloride
Magnesium sulfate
Allyl bromide*
Acetone
Potassium carbonate
10% Sulfuric acid
Ether
Sodium hydroxide
Dimethyl sulfoxide*
Methylene chloride
Magnesium sulfate

Supplementary 3. Synthesis of Safrole from Eugenol

1. Making 4-allylbenzene-1,2-diol. This is done by reacting eugenol with aluminum iodide under reflux in the presence of a catalyst, followed by filtration, purification, and further purification with pyridine.
2. Conversion of 4-allylbenzene-1,2-diol to safrole. React with potassium fluoride and methylene chloride under high

temperature. Quench with water and extract with ether. Vacuum distill with solvent purification or low-temperature fractional crystallization.

Chemicals required

Aluminum foil
Iodine
Eugenol
n-Butyl ammonium iodide
Ethyl acetate
Toluene
Pyridine
Ether
Hydrogen chloride gas
Sodium sulfate
Potassium fluoride
Dimethylformamide
Methylene chloride
Sodium hydroxide

Supplementary 4. Making Eugenol from Cloves

1. Steam distill cloves.
2. The oil/water mixture is then extracted with methylene chloride and a potassium hydroxide solution.
3. Treat with acid, and re-extract with methylene chloride.
4. Evaporate methylene chloride to recover the eugenol.

Chemicals required

Cloves*
Methylene chloride
Potassium hydroxide
5% Hydrochloric acid.
Sodium chloride
Sodium sulfate

Supplementary 5. Making Benzyl Chloride

Benzyl chloride can be made by bubbling chlorine gas through toluene in the sunlight. This can be done easily by adding an acid to calcium hypochlorite in one container and bubbling the gas created through a second container.

Chemicals required

Toluene
Chlorine gas
Calcium hypochlorite
Hydrochloric acid

Synthesis 18-A. P-2-P from Methyl Styrene

This is a separate synthesis and is not related to Synthesis 18. A solution of thallium(III) nitrate in methanol is added to a stirred solution of methyl styrene at room temperature. The end product dimethyketol is hydrolyzed to P-2-P in a reparatory funnel, with the addition of sulfuric acid. The P-2-P is then distilled.

Chemicals required

Methyl styrene*
Thallium(III) nitrate*
Methanol
Sulfuric acid

Synthesis 19. Making Methylamine·HCl

Add formaldehyde to ammonium chloride and let cook in the presence of chloroform and ethanol to produce methylamine hydrochloride.

Methylamine hydrochloride crystals are very hydrophilic and must be kept dry. Look for a sealed Mason jar filled with clear/white crystals.

Chemicals required

Formaldehyde*
Ammonium chloride*
Chloroform
Ethanol

Synthesis 20. Benzyl Cyanide to Make Phenylacetic Acid

Benzyl cyanide is refluxed with dilute sulfuric acid, which produces phenylacetic acid. Also see Synthesis 16.

As stated at the beginning of the book, organic chemistry provides a set of building blocks, and the number of ways that these building blocks can be attached to one another is endless. Phenylacetic acid is an important compound in organic chemistry. As such, there is an endless number of ways that this key component to the manufacture of P-2-P can be made. Eight are listed below as Supplementaries 1 through 8.

Chemicals required

Benzyl cyanide
Sulfuric acid

Supplementary 1

Benzene
Chloroacetic acid *
Ferric chloride

Supplementary 2

Benzoylformic acid
Hydrazine hydrate
Potassium hydroxide

Supplementary 3

Styrene
Morpholine
Sulfur
50% Sulfuric acid

Supplementary 4

Mandelic acid
Hydriodic acid
Red phosphorus

Supplementary 5

Allyl benzene
Ozone
Hydrogen peroxide
Sulfuric acid
Phenyl ethyl alcohol
Dinitrogen tetroxide

Supplementary 6

Toluene
Potassium metal
Sodium peroxide

n-Heptane
Dry ice

Supplementary 7

Acetophenone
Sodium hydroxide
Hydroxylamine
Hydrogen chloride
Sulfur

Supplementary 8

Benzyl chloride
Magnesium
Ether
Carbon dioxide gas

METHOD 3. Addition

Any librarian can tell you that these methods are difficult to find because they are frequently torn out of the literature. It is possible that one of these methods may suddenly appear in the field, depending on the ease of obtaining or making the required ingredients.

Synthesis 21. Grignard Reagent

Variation 1. Uncle Fester

This is an addition reaction. First the Grignard reagent is made by mixing benzyl chloride with magnesium metal and diethyl ether. Separately, acetaldehyde is added to methylamine to make methylaminoethane. The Grignard reagent (benzyl magnesium chloride) is added to the methylaminoethane in the presence of dilute sulfuric acid to produce methamphetamine. Also required are iodine and a 40% concentration of methylamine in water or the hydrochloride salt. If the magnesium is too fine, the reaction may explode.

Chemicals required

Step 1
Benzyl chloride*
Magnesium metal
Ether

Step 2
Acealdehyde*
Methylamine
Sulfuric acid

Variation 2. Chewbacca Darth

Step 1
Nitrogen gas
Calcium chloride
Benzyl chloride*
Magnesium turnings
Ether
Iodine
Acetaldehyde*

Step 2
40% Methylamine
Dilute sulfuric acid
Dry ice
Methanol

Salting

Hydrochloric acid
Table salt

Synthesis 22. Friedel–Krafts Reaction

Benzene and allyl chloride are used to make the intermediate 1-phenyl-2-chloropropane. Methylamine is added to make the methamphetamine. In many ways this looks like a P-2-P addition method and could probably be placed in that section. However, there are some significant differences. Note that the manufacture of allyl chloride is also discussed in Synthesis 18. The intermediate 1-phenyl-2-chloropropane is found in some reduction and addition methods. Methylamine is used and made in many of the P-2-P methods. The second part of this synthesis requires a unique reaction vessel, metal pipes in a pressure cooker.

Step 1 is done in a round-bottomed reaction vessel. The second part of this reaction is done by putting the purified solution from step 1 into metal pipes and then placing the pipes into a pressure cooker to react.

Chemicals required

Benzene
Allyl chloride*
Ferric chloride*
Methanol
Hydrochloric acid

DATE RAPE

γ-Hydroxybutyate (GHB), or sometimes called *Liquid E* or *Liquid G*, is popular at raves and parties because many users compare its Effects to that of Ecstasy. There are many ways of doing Variation 1 below, but all of them use the same three chemicals.

Making γ-Hydroxybutyrate
302-17-0

Variation 1

γ-Butrolactone
Sodium hydroxide
Ethanol

Variation 2

Tetrahydrofuran
Sodium bromate
Methylene chloride

Making Chloral hydrate (Mickey Finn)

Ethanol
Chlorine gas
Sulfuric acid

DESIGNER DRUGS

This section represents is the battle of the chemists. It is reminiscent of the battles between Chewbacca Darth, Uncle Fester, and the DEA. Although there is definitely activity here, the numbers of people making these at present is very small, perhaps hundreds in the entire country. Most, although not all of these syntheses require a very good chemist and very decent laboratory facilities. Despite this, the DEA laboratory personnel have spent considerable time in this area, and the Web is full of formulas to make these drugs.

Designer drugs mimic amphetamine, illegal drugs, and the chemical weapon incapacitating agents because they are nothing more than modifications of those drugs. Before 1987, chemists/dealers were hoping to sell these modifications before the DEA changed the law to include the modification. As a result, the DEA revised the law in 1987 to use the effects of the drug, not the formula, as the method of determining if it is illegal. This did not stop the sale of designer drugs. (The law now calls them *controlled substances analogs* (CSAs). During the period of experimentation, it was found that many of the CSAs were more potent than the original drugs. One modification of fentanyl is 1000 times as potent as heroin, which means that a person using $600 worth of chemicals can make a kilogram of the drug, which would be the equivalent of a ton of heroin worth millions of dollars. The benefit of that is obvious. Also, many drugs that started out as CSAs are now considered to be standard drugs. Examples include Cat (methcatanone) and Ecstasy. Little if any, research has been done on the toxicology or effects of most of these drugs.

In this section we often do not even try to give formulas, because, for example, there are over 1000 variations of fentanyl alone. Some of these would have no effect, where others might cause death. Most are very unlikely to ever be seen, or may only be tried once. Instead, we include only key formulas or chemicals, which would always be expected to be there. Our only purpose in this book is to avoid a designer drug lab being passed off as just another meth lab and turned over to a cleanup company. The DEA has chemists who would understand the modifications if the chemicals were brought to their attention. We have broken down these chemicals into four groups.

Group 1. Piperidyl Benzilate Esters

These compounds are closely related to BZ. They have effects, which are described as deliriums, similar to those caused by datura,

belladonna, and atropine. Although classified as hallucinogens, they are very effective as incapacitating agents. As a designer drug, BZ seems to have no following, due in part to the very long time effects of BZ. Many of the analogs have effects that last for 8 to 12 hours. You should see Facility 13 for the full formula for BZ. We have included some key chemicals in Variation 3 below. The antidote for piperidyl benzilate esters is tetrahydroacridine.

Variation 1. Ditran (DMZ)

Methyl benzilate
Toluene
Sodium metal
3-Quinuclidinol
Hydrochloric acid
Sodium hydroxide
Chloroform
Sodium sulfate
Acetone

Variation 2. TWA (Benacyzine)

Benzillic acid
3-Quinuclidinol

Variation 3. BZ

Benzaldehyde
Potassium cyanide
Lead oxide
Benzilic acid
3-Hydroxy-1-methylpiperidine
Methyl benzilate
3-Quinuclidinol

Group 2. Fentanyl-Based Analogs

Fentanyl is a Russian incapacitating drug which received much publicity when their alpha team killed several people during a siege on a theater in Moscow by gassing the theater with fentanyl. Fentanyl or meperidine (Demerol®) is in effect very analogous to heroin. It has a built-in market (1,000,000 hardcore and casual heroin users), which increases with the decreased availability of heroin. The modifications of fentanyl generally produce shorter effect times but greater potency than the effects of fentanyl itself, which can last for days.

This group of drugs, which include MPPP, alfentanil (Alfenta®), sufentanil (Sufenta®), remifentanil (Ultiva®), carfentanil (Wildnil®), and α-methylfentanyl, which were all sold as China White. Fentanyl-based drugs have caused many deaths,s due in part to the relative potency of the analogs compared to heroin. An effective dose is smaller than a grain of salt. A person not familiar with the product is prone to overdose. Improper preparation of MPPP (new heroin); (1-methyl-4-phenyl-4-propionoxylpiperidine) can produce MPTP (1-methyl-4-phenyl-1,2,3,6,-tetrahydropyridine), which is a neurotoxin causing immediate symptoms much like those of Parkinson's syndrome. MPTP was suggested initially as a pesticide; however, it was withheld when workers began showing up with Parkinson-like symptoms.

Fentanyl is sold legally in the United States as a patch for people (e.g., cancer patients) in extreme pain, and as a lollipop-type device also used for the relief of pain. A quick antidote for fentanyl and all its analogs is naloxone.

Variation 1. Fentanyl

Step 1
N-(4-Piperidyl)propionanilide
Sodium carbonate
Potassium iodide
Hexane

Step 2
Phenylethyl chloride
4-Methyl-2-pentanone

Step 3
Diisopropyl ether

Variation 2. Fentanyl·HCl

Step 1
Piperidone
Phenethyltosylate or Phenethylbromide

Step 2
N-Phenethylpiperidone (NPP) from step 1
Aniline
Tetrahydrofurane

Methanol
Sodium borohydride

Step 3
4-Anilino-N-phenethylpiperidine (4-ANPP)
from step 2
Propionyl chloride
Sodium chloride
Sodium hydroxide

Group 3. Indole-Related Compounds

Perhaps the best examples of these are found under "Illegal Drugs, Psilocybin." Found here also is LSD, although experimentation has shown that so far, analogs of LSD are all ineffective and are not likely to be found being produced as designer drugs.

Variation 1

All the chemicals used to make psilocin from scratch are also listed in Cook 5. We have only listed the unusual chemicals used to make psilocin in Variation 1 below.

Key chemicals used to make psilocin
2,6-Dinitrotoluene*
Ammonium sulfide
Sodium nitrite
Dimethyl sulfate*
Potassium*
Ether
Diethyl oxalate
Ferrous sulfate
Oxalyl chloride*

Supplementary 1. Fischer Indole Synthesis

This synthesis is very likely to be encountered at a lab making indole-based designer drugs. Unfortunately, there are a number of ketones that could be used, so many that it would be useless to try to list them all. The primary indicator of this reaction is phenylhydrazine.

Supplementary 2

Indole can be used in hundreds of compounds, which also have hallucinary effects.

Leimgruber–Batcho Synthesis

o-Nitrotoluene
N,N-Dimethylforamide
Pyrrolidine
Raney nickel and hydrazine, or palladium on carbon and hydrogen, or stannous chloride and sodium dithionite

Chemicals that could be expected to be used to formulate indole designer drugs

(ET) 3-(2-aminobutyl)indole

Group 4. Stimulant Compounds

Stimulants strengthen the action of the heart, increase vitality, and promote a sense of well-being. Abuse may cause hallucinations and euphoria. This group includes the amphetamines and cocaine.

Euphoria

4-Methylaminorex, which is commonly called Euphoria, U4EA U4Euh, Ice, and others, has effects comparable to those of methamphetamine, but with a much longer duration and supposedly fewer side effects. This Schedule 1 compound is made from other scheduled compounds, something that appears to be common with designer drugs.

Variation 1. Cyclization; One Step

d, l-Norephedrine
Cyanogen bromide

Variation 2. Cyclization: More Steps

Sodium or potassium cyanide
Concentrated hydrochloric acid
d,l-Norpseudoephedrine

Chemicals that could replace P-2-P in reactions

In Amphetamines, Synthesis 18, allyl chloride is suggested as a substitute for P-2-P. Listed below are several other chemicals that would produce either methamphetamine or analogs of that compound.

1-Phenylbutanone
1-(4-Fluorophenyl)propan-2-one
Methylphenidate
Phenmetrazine
4-Methylaminorex
Phenylisopropylamine
Tranylcypromine
    (2-phenylcyclopropylamine)
2-Phenylcyclopylamine
2-Phenylcyclohexylamine
2-Amino-3-phenyl-trans-decalin
2-Aminotetraline

Schedule 2 Drugs

These drugs are available only by prescription, and distribution is controlled and monitored carefully by the DEA. Oral prescriptions are allowed, except that the prescription is limited to 30 doses, although exceptions are made for cancer patients, burn victims, and others. No refills are allowed. These are subject to production quotas set by the DEA.

Cocaine (used as a topical anesthetic)
Methylphenidate (Ritalin®)
Phencyclidine (PCP)
Opioids
Pethidine
Meperidine
Fentanyl
Hydromorphone
Opium
Oxycodone (Percocet®)
OxyContin®
Morphine
Short-acting barbiturates
Secobarbital
Amphetamine
Methamphetamine (injectable)

ILLEGAL DRUGS

Most of the first seven or eight of these cooks are very reminiscent of the 1960s. Although today there is not the emphasis on some of these types of drugs, availability and access to information for making these drugs (both in legitimate bookstores and on the Web) is a strong indication that people are still interested in or making them.

Cook 1. Cocaine
$C_{17}H_{21}NO_4$
50-36-2

Schedule 1

Despite the fact that we are starting off with a recipe to make cocaine from scratch, it is unlikely that anyone will encounter a person making cocaine. We also doubt that cocaine leaves would be smuggled in to process in the United States. However, these recipes were available and are therefore included here. The last recipes for crack cocaine are very likely finds. The first one is described on the Web as simple enough that "if your IQ is higher than that of a houseplant, you will be able to do this." The LD_{50} of cocaine is 95 mg/kg. At this level, seizures occur followed by respiratory arrest.

Variation 1. Manufacturing Cocaine from Scratch

Furan*
Methylene chloride
Ferric chloride
Methanol
Chloroform
Ice
Ether
Calcium chloride
Norite
Ammonia gas
Benzoyl chloride*
Ethanol
Raney nickel*
Ammonia hydroxide
Hydrogen gas
Ethanol
Hydrochloric acid
Methylamine hydrochloride*
Potassium hydroxide
Benzene
Sulfuric acid
Sodium
Mercury*

Variation 2. Purifying Cocaine from Cut

Sulfuric acid
Ammonium hydroxide
Chloroform
Alcohol
Water
Petroleum ether

Variation 3. Crack Cocaine

Baking soda
Cocaine*

Variation 4. "Best Damned Crack Recipe Ever"

Cocaine*
Grand Marnier® liqueur
Famous Grouse® scotch
Pellegrino® sparkling water
Coriander
Rexal Formula III® baking soda

Supplementary 1. Extraction of Cocaine from Coca Leaves

Coca leaves*
Water
Dilute sulfuric acid or hydrochloric acid
Sodium carbonate
Kerosene
Benzene
Methyl alcohol
Chloroform
Benzoyl chloride
Benzoic anhydride
Sodium hydroxide
Ether
Acetone

Cook 2. LSD
$C_{20}H_{25}N$
50-37-3

We apologize for the proliferation of formulas here, but the 1960s were a wild time. There were three main sources for this information: Chewbacca Darth, Valentine Smith, and Robert E. Brown. I believe Uncle Fester also wrote a book on the subject. They seem to have read each other, as there is some similarity in these formulas. It would take a very sophisticated person to try one of these formulas. If an amateur is doing one of these, **watch out**. Some parts are hazardous and the products they produce can be nasty.

Use of LSD is dropping off from a high of around 2% of the population who have tried it. By mass, LSD is one of the most potent drugs discovered. Dose for effects: 20–30 μg (1/1,000,000 of a gram). Most drugs are taken at the mg (1/1000 gram) level. A lethal dose of LSD is 200 μg/kg.

Variation 1

Morning glory seeds* or pearly gates, wedding bells, or heavenly blue seeds* or Hawaiian baby wood rose* or rye grass
Petroleum ether, ligroine, or lighter fluid
Tartaric acid*
Ethanol
Ammonium hydroxide
Chloroform isobutanol
Chloroform

Variation 2

Morning glory seeds (pearly gates, wedding bells, or heavenly blue)*
Petroleum ether
Ligroine
Chloroform
Ammonium hydroxide
Black light*
Tartaric acid*
Sodium bicarbonate
Menthol
Potassium hydroxide
Nitrogen gas
Hydrochloric acid
Congo Red

Variation 3

Aluminum foil
Yellow light*
Red light*
Black light*
Hydrazine (anhydrous)*
Ice
Hydrochloric acid

Sodium nitrate
Sodium bicarbonate
Ether
Diethylamine*
Ethanol
Potassium hydroxide
Chloroform
Benzene
Alumina
Tartaric acid*
Methanol

Variation 4

Ergot
Acetonitrile
Dry ice
Acetone
Trifluoroacetic (anhydrous)*
Diethylamine*
Chloroform

Variation 5

Sulfur trioxide (anhydrous)*
Ferric sulfate (anhydrous)
Dimethylformamide*
Sodium hydroxide
Lithium hydroxide*
Methanol
Ethylene dichloride

Variation 6

Yellow light*
Red light*
Ergot*
Any alkali
Hydrazine*
Ice
Hydrochloric acid
Sodium nitrite*
Ethyl ether
Bicarbonate of soda
Diethylamine*

Variation 7

d-Lysergic acid
Acetonitrile
Acetone
Trifluoroacetic anhydride*
Diethylamine*

Chloroform
Ice
Sodium sulfate

Variation 8

Red light*
Ferric sulfate
Dimethylformamide*
Water
Sodium hydroxide
Phenolphthalein
d-Lysergic acid
Lithium hydroxide hydrate*
Diethylamine*
Saline solution
Ethylene dichloride

Supplementary 1. Purifying LSD from Any of the Formulas Above

Black light*
Red light*
Any product from Variations 1 to 8
Benzene or chloroform
Alumina
Methanol
Tartaric acid*
Potassium hydroxide
Ammonium hydroxide
Chloroform

Supplementary 2. Extraction of Lysergic Acid from Plants

Red light*
Yellow light*
Black light*
Ergot, morning glory seeds, rye grass or grain*
Methanol
Potassium hydroxide
Nitrogen gas*
Congo Red
Tartaric acid*
Aluminum foil

Supplementary 3. Growing Ergot

Cotton
Black light*
Sucrose

Chickpea meal
Calcium nitrate
Potassium phosphate
Potassium chloride
Ferric sulfate hydrated
Zinc sulfate hydrated
Sodium hypochlorite
95% Ethanol
Tartaric acid*
Ammonium hydroxide
Chloroform isobutanol
Benzene

Supplementary 4. Extraction from Ergot or Morning Glory Seeds

Lighter fluid
Blender
Hair dryer
Bicarbonate of soda
Ascorbic acid*

Storage

Lysergic acid compounds are unstable to heat, light, and oxygen. In any form it helps to add ascorbic acid as an antioxidant, keeping the container tightly closed, light-tight with aluminum foil, and in a refrigerator.

Cook 3. PCP
$C_{17}H_{25}N$

PCP is available as a powder, crystal, tablet, capsule, and liquid. It is injected, snorted, swallowed, or smoked on leafy materials such as tobacco, marijuana, parsley, mint, and oregano. Most often seen in large metropolitan areas, such as Baltimore, Chicago, Dallas, Houston, New York, Philadelphia, and Washington, DC. PCP increases heart rate, blood pressure, breathing rate, and urinary output. Induces sweating and vomiting. Causes jerky eye movement that can last for months after a single dose. Body heat increases, which is why so often users will strip.

The street product is usually the water-soluble sulfate salt. This is usually yellow to dark brown, due to contamination during the cook. Base has also been seen on the street.

Variation 1. Chewbacca Method

Piperidine*
Cyclohexanone*

Step 1
Benzene
p-Toluenesulfonic acid*
Toluene

Step 2. Grignard
Ether
Bromobenzene*
Magnesium
Ammonium chloride
Ammonium hydroxide
Potassium carbonate

Step 3. Salting
Hydrogen chloride gas
Methanol

Variation 2. Bucket Method

It appears that the phenyl portion of this compound will always be added with a Grignard reagent. It is surprising that there are not more fires. Apparently, this operation is usually done in the back of a van while driving around. It is also surprising that more cooks do not just fall asleep. Apparently, the limiting factor in this operation is the piperidine, which probably will be about 500 g. The parentheses indicate the variability of this formulation. Not all of these will be present.

Step 1
Cyclohexanone*
Piperidine*
(Potassium cyanide)
(p-Toluenesulfonic acid)*
(Hydrobromic acid)

Step 2. Grignard
Magnesium shavings
Ether
Bromobenzene*
(Iodine)

Step 3. Salting
White gasoline
Hydrochloric acid

Cook 4. Tetrahydrocannabinol

Variation 1. Making Tetrahydrocannabinol

Citral
Olivetol
Boron trifluoride etherate
Sodium hydroxide
Magnesium sulfate
Fluorisil column
Hexane
Ether

Variation 2. Making Tetrahydrocannabinol

Citral
Olivetol
Boron trifluoride etherate
Sodium hydroxide
Magnesium sulfate
Fluorisil column
Hexane
Ether

Variation 3. Making Tetrahydrocannabinol

Lithium
n-Butyl chloride
Olivetol
Dimethyl ether
Nitrogen gas
1-Methyl-4-isopropylidine*
3-Cyclohexanone *
Ether
Ice
Ethanol
Hydriodic acid

Variation 4. Making Honey oil (Dark Yellow)

Marijuana*
Butane
Isopropyl alcohol
(Plastic pipe)*

Variation 5. Extraction of Oil from Marijuana (Green)

Marijuana*
Ethanol (Everclear®)
(Electric stove)

Variation 6. Extraction of Oil from Marijuana (Yellow)

Marijuana*
Petroleum ether
Tobacco
(Distillation)

Supplementary 1. Making Olivetol

Calcium chloride
1-Chlorobutane*
Magnesium turnings
Iodine crystal
Cadmium chloride (anhydrous)*
Ether
3,5-Dinitrobenzoyl chloride*
Benzene
Mercuric chloride
Zinc
Hydrochloric acid
Sulfuric acid
Sodium nitrite
Ice

Supplementary 2. Making Heptylresorcinol

Hydrochloric acid
Thionyl chloride*
Water
2-Chloroheptane*
Powdered iron
3,5-Dinitrobenzoic acid*
Magnesium turnings
Sulfuric acid
Iodine
Sodium nitrite
Dry ice
Ether
Acetone
Ethanol
Ferric chloride (anhydrous)
Sodium hydroxide

Sodium sulfate
Sodium sulfite
Methyl bromide*
Ice
Sodium
Dimethyl sulfate*

Supplementary 3. Making Boron Trifluoride Etherate

Boron trifluoride
Ether

Cook 5. Psilocybin
$C_{12}H_{17}N_2O_4P$
520-52-5

Schedule 1 drug
The toxicity of psilocybin (psilocin becomes psilocybin) in the body, where it is metabolized by the liver), is relatively low; no human over-doses have been reported, animal data is between LD_{50} 12 and 150 mg/kg. This is almost one and a half times that of caffeine. The mushroom dose is 2 to 5 g of dried mushrooms. Psilocybin is from 1 mg to 1 g/L of fresh mushrooms and 5 mg of dried mushroom. There is a great deal of ambiguity about the legal status of psilocybin mushrooms and the spores of these mushrooms, as well as a strong element of selective enforcement in some places. Because of the ease of cultivating psilocybin mushrooms or gathering wild species, purified psilocybin is practically nonexistent on the illegal drug market.

Variation 1. Making Psilocin from Scratch

This is a lot of work for something you can get just by drying mushrooms. This is a difficult, lengthy, and dangerous synthesis. A person doing this should be very knowledgeable.

Step 1. Ethanol
2,6-Dinitrotoluene*
Ammonium sulfide
Hydrochloric acid
Ammonium hydroxide

Step 2. Concentrated sulfuric acid
Water
6-Nitro-*o*-toluidine (from step 1)
Sodium nitrite
Water
Ether

Step 3
Sodium hydroxide
6-Nitro-*o*-cresol (from step 2)
Sodium sulfite
Dimethyl sulfate*
Benzene
Sodium sulfate

Step 4. Making potassium ethylate
Potassium*
Xylene
Ether
Ethanol

Step 5
Diethyl oxalate
Potassium ethylate (from step 4)
2-Nitro-6-methoxytoluene (from step 3)
Water
Hydrochloric acid

Step 6
Ammonium hydroxide
2-Nitro-6-methoxyphenylpyruvic acid (from step 5)
Ferrous sulfate
Hydrochloric acid
Acetone

Step 7
4-methoxyindole-2-carboxylic acid (from step 6)
Potassium carbonate
Concentrated sulfuric acid
Petroleum ether

Step 8
Ether
4-Methoxyindole (from step 7)
Oxalyl chloride*

Step 9
Sodium chloride
Ice
Ether

4-Methoxyindole oxalyl chloride (from step 8)
Dimethylamine

Step 10
4-Methoxyindole glyoxal amide (from step 9)
Ether
Water
Methanol
Benzene
Aluminum foil
Lithium aluminum hydride*
Sodium sulfate
Hydrochloric acid
Sodium hydroxide
Chloroform
Sodium sulfate
Petroleum ether

Step 11
4-Methoxydimethyltryptamine (from step 10)
Hydriodic acid
Ammonium hydroxide
Petroleum ether

Variation 2. Psilocybin (Naturally Occurring) from Mushrooms

Step 1
Mushrooms
Water PDY agar*

Step 2 (a test)
PDY broth
Methanol

Step 3. Keller's reagent
Media from step 1
PDY agar on a slant
Refrigerator
Rye grain media

Step 4
Pressure cooker
Mason jars
Inoculation loops
Saccharimeter
Fungus from step 1, positive in steps 2 and 3

Step 5
Mortar and pestle

Hair dryer
Keller's reagent
Methanol

Supplementary 1. Making PDY Agar

Potatoes
Water
Agar
Dextrose
Yeast extract

Supplementary 2. Making PDY Broth

Potatoes
Water
Agar
Dextrose
Yeast extract

Supplementary 3. Making Rye Broth

Rye grain
Calcium carbonate
Water

Supplementary 4. Making Keller's Reagent

Ferric chloride
Glacial acetic acid

Cook 6. Mescaline
$C_{11}H_{17}NO_3$
128-04-6

LSD is 100 times as potent as psilocybin and 1000 times more potent than mescaline. However, not counting making this from scratch, this is 1000 times more easily acquired, especially if in Texas. Since it is sold extensively as a decorative plant, it is available off the Net or at nursery stores without notice any time. This is also included in recreational drugs despite the fact that it is a Schedule 1 drug. The cactus is chewed, or soaked in water for a beverage, which can have effects lasting up to 12 hours. The effective dose is 0.3 to 0.5 g. Note that the water extraction will also extract many other not-so-nice alkaloids.

The euphoric effects may be mixed with nausea and vomiting.

Variation 1. Making Mescaline from Scratch

Step 1
Sodium hydroxide
Sodium sulfate
3,4,5-Trihydroxybenzoic acid*
Dimethyl sulfate*

Step 2
3,4,5 Trimethoxybenzoic acid (from step 1)
Sodium hydroxide
Sodium carbonate
Dimethyl sulfate

Step 3
Lithium aluminum hydride
Ether
Methyl ester trimethoxybenzoic acid (from step 2)
Ice water
10% Sulfuric acid
Sodium sulfate

Step 4
Trimethoxybenzyl alcohol (from step 3)
Concentrated hydrochloric acid
Benzene
Ether

Step 5
Ethanol
Trimethoxybenzyl chloride (from step 4)
Hexamethylenetetramine (from Supplementary 1)

Step 6
Trimethoxybenzaldehyde (from step 5)
Ethanol
Nitromethane
Potassium hydroxide
Methanol

Step 7
Ether
Lithium aluminum hydride*
Trimethoxynitrostyrene (from step 6)
Ice water
Sodium hydroxide

Benzene
2 N Sulfuric acid

Variation 2. Extracting Mescaline from Peyote

This will probably look very much like a meth lab, as it follows many of the same steps, such as extractions, reaction, making base, and salting. In the extraction, benzene must be used, as it is specific to mescaline. There are many other alkaloids in peyote, which will also be extracted if other solvents are used.

Step 1. Extraction of mescaline
Peyote cactus*
Ethanol
Ammonia

Step 2. Reacting
Citric acid
Boiling water
Benzene

Step 3. Making mescaline base
Sodium hydroxide
Unspecified acid

Step 4. Extraction and salting
Benzene
Sulfuric acid

Step 5. Wash
Acetone

Variation 3

Peyote buds
Water
By extracting with water the person will also extract the following alkaloids, some of which are toxic. A person using other solvents in this method will probably get sick prior to the desired effects taking place.
N-Methylmescaline
N-Acetylmescaline*
Anhalonidine
Anhalonine
Pellotine
Lophophorine
Anhalamine
Anhaline
Anhalidine
O-Methyl-d-enhalonidine

Supplementary 1. Making Hexamethylene tetramine

Formuline*
Concentrated ammonium hydroxide

Cook 7. Dimethyltryptamine

Ether
Indole*
Oxalyl chloride*
Sodium chloride
Ice
Dimethylamine
Methanol
Benzene
Aluminum foil
Lithium aluminum hydride
Methanol
Purification
Sodium sulfate
Hydrochloric acid
Sodium hydroxide
Chloroform
Petroleum ether

Cook 8. Heroin

I could find no formula for making heroin from scratch. The methods below simply make heroin from morphine. The morphine is a Schedule 2 drug. Heroin is a Schedule 1 drug. Morphine is often stolen from hospitals and emergency vehicles.

Variation 1

Morphine hydrate*
Methanol
Acetic anhydride
Ethyl acetate
Carbon

Variation 2

Morphine*
Acetic anhydride
Benzene
Sodium carbonate
Sodium sulfate

Supplementary 1. Extracting Opium from Poppies

Opium poppies*
Water

Supplementary 2. Extracting Morphine from Opium Kabay Reaction

Opium poppies*
IN sulfuric acid
Ammonium hydroxide

Cook 9. 2-Phenyl-3-aminobutane (freebase)

This is similar to amphetamine, but it is more effective in fighting symptoms of fatigue. It is a colorless mobile liquid with an amine-like odor.

Sodium bisulfite
Methyl ethyl ketone
Potassium cyanide
Hydrochloric acid
Acetone
Acetic anhydride
95% Ethanol
Benzene
Aluminum trichloride
Ether
Chloroform
98% Sulfuric acid
Sodium azide
Sodium carbonate

RECREATIONAL DRUGS

Variation 1. Absinthe

True absinthe is distilled. The final product is green. Absinthe is illegal in most states but does not come up to the status of illegal drug; however, the popularity has been growing since 1990. This is probably a yuppy thing, as there is considerable ritual in drinking absinthe.

Absinthe

Wormwood
Hyssop
Melissa
Florence fennel
Green anis

Sometimes

Angelica root, sweet flag, dittany leaves, coriander, veronica, juniper, nutmeg

Variation 2. Fake Absinthe

There are numerous recipes on the Web for homemade absinthe. There are also kits, which somewhat incorporate the formula below. There would be some question as to whether this was illegal. This is not considered absinthe since it is not distilled. Since wormwood, the primary ingredient, is poisonous, overzealous cooks may create a poisonous drink.

Any combination of store-bought herbs
Wormwood
Vodka or Everclear®

Bananadine

Banadine is a fictional psychoactive substance, allegedly extracted from banana peels. A recipe for its extraction from banana peel was originally published as a hoax in the *Berkeley Barb* in March 1967. It became more widely known when William Powell, believing it to be true, reproduced the method in *The Anarchist Cookbook*. The recipe is given below. We have listed it under recreational drugs, but the only recreation is cooking up the "bananadine."

Microwave oven
10 lb bananas
Methylene chloride or Ronson® lighter fluid
Ether
Ethyl alcohol

Belladonna

Occasionally, the plant is used for recreational purposes: it is consumed in the form of either a tea or simply raw, which can produce vivid hallucinations, described by many as a "living dream." The Manson Family used it.

Khat

Khat is a stimulant, composed of the active ingredients cathinone and cathine, which causes excitement and euphoria. In 1980 the World Health Organization classified Khat as a drug of abuse that can produce mild to moderate psychological dependence. It is taken as a tea, or chewed. It is used mostly by people from eastern Africa.

Jimsonweed

Forms of ingestion: eating jimson weed seeds or drinking tea made of the plant's leaves and stems. Teenagers occasionally use jimson weed as a cheap alternative to illegal drugs. It can also be eaten or smoked. The effects of Datura have been described as a living dream: consciousness falls in and out, people who don't exist or are miles away are conversed with; and so on. The effects can last for days. Tropane alkaloids are some of the few substances that cause true hallucinations which cannot be distinguished from reality. This is unlike psylocybin or LSD, which cause only sensory distortions.

The doses that cause noticeable effects and the doses that can kill are very close. This makes overdosing on datura very easy. It can be fatal; it can cause fevers in the range 105 to 110°F, which is a range that can kill brain cells and lead to brain damage. There are many instances of teenagers looking for a cheap high who poison themselves to death on datura. If someone overdoses on datura, it is advised to cause vomiting and to wash out the stomach.

If taken recreationally, the datura experience seems almost identical from person to person. After ingestion the user does not notice any conscious effects, no matter how bizarre, for quite a while, which is why overdoses are so common; most people will take more, thinking it's not working, when in fact they're passing an imaginary cigarette to an imaginary friend. At the peak of such experiences, users often enter a true psychotomimetic state in which they lose touch with reality altogether. The majority of users report their experiences as unpleasant and often terrifying. Overall, it

is in very low demand as a recreational drug, due to the unpleasant high.

Laughing gas: Although nitrous oxide is pretty innocuous and the effects are very short lived, the method of making it start with fairly hazardous materials, and a mistake in setting up the apparatus could be lethal.

1.
Aluminum foil
Sand
Ammonium nitrate
Sodium carbonate
Water
Steel wool
2.
Ammonium nitrate
Water
Alcohol lamp

Peyote: Although this is a Schedule 1 drug, it is quite possible for a person to walk out in the dessert in Texas, pick up a cactus bud (peyote is quite common), and chew it. This would probably be considered more recreational. A person doing this will experience nausea long before they experience euphoria (because there are many other alkaloids in peyote); however, it is still done.

Rush®: Amyl nitrite (poppers) is popular with homosexuals as a sexual enhancer. We know of three events in San Francisco alone involving the manufacture of this compound. A fire in one amyl nitrite facility put several firemen into the hospital. We have presented the manufacture of this product in two ways: one in the Amphetamines section (Synthesis 18), where it is only a slight variation from making MDA. The two methods for making this product below come from chewbacca Darth, *Whole Drug Manufacturers' Catalog.* While the sale of amyl nitrite for use as an inhalant is illegal, it is still sold extensively as a Room odorizer or as a tape deck head cleaner.

Variation 1

Variation 1 is a very dangerous synthesis. Persons making this product are serious about what they are doing. If the nitric acid is too concentrated, they are making a primary explosive. If you encounter this operation, be aware that if the person was not very careful, especially with the nitric acid, there could be an explosion. Too much nitric acid would create a primary explosive. If you see this, be aware that it could be explosive. **This is a dangerous synthesis.**

Amyl alcohol
Strong sulfuric acid
Nitric acid
Water
Sodium hydroxide
Potassium carbonate

Variation 2

Isoamyl alcohol
Boric anhydride
70% Nitric acid

Shrooms

This is one of the many psilocybe mushrooms. Although possession of the mushroom is generally illegal, it is difficult to control eating them. Ingestion of 2 to 5 g of dried mushroom is sufficient to have noticeable effects. Although high doses can kill, Death has NOT been reported in recreational drug users.

4

PYROTECHNICS

AMMUNITION

Shells

Smokeless powders, black powder, rim fire primer fuses

Black powder is more a hobby than anything else. There are people who still shoot muzzle-loading rifles and pistols, firing cap and ball weapons, and even cannons.

Load 1. Black Powder

Potassium nitrate
Charcoal
Sulfur

Load 2. Gun Powder

Variation 1

Nitroglycerin
Cellulose

Variation 2

Mercury fulmanate
Pulverized soapstone
Gum arabic

Load 3. Smokeless Gun Powder

One or two items from each section below:

1. Propellents
Nitroglycerin
Nitroguanidine

2. Plasticizers
Dibutylpthalate
Polyester adipate
Rosin
Ethyl acetate

3. Binders
Diphenylamine
2-Nitrophenylamine

4. Stabilizers
4-Nitrodiphenylamine
N-Nisotrodiphenylamine
N-Methyl p-nitroaniline

5. Decopperizing
Tin dioxide
Bismuth
Bismuth trioxide
Bismuth nitrate

6. Flash Reducers
Potassium nitrate
Potassium sulfate

7. Barrel Wear Reducers
Wax
Talc
Titanium dioxide

Variation 1

Graphite
Calcium carbonate

Variation 2

14% Nitrocellulose
Water

Chemicals Used for Illegal Purposes: A Guide for First Responders to Identify Explosives, Recreational Drugs, and Poisons, By Robert Turkington
Copyright © 2010 John Wiley & Sons, Inc.

Ether
95% Ethanol
Diphenylamine
Potassium nitrate

Variation 3

Water
14% Nitrocellulose
Ether
95% Ethanol
Potassium nitrate
Diphenylamine

Load 4. High-Performance Gun Propellant

Step 1
Butyne-1,4-diol
Paraformaldehyde
p-Toluenesulfonic acid
Toluene
Water

Step 2
Polymer (from step 1)
HMX (*see* Explosives)

Variation 1. Military-Grade Gunpowder (double-based)

14% Nitrocellulose
Potassium sulfate
NENA
DIANP
Diethydiphenylurea
Ether
95% Ethanol

Rim Fire

Rim-fire ammunition came out just before the Civil War. It was the first attempt at multifire weapons. Rim-fire bullets were really small rockets. The advantage was that they were preloaded; the big disadvantage was that they were greatly affected by moisture and lost their impact after a relatively short distance. Firing rim-fire weapons is still a hobby. The modern meaning of rim fire is 22-caliber ammunition.

Load 5. Mercury Fulmanate Rim-Fire Mix

Barium peroxide
Antimony sulfide
TNT (*see* Explosives)
Mercury fulmanate

Load 6. Mercury Fulmanate Rim-Fire Primer

Potassium chlorate
Mercury fulmanate
Antimony sulfide
Styphnic acid

Load 7. Primary Mixture

Sulfur
Tin
Mercury fulmanate
Flour
Charcoal
Gum arabic

Load 8. CNTA Priming Mix

For ammo from 22 to 50 caliber
Antimony pentasulfide
Lead thiocyanate
Potassium chlorate
CNTA
Steel balls

Load 9. Silver Fulminate Priming Mix

Sulfur
Tin (powdered)
Silver fulminate
Flour
Charcoal
Gum arabic

Load 10. Mercury Fulminate Priming Mix

Sulfur
Tin (powdered)
Mercury fulminate
Flour
Charcoal
Gum arabic

McGyver Gun Powder

Potassium nitrate
Sugar
Ferric oxide

Bullets

Exploding bullets:

1. The bullet is hollowed out and filled with mercury, then sealed with candle wax.
2. The bullet can be hollowed out and filled with impact-ignited explosive, then sealed with candle wax.

Armor-piercing bullets:

A drill bit is inserted into the tip of the bullet.

Bismuth

Along the east coast of the United States, bismuth is used in place of lead for bullets, as it is safer for the environment.

EXPLOSIVES

Definitions

Deflagration: A sudden or rapid burning, as opposed to a detonation or explosion.

Detonation: A rapid chemical reaction that propagates at a supersonic velocity. (*Detonation wave*: a shock wave in a combustible mixture that originates as a combustion wave)

Explosion: The sudden production of a large quantity of gas, usually hot, from a much smaller amount of a gas, liquid, or solid (specifically, an explosion produced by combustion of a fuel and an oxidizer).

The distinction between an explosion and a detonation is that in an explosion the heat release rate and the number of molecules per unit volume increase with time more or less uniformly, whereas a detonation is propagated by an advancing shock front behind which exothermic reactions take place and thus is (spatially) nonuniform.

High Explosives:

HMX, RDX, TNT, tetryl, HNS, picric acid, TATB, PETN, and nitroglycerin

Astrolite, a liquid explosive, is considered to be the most powerful nonnuclear explosive, about twice as powerful as TNT. It is also easier to handle.

Process 1. Making Sodium Azide

Ammonia gas
Sodium metal
Nitrogen dioxide gas

Process 2. Making 99% Nitric Acid (see figure)

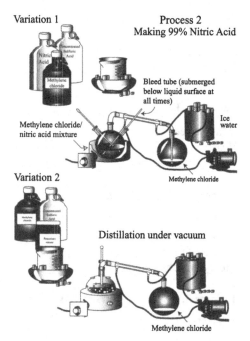

Variation 1

Process 2
Making 99% Nitric Acid

Variation 1

Potassium or sodium nitrate
Sulfuric acid
Water
Methylene chloride

Variation 2. Making 99% Nitric Acid

70% Nitric acid
98% Sulfuric acid
Methylene chloride

Variation 3. Making 70% Nitric Acid

Potassium or sodium nitrate
Sulfuric acid
Water
Methylene chloride

Process 3. Making 30% Sulfuric Acid

98% Sulfuric acid
Sulfur trioxide

Supplementary 1. Making Sulfur Trioxide

Variation 1

Sulfur dioxide
Oxygen
Platinum

Variation 2

Sulfur dioxide
Oxygen
Vanadium pentoxide

Variation 3

(May be seen in drug labs)
Heat
Ferric sulfate

Process 4. Making Trioxane

Paraffin oil
70% Formaldehyde
98% Sulfuric acid
Methylene chloride

Process 5. Making Chlorine Gas

Variation 1

Down's cell
Sodium chloride

Variation 2

10% Hydrochloric acid
Calcium hypochloride
Water

Calcium chloride

Process 6. Making Ammonium Gas

Variation 1
Any ammonium salt
Sodium hydroxide

Variation 2

10% Ammonium hydroxide
35% Hydrochloric acid
98% Sulfuric acid
Sodium hydroxide

Process 7. Making Bromine

Chlorine gas
Sodium bromide

Acetone Peroxide (Primary Explosive)

Detonates when ignited

Variation 1

Acetone
98% Sulfuric acid
30% Hydrogen peroxide
Ether

Variation 2

Acetone
98% Sulfuric acid
30% Hydrogen peroxide
Ether
Acetone
Smokeless gun powder

Variation 3

Acetone
Concentrated hydrochloric acid
30% Hydrogen peroxide
Ether
Acetone
Smokeless gun powder

ADBN
(4-Azido-4,4'-dinitro-1-butyl nitrate)
(Secondary explosive)

Hydrochloric acid
Methanol

ADNB
Methylene chloride
Nitric acid
10% Baking soda
Magnesium sulfate (anhydrous)

ADN
(Ammonium dinitramide)
(Secondary explosive)

1.
Nitric acid
Urea
98% Sulfuric acid
Nitronium tetrafluoroborate
Ammonia gas
Acetonitrile
Ethyl acetate
Chloroform
2.
Acetonitrile
Nitronium tetrafluoroborate
Ammonium carbonate
Isopropyl alcohol
Ammonia gas
Ether
Acetone
Ethyl acetate
Butanol
3.
KDN
Ammonium sulfate
Isopropyl alcohol
Petroleum ether

ADNB
(4-Azido-4,4′-dinitro-1-butyl acetate)

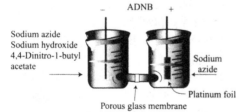

4,4′-Dinitrobutyl acetate
Methylene chloride
Magnesium sulfate (anhydrous)
Sodium azide
Sodium hydroxide

Acetyl chloride
Ethyl acetate
Hexane

ADNBF
(7-Amino-4,6-dinitrobenzofuroxan)
(Secondary explosive)
May detonate upon ignition

Tetraniline
Glacial acetic acid
Sodium azide

Ammonium azide
(Primary explosive)

Sodium azide
Ammonium gas
Carbon dioxide gas

Ammonium chlorate
(Secondary explosive)
Nonflammable; may flash when ignited

30% Ammonium hydroxide
Carbon dioxide gas or dry ice
Sodium chlorate

Ammonium perchlorate
(Secondary explosive)
Nonflammable; decomposes when heated

30% Ammonium hydroxide
Carbon dioxide gas or dry ice
Sodium perchlorate

Ammonium picramate
(Primary explosive)
Flammable; deflagrates

Picric acid
13% Ammonium hydroxide
Ammonium bisulfide
Carbon disulfide

Ammonium picrate
(Primary explosive)
Flammable; may flash

Benzene
Ammonia gas
Picric acid

Ammonium triiodide
(Primary explosive)
VERY UNSTABLE

(Probably useless)
Household ammonia
Iodine

AN
(Ammonium nitrate)
(Secondary explosive)

1.
70% Nitric acid
Ammonium gas
Methanol
30% Ammonium hydroxide
2.
Sodium nitrate
Ammonia gas or ammonium hydroxide
solution
Methanol
Carbon dioxide gas or dry ice
3.
Household ammonia
Nitric acid

ANFO
See Blasting agents

A-NPNT
(4-Amino-*N*,2,3,5,6-pentanitrotoluene
(Secondary explosive)

TNT
Dioxane
Hydrogen disulfide
30% Ammonium hydroxide
Hydrochloric acid
Potassium iodide
98% Sulfuric acid
Paraformaldehyde
90% Nitric acid
Methylene chloride
Methanol
Magnesium sulfate
Chloroform

Astrolite
(Liquid explosive)

Variation 1. Astrolite A

Ammonium nitrate
Hydrazine

Variation 2. Astrolite G

Ammonium nitrate
Hydrazine
Aluminum

Making Hydrazine

Supplementary 1. Olin Raschig Process

Ammonium hydroxide
Sodium hypochlorite
Ammonia gas

Supplementary 2. Atofina-PCUK

Acetone
Ammonium hydroxide
Hydrogen peroxide
Ammonia gas

Supplementary 3. Ketazine Peroxide
Process

Ammonia gas
Chlorine gas or chloramine
Acetone
Gelatin or glue

Azido Ethyl
(Primary explosive)

Trimethylformamide
Sodium azide
2,2′,2″-Trichloroethylamine
(Nitrogen mustard!!!!!)
Chloroform
Sodium sulfide

Barium Styphnate
(Primary explosive)
Flashes when ignited

Styphnic acid
30% Ammonium hydroxide
Barium chloride
Acetone

BDC
(Biguanide diperchlorate)
(Secondary explosive)
Flammable; may deflagrate

Biguanide
Ethanol

Perchloric acid
Ether

BDPF
(Primary explosive)

1,3-Dichloro-2-propanol
Troxane (as formaldehyde)
1,2-Dichloroethane
Sulfuric acid
1% Baking soda
Dimethyl sulfoxide
Sodium azide
Methylene chloride
Magnesium sulfate

Benzoyl Peroxide
SADT of 114°F

Benzoyl chloride
Sodium peroxide

Supplementary 1. Making Benzoyl

Chloride
Benzoic acid
Sulfuyl chloride

Supplementary 2. Making Benzoyl

Chloride
Benzoic trichloride
Water
Zinc chloride

BINARY EXPLOSIVES

Liquid Explosives

Binary explosives are used for ease of shipping. While the final product requires considerable preparation and expense in shipping, each of the binary parts can be shipped under considerably less restrictive conditions: in some cases, with no hazmat regulation at all. Although most of these require blasting caps, some can be detonated using just a booster. There are unspecified formulations that can be ignited with a black powder fuse. While liquid explosives are versatile, because they are liquid, a gelling agent is often added, as it makes the compound more explosive. Nitromethane to be used as an explosive is often dyed RED.

Variation 1. Kinepak Kenepouch

Requires blasting cap

Nitromethane
Ammonium nitrate

Variation 2. Helex

Booster sensitive

Looks like runny silver toothpaste
Nitromethane
Flaked aluminum
(Steric acid)

Variation 3

Nitromethane
Amine (*see* Supplementary 1)

Variation 4

Nitromethane
Nitroethane
Aluminum

Variation 5. Making Nitromethane

Sodium nitrite
Monochloroacetic acid
Sodium hydroxide

Variation 6. Making Nitromethane

Methane or propane
Nitric acid

Variation 7. Binex
Requires blasting cap

Sodium perchlorate (aqueous)
Aluminum powder

Supplementary 1. Nitromethane
Sensitizers

(Amines that sensitize nitromethane)
Variation 3 is a combination of nitromethane and an amine.
Diethylamine
Triethylamine
Ethanolamine
Ethylenediamine (most sensitive)
Morpholine
(Nonchemical entrappers)
Microballoons
Polymeric foam

Supplementary 2. Other Nitro Paraffins That Form Explosives

These become less sensitive as the carbon number increases.

Nitroethane
Nitropropanes
Etc.

BLASTING AGENTS
Require boosters for detonation

ANFO
(Tertiary explosive or a blasting agent)
Requires high-explosive booster

Variation 1

Ammonium nitrate
Diesel fuel

Variation 2

Ammonium nitrate
Black powder

Variation 3

Ammonium nitrate
Sulfur
Charcoal
Sodium nitrate
Calcium carbonate
Urea
Kerosene

Variation 4

Ammonium nitrate
Gar gum
Aluminum powder
Polytetrafluoroethylene
Water

Tovan
(Tertiary explosive or blasting agent)
Requires high-explosive booster

Ammonium nitrate
Mineral oil

Blasting Caps and Primers

Blasting Caps (see figure next page)

RDX
Lead azide
Sulfur
Wires
Aluminum tube
Styphnic acid
Barium chromate

Black Powder

Potassium nitrate
Charcoal
Sulfur

Casein Nitrate
See Nitrated milk powder

CDBTA
(3,5-Dinitro-1,2,4-triazole-copper salt)
(Primary explosive)
Detonates when ignited

Hydrazine
Cyanogen bromide
Isopropyl alcohol
Sodium nitrite
Baking soda
Copper nitrate

CNTA
(Copper salt of 5-nitrotetrazole)
(Primary explosive)
Detonates when ignited
1.
Cupric sulfate
5-Aminotetrazole
98% Sulfuric acid
Sodium nitrite
2.
Cupric sulfate
5-Aminotetrazole monohydrate
98% Sulfuric acid
25% Sodium nitrite
Sulfuric acid

Copper Acetylide

Copper sulfate
Sodium hydroxide
Acetylene

Blasting Caps

Typical blasting cap

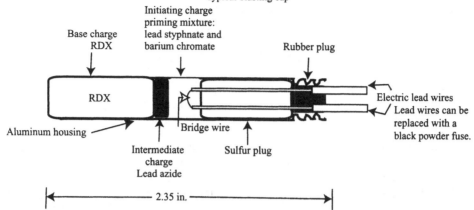

Explosive Train Materials
1. Ignition
   Black powder fuse
   Electrical bridge
2. Initiators
   Metal compounds
   Lead styphnate
   Fulmanates
   Salts of nitrotriazoles
   Tetrazoles
   Ammonium nitrophenyl compounds
   Diazodinitrophenol
3. Booster charge
   Lead azide
   Azides
   Metal or ammonium salts of nitrophenyls
   Salts of nitrotriazoles
   Salts of tetrazoles
   Azonitro compounds
4. Base charge
   RDX
   HMX
   Methylnitramine
   Other nitramine compounds

Copper Fulminate
(Primary explosive)

Nitric acid
Copper(II) nitrate
Ethanol

Cyclotrimethylenetrinitramine
Hexamine
Nitric acid

CZ
(Copper azide)
(Primary explosive)
Detonates when ignited
UNSTABLE

Sodium azide
Copper sulfate pentahydrate
Ice

DANP
(1,5-Diazido-3-nitrazapentane)
(Secondary explosive)

Acetic anhydride
99% Nitric acid
Hexamine
Glacial acetic acid
Methylene chloride
Baking soda
Magnesium sulfate

Dioxane
Hydrogen chloride gas
Acetone
Sodium azide

DATB
(1,3-Diamino-2,4,6-trinitrobenzene)
(Secondary explosive)

m-Phenyleneamine
Methanol
Sodium carbonate
Ethyl chloroformate
Ethylene glycol dimethyl ether
98% Sulfuric acid
90% Nitric acid
30% Fuming Sulfuric acid

DATBA
(5-Carboxyl-1,3-diamino-2,4,6-
trinitrobenzene)
(Secondary explosive)

2-Amino-2-methylpropane
Methylene chloride
1,3,5-Trifluoro-2-4,6-trinitrobenzene
Trifluoroacetic acid
Cyanotrimethyl silane
Nitromethane
Acetonitrile
98% Sulfuric acid

DDD
(5,7-Dinitro-5,7-diaza-1,3-
dioxabicyclooctan-2-one)
(Secondary explosive)
Requires blasting cap

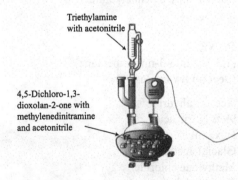

Triethylamine
with acetonitrile

4,5-Dichloro-1,3-
dioxolan-2-one with
methylenedinitramine
and acetonitrile

Triethylamine
Acetonitrile

Methylenedinitramine
4,5-Dichloro-11,3-dioxolan-2-one
Silica gel
Benzene
Hydrochloric acid
Sodium nitrite

DDNP
(4,6-Dinitrobenzene-2-diazo-1-oxide;
diazodinitrophenol)

Picric acid
Sodium or potassium nitrite
85% Sulfuric acid
Sodium hydroxide
Sulfur

Detonator Cord
(Secondary explosive)
Requires blasting cap

PETN
Cotton tube
Asphalt
Plastic coating

Diaminesilver Chlorate
(Shock sensitive)

Silver nitrate
Sodium chlorate
25% Ammonium hydroxide

DIANP
(1,5-Diazido-3-nitrazapentane)
(Liquid explosive)
(Secondary explosive)

Dimethyl sulfoxide
Sodium azide
Dinitroxydiethylnitramine
Methylene chloride
Sodium sulfate
Alumina

Diazodinitrophenol
(Secondary explosive)

1.
Sodium picramate
Hydrochloric aicd
Sodium nitrite

2.
99% Nitric acid
Kerosene
Ammonium picramate
Ethanol
3.
Methyl green
Sodium picramate
Hydrochloric acid
Sodium nitrite

DINA
(Dinitroxydiethylnitramine)
(Secondary explosive)
Burns with little smoke
MAY DETONATE

1.
Acetic anhydride
38% Hydrochloric acid
Diethanolamine
Methylene chloride
70% Nitric acid
5% Sodium bicarbonate
2.
99% Nitric acid
Diethanolamine
Acetic anhydride
Acetyl chloride
Acetone
Potassium carbonate
3.
99% Nitric acid
Diethanolamine
Hydrogen chloride gas
10% Baking soda

DITN
(Diisopropylamine trinitrate)
(Secondary explosive)

Diisopropylamine
70% Nitric acid
99% Nitric acid

DMMD
2,4-Dinitro-2,4-azapentane)
(Secondary explosive)

Formaldehyde
Methylnitramine

Methylene chloride
Magnesium sulfate
98% Sulfuric acid
Chloroform
Hexane

DNAN
(Dinitroxydiethylamine nitrate)
(Secondary explosive)
Flashes when ignited

90% Nitric acid
Diethanolamine
Ethanol
Ether
Acetone

DNAT
(1,1'-Dinitro-3,3'-azo-1,2,4-triazo)
(Secondary explosive)

Potassium permanganate
3-Amino-1,2,4-triazole
Sodium hydroxide
Sodium bisulfide
Hydrochloric acid
99% Nitric acid
Acetic anhydride
Acetone

4,4-DNB
(4,4-Dinitro-1-butanol)
(Secondary explosive)
May burn when ignited

TNB
Methanol
Potassium iodide
38% Hydrochloric acid
Sodium bisulfite
Methylene chloride
Magnesium sulfate

DNFA-P
(1,4-Dinitrofurazano[3,4-*b*] piperazine)
(Secondary explosive)
May burn when ignited
Requires blasting cap

Hydroxylamine·HCl
Glyoxal

Sodium carbinate
Ethanol
Chlorine gas
Methanol
Ethylenediamine
Ethylene glycol
Sodium hydroxide
Trifluoroacetic anhydride
99% Nitric acid
Acetone

DNP
(2,4-Dinitrophenol)
(Secondary explosive)
Burns with smoky flame

Phenol
Sodium nitrite
Sodium hydroxide
Nitric acid
60% Nitric acid

DNPU
(2,4-Dinitrophenylurea)
(Secondary explosive)

Phenyl urea
Methylene chloride
98% Sulfuric acid
90% Nitric acid
5% Sodium bicarbonate

DNR
(4,6-Dinitroresorcinol)
(Secondary explosive)
Burns with smoky flame

70% Nitric acid
90% Nitric acid
98% Sulfuric acid
Resorcinol diacetate
Urea

Double Salts
See Silver acetylide

DPT
(1,5-Methylene-3,7-Dinitro-1,3,5,7-
tetracyclooctane)
(Secondary explosive)

Requires primer

Nitrourea
Formaldehyde
Sodium hydroxide
Ammonium hydroxide

Dynamite
(Secondary explosive)
Requires blasting cap

Diatomaceous earth
Sodium carbonate
Nitroglycerin
Cardboard or plastic tube

EDDN
(Ethylenediamine dinitrate)
(Secondary explosive)
Flammable, may flash

Ethylenediamine
70% Nitric acid
Ethanol

EDT
(N,N'-Di(2-ethanol)ethylenediamine
tetranitrate)
(Secondary explosive)
May flash upon ignition

1.
99% Nitric acid
N,N'-Diethylanolethylenediamine
Ethanol
2.
Ethylenediamine
Ethylene oxide
Phosphorus pentoxide
99% Nitric acid
10% Ammonium hydroxide
Ether

EGDN
(Ethylene glycol dinitrate)
(Secondary explosive)
May detonate upon ignition

90% Nitric acid
98% Sulfuric acid
Ethylene glycol
5% Sodium carbonate
Sodium sulfate

ETN
(Erythritol tetranitrate)
(Secondary explosive)
May flash upon ignition

98% Sulfuric acid
Erythritol
90% Nitric acid
10% Sodium carbonate
Ethanol

F-TNB
(Secondary explosive)
Burns with smoky flame
Used in hand grenades

30% Fuming sulfuric acid
Potassium nitrate
1,3,5-Trifluorobenzene
Methylene chloride
Hexane
Charcoal
Sodium sulfate

Gun Cotton
See Nitrocellulose

Helex
See Binary explosives

Hexaditon
(2,2′4,4′6,6-Hexanitrodiphenylmethane)
(Secondary explosive)
May flash upon ignition

TNT
Tetrahydrofuran
Potassium hydroxide/methanol
Picryl chloride
DMSO
Hydrochloric acid
Methanol
Acetone

Hexanitrate
(Secondary explosive)
May flash upon ignition

90% Nitric acid
98% Sulfuric acid
Sorbitol

Ethanol
10% Sodium carbonate

Hexol
(Secondary explosive)
Burns with smoky flame

Benzene
Methanol
TNT
Sodium hydroxide
5% Sodium hypochlorite

HgNTA
(Mercury nitrotetrazole)
(Primary explosive)
Detonates upon ignition

Sodium hydroxide
CNTA
Mercury(II) nitrate
70% Nitric acid

HMTD

Variation 1

Hydrogen peroxide
Hexamine
Citric acid

Variation 2

37% Formaldehyde
3% Hydrogen peroxide
Ammonium sulfate

HMX
(Otgen: 1,2,5,7-tetranitro-1,3,5,7-tetrazacyclooctane)
(Secondary explosive)
Requires a primary

Phosphorus pentoxide
Nitric acid
Solex

β-HMX
(Secondary explosive)
Requires a primary

Nitric acid
Ammonium nitrate
Paraformaldehyde

Hexamine
Acetone
Glacial acetic acid
Acetic anhydride

HNB
(Secondary explosive)
Burns with smoky flame

Benzene
Methanol
TNT
Sodium hydroxide
5% Sodium hypochlorite solution

HNBP
(Hexon: trobiphenyl)
(Secondary explosive)
Burns with smoky flame

Picryl chloride
Ethylene dichloride
Copper powder
Acetone

HNF
(Hydrazine nitroform)
(Secondary explosive)
May deflagrate upon ignition

1.
Nitroform
Ether
Hydrazine
2.
Nitroform
Methanol
Hydrazine
Carbon tetrachloride

HNH-3
(1,1,1,6,6,6-Hexanitrohexyne-3)
(Secondary explosive)
May deflagrate upon ignition

Nitroform
Methyl acetate
Silver oxide
1,4-Dibromobutyne-2
Chloroform

HNIW
(2,4,6,8,10,12-Hexanitro-2,4,6,8,10,12-
hexaazaisowurtzitane)
(Secondary explosive)
Requires RDX detonator

Glyoxal
Benzyl amine
88% Formic acid
Acetonitrile
Acetic anhydride
Pearlman's catalyst
Bromobenzene
Chloroform
Sulfolane
Nitrosonium tetrafluoroborate
Ethyl acetate

HNS
(Hexanitrostilbene)
(Secondary explosive)
Burns with smoky flame

1.
Trinitrobenzyl chloride
Triethylbenzyl ammonium chloride
Methylene chloride
Sodium hydroxide
Methanol
2.
TNT
DMSO
Oxygen gas
Sodium benzoate
Hydrochloric acid
Methanol

HNTCAB
(Hexanitrotetrachloroazobenzene)
(Secondary explosive)
Requires blasting cap

90% Nitric acid
30% Fuming sulfuric acid
3,5-Dichloroaniline
Sulfuric acid
Acetone
Hexane

Isoitol Nitrate
(Secondary explosive)
Flashes upon ignition

Quebrachitol
Hydriodic acid
Ethanol
Ether
99% Nitric acid
98% Sulfuric acid
10% Baking soda
Sodium sulfate

KDN
(Potassium dinitramide)
(Secondary explosive)

1.
Nitronium tetrafluoroborate
Acetonitrile
Ammonium carbonate
Potassium carbonate
Ether
Acetone
Ethyl acetate
Butanol
2.
Sulfamic acid
Potassium hydroxide
Ethanol
90% Nitric acid
98% Sulfuric acid
Acetone
Isopropyl alcohol

KNF
(Potassium nitroform)
(Secondary explosive)
Highly flammable; deflagrates

Dioxane
Potassium nitrite
Potassium bicarbonate
Tetranitromethane

Kinepac
See Binary explosives

Lead Nitroanilate
(Detonates easily)

Lead nitrate
Salicylic aicd
Hydrochloric acid
Potassium chlorate
Ethanol

Lead Picrate
(Primary explosive)
Detonates upon ignition

Variation 1

Picric acid
Sodium hydroxide
Lead nitrate

Variation 2 (military)

Methanol
Lead monoxide
Picric acid

Supplementary 1. Lead Monoxide

Sodium or potassium nitrite
Lead
Propane torch
Metal pipe

Lead Styphnate
(Primary explosive)
Detonates upon ignition

1.
Styphnic acid
Sodium hydroxide
Lead(II) nitrate
2.
Styphnic acid
Sodium hydroxide
Lead(II) nitrate
3.
Styphnic acid
Magnesium carbonate
Lead(II) nitrate
70% Nitric acid

Lead TNP
(Primary explosive)
Detonates upon ignition

Phloroglucinol
Glacial acetic acid

Sodium nitrite
Lead nitrate

LNTA
(Lead nitrotetrazole, basic salt)
(Primary explosive)
Detonates upon ignition
NOT STABLE

CNTA
Hydrogen sulfide
Benzene
Lead(II) hydroxide

LZ
(Lead azide)
(Primary explosive)

Variation 1

Lead(II) acetate
Sodium azide
Water

Variation 2

Sodium azide
Dextrin
Sodium hydroxide
Lead nitrate

McGyver (Improvised) Explosives

Plastic Explosive
Requires blasting cap
Vaseline®
Potassium chlorate

Armstrong's Powder
Impact explosive
(Contact explosive)

Variation 1

Potassium chlorate salt
Red phosphorus

Variation 2

Any chlorate salt
Red phosphorus
Sulfur

Rack-a-Rock

Potassium chlorate
Nitrobenzene

Generic Explosive

Potassium chlorate
Powdered sugar
Sulfur

Solidox Bomb

Solidox
Sugar

Variation 1

Match heads
Steel pipe

Variation 2

Perchlorate
Sugar
Steel pipe

Variation 3. Black Powder

Steel pipe
Oxidizing fertilizer
Other ingredients could be:
Charcoal
Aluminum
Sulfur

Pop Bottle (Plastic Pressurized Bottles)

1.
Dry ice
2.
Aluminum foil
Sodium hydroxide solution
3.
Aluminum foil
Hydrochloric acid

Depth Charge

Bangsite
Tin cans

Oxidizer–Fuel Mixtures

The table in the following page is a published
list of mixtures that can be used to make

low-order explosives (deflagrates). That some of the mixtures will work is doubtful; some would work, however. A person making one of these mixtures is most likely driven by availability of product. Do not discount the fact that low-order explosives can still cause severe damage.

Mercury Fulminate
(Primary explosive)
Detonates upon ignition
TOXIC
UNSTABLE

Nitric acid
Mercury
Ethanol

OXIDIZER-FUEL MIXTURES

Oxidizer	*Fuel*	*Notes*
Potassium chlorate	Sulfur	Unstable
	Sugar, charcoal	Unstable, slow
	Sulfur, Mg, or Al	Very unstable
	Charcoal sulfur	Unstable
	Phosphorus sesquisulfide	
Potassium chlorate & calcium carbonate	Red phosphorus, sulfur	Very unstable Impact sensitive
Sodium nitrate	Mg, sulfur	Unpredictable
Potassium permanganate	Glycerin	Depends on grain size
	Sulfur	Unstable
	Sulfur, Mg, or Al	Unstable
	Sugar	
	Powdered sugar, Al, or Mg	Ignites if gets wet Unstable
Potassium nitrate	Charcoal sulfur	Black powder
Ammonium perchlorate	Fe and Mg	Solid rocket fuel
K or Na perchlorate	Al or Mg	Flash powder
	Al or Mg and sulfur	Flash powder
Barium nitrate & potassium perchlorate	Al	Flash powder
Barium peroxide	Al and Mg	Flash powder
Potassium perchlorate	Mg or Al and sulfur	Slightly unstable

Mercury Nitride
(Primary explosive)
Detonates upon ignition
TOXIC
UNSTABLE

Ammonia gas
Mercury(II) oxide
Nitric acid

Methylene Dinitroamine
(Secondary explosive)

99% Nitric acid
Methylene diformamide
Acetic anhydride
99% Formic acid
Benzene
Ice
Sulfuric acid

Methyl Ethyl Ketone Peroxide
(Liquid explosive)

Methyl ethyl ketone
30% Hydrogen peroxide
Sulfuric acid
Ice

Methyl Ethyl Ketone Oxyacetylene
(Liquid explosive)

Methyl ethyl ketone
Oxyacetylene
One of the following gases: butane, ethane, ethylene oxide

Methylpicric Acid
(Secondary explosive)
Burns with smoky flame

m-Cresol
Sodium nitrite
Sodium hydroxide
90% Nitric acid
60% Nitric acid

Metriol
(Secondary explosive)
Nonflammable, but may detonate upon ignition

90% Nitric acid
98% Sulfuric acid

Mentriol
Ether
5% Sodium carbonate
Sodium sulfate

MGP
(*N*-Methylgluconamide pentanitrate)
(Secondary explosive)
May flash upon ignition

99% Nitric acid
N-Methyl gluconamide
Acetic anhydride
10% Baking soda
Methanol

Milk Booster
See Nitrated milk powder

MNA
(Methylnitramine)
(Secondary explosive)

Dimethylurea
70% Nitric acid
30% Fuming sulfuric acid
Methylene chloride
Sodium carbonate

MNTA
(1-Methyl-3,5-nitro-1,2,4-triazole)
(Secondary explosive)
Burns with mild flame

Hydrazine
Cyanogen bromide
Isopropyl alcohol
Sodium nitrite
Copper nitrate
70% Nitric acid
Ether
Dimethyl sulfate
Sodium carbonate

MON
(Maltose octanitrate)
(Secondary explosive)
Flammable; may flash upon ignition

Maltose (ahnydrous)
Urea
99% Nitric acid
30% Fuming sulfuric acid

5% Sodium carbonate
Ether
Ethanol

MX
(Nitromannite)
(Secondary explosive)
Flammable; may flash upon ignition

99% Nitric acid
98% Sulfuric acid
D-Mannitol
5% Sodium bicarbonate

MZ
(Mercury azide)
(Primary explosive)

Mercury nitrate
Sodium azide

NDTT
(5-Nitro-2(3,5-diamino-2,4,6-
trinitrophenyl)-1,2,4-triazole)
(Secondary explosive)
Burns with smoky flame

Potassium nitrate
30% Fuming sulfuric acid
1,3,5-Trifluorobenzene
Methylene chloride
Hexane
tert-Butylamine
Trifluoroacetic acid
1,2-Dichloroethane
3-Amino-1,2,4-triazole
Glacial acetric acid
Sodium nitrite
98% Sufuric acid
Urea
Ethyl acetate
Dimethylformamide
Ether
Sodium sulfate
Methanol

NENA
(*N*-(2-Nitroxyethyl)nitramine)
(Secondary explosive)
May burn with smoky flame

Ethanol amine
Ether

Ethyl chlorocarbonate
Sodium hydroxide
Magnesium sulfate
99% Nitric acid
Ammonia gas
98% Sulfuric acid

NINHT
(2-Nitroimino-5-nitro-hexahydro-1,3,5-
triazine)
(Secondary explosive)
Requires primer

39% Hydrochloric acid
Nitroguanidine
Hexamine
Methanol
Sodium nitrite
99% Nitric acid

Nitrated Milk Powder

Milk
70% Nitric acid
Sulfuric acid
Vinegar
Bicarbonate of soda

Nitrocellulose
(Secondary explosive)
Highly flammable; deflagrates

1.
Phosphorus pentoxide
Methylene chloride
99% Nitric acid
Cotton balls
10% Baking soda
2.
99% Nitric acid
98% Sulfuric acid
Wood pulp
10% Baking soda
Sodium hydroxide
Carbon disulfide
3.
60% Nitric acid
98% Sulfuric acid
Cellophane
Cotton
10% Baking soda
4.
98% Sulfuric acid

70% Nitric acid
Cotton
10% Baking soda
5.
98% Sulfuric acid
Newspaper (not recycled)
70% Nitric acid
10% Baking soda
6.
99% Nitric acid
98% Sulfuric acid
Wood cellulose
10% Baking soda
Depending on the concentration of nitric acid, cellulose mononitrate, dinitrate, or trinitrate is formed. Trinitrate is used in explosives.

WARNING: Nitrocellulose decomposes very slowly on storage. The decomposition is auto-catalyzing and can result in spontaneous explosion if the material is kept confined over a period of time.

Nitroform
(Secondary explosive)
Difficult to ignite

1.
99% Nitric acid
Isopropyl alcohol
Metyhylene chloride
Calcium chloride
2.
Methanol
Sodium nitrite
Baking soda
Tetraaminomethane
38% Hydrochloric acid
Methylene chloride
Calcium sulfate
3.
Methanol
Potassium carbonate
Hydrogen chloride gas
Ether
Potassium nitrite
Tetranitromethane
Methylene chloride

Nitrogen Sulfide

Sulfur
Chlorine
Manganese dioxide
Benzene
Ammonia gas

Nitrogen Trichloride
(Liquid explosive)
Explodes violently when heated above
60°C or exposed to open flame
Explodes on contact with dust, organic
material

Variation 1

Battery charger
Ammonium chloride
Ammonia
Hydrogen chloride gas
Carbon

Variation 2

Battery charger
Ammonium chloride
Ammonia
Hydrochloric acid
Lead rods from battery

Variation 3

Ammonium nitrate
Water
Hydrochloric acid
Potassium permanganate

Nitroglycerin
(Liquid explosive)
(Secondary explosive)
Causes severe headaches from contact or
inhalation
Detonates above 50°C

1.
Glycerin
Nitric acid
Sulfuric acid
Ice
2.
Sodium chloride
Bicarbonate of soda

Concentrated nitric acid
Concentrated sulfuric acid
Glycerin1

Nitroguanidine
(Secondary explosive)
Burns with smoky flame
May flash

Guanidine
98% Sulfuric acid
70% Nitric acid

Nitromannitol
Nitromannite
Mannitol hexanitrate

Variation 1

Mannitol
Nitric acid
Sulfuric acid

Variation 2

Mannitol
Nitric acid
Sulfuric acid
Ethanol
Sodium chloride
Sodium bicarbonate

Nitro PCB
(Secondary explosive)
Flammable; may flash

70% Nitric acid
98% Sulfuric acid
Perchlorobenzene
Benzene
Petroleum ether

Nitrostarch
(Secondary explosive)
Highly flammable; deflagrates

99% Nitric acid
98% Sulfuric acid
Cornstarch
55% Nitric acid
1% Ammonium hydroxide

Nitrourea
(Secondary explosive)
Burns with smoky flame
May flash

1. Making urea nitrate
Concentrated nitric acid
Urea
2.
Urea nitrate
Concentrated sulfuric acid

NMHAN
(*N*-Nitro-*N*-methylhydroxyacetamideni-
trate)
(Secondary explosive)
May flash upon ignition

N-Methylhydroxyacetamid
99% Nitric acid
98% Sulfuric acid

NQ
See Nitroglycerin

NTA
(3,5-Dinitro-1,2,4-triazole)
(Secondary explosive)
May flash when ignited

Hydrazine
Cyanogen bromide
Isopropyl alcohol
Sodium nitrite
Baking soda
Copper nitrate
70% Nitric acid
Ether

NTND
(2-Methyl-2(*N*-nitro-*N*-trinitroethylamino)-
1,3-propyldinitrate)
(Secondary explosive)
Flammable

2-Methyl-2-amino-1,3-propanediol
Nitroform
37% Formaldehyde solution
Ethanol
Magnesium sulfate
Acetic anhydride

90% Nitric acid
Baking soda

NTO
(3-Nitro-1,2,4-triazol-5-one)
(Secondary explosive)
May burn

Simicarbazide hydrochloride
85% Formic acid
70% Nitric acid

NU
See Nitrourea

PCB
(Perchlorobenzene)
 (Secondary explosive)
Flammable; may flash upon ignition

Aluminum trichloride
Benzene
Perchloryl fluoride

PEN
(Pentaerythritol trinitrate)
(Secondary explosive)
May flash when ignited

Pentaerythritol
99% Nitric acid
98% Sulfuric acid
Methylene chloride
Urea
Baking soda
Ether

PETN
(Pentaeythritol trinitrate-2,2-bis[(nitrooxy)-methyl]-1,3-propanadiol dinitrate)
(Secondary explosive)
May flash when ignited
Also see Detonator cord

1.
Pentaerythritol
90% Nitric acid
Acetone
Ethanol
2.
Pentaerythritol

99% Nitric acid
Acetone
Ethanol
3.
Pentaerythritol
90% Nitric acid
Acetone
98% Sulfuric acid
10% Sodium carbonate
4.
PETN bonded explosive charge
PETN
Polyethylene
Dimethyl glycol
Phthalate

PGDN
(Propylene glycol dinitrate)
(Secondary explosive)
Flammable; detonates upon ignition

90% Nitric acid
98% Sulfuric acid
5% Sodium cabonate
Propylene glycol
Sodium sulfate

Picaramic Acid
(Secondary explosive)
Burns with smoky flame

Glacial acetic acid
Ammonium picramate

Picric Acid
(Secondary explosive)
Burns with smoky flame

1.
Phenol
Sodium hydroxide
Sodium nitrite
26% Nitric acid
90% Nitric acid
2.
Aspirin
Ethanol
98% Sulfuric acid
Sodium nitrite

**Picryl Chloride
(Secondary explosive)
Burns with smoky flame**

Chlorobenzene
Potassium nitrate
30% Fuming sulfuric acid
Acetone
Methanol

Plastic Explosives

Variation 1. TNTC Plastic Explosive

TNTC
Water
Polypropylene glycol

Variation 2. PBX, RDX Plastic Explosive

RDX
Water
Oxalin
1-Chloroethane
Polystyrene
Dioctylpthalate
Methyl ethyl ketone

**Variation 3. C3
Requires blasting cap**

Poly urethane
Water
Polyethylene glycol
RDX

**Variation 4. C4
Requires blasting cap**

RDX
Polypropylene glycol
Water

Variation 5. HMX Plastic

Ethyl silicate resin
(Sylgard)
HMX

Variation 6. PBX HMX Plastic Explosive

HMX
Water
Polytetrafluoroethylene resin

Nonionic wetting agent
Acetone

**Variation 7. TNT Wax Castable
Explosive**

1.
Ozokerite
Lecithin
Nitrocellulose
Methanol
2.
TNT
Aluminum powder

Variation 8

PETN bonded explosive charge
PETN
Polyethylene
Dimethyl glycol
Phthalate

PNT
(2,3,4,5,6-Pentanitrotoluene)
(Secondary explosive)
Burns with smoky flame

TNT
Glacial acetic acid
Iron powder
80% Sulfuric acid
99% Nitric acid
98% Sulfuric acid
Anisole
Methylene chloride
30% Fuming sulfuric acid
Magnesium sulfate
Chloroform

**Potassium Chlorate Primer
(Friction primer)**

Potassium chlorate
Antimony sulfate
Gum arabic
Sulfur
Ground glass
Calcium carbonate
Meal powder

PVN
(Poly (vinyl nitrate))
(Secondary explosive)
Burns with smoky flame

99% Nitric acid
Methylene chloride
Acetic anhydride
Poly(vinyl alcohol) (powder)
5% Sodium Bicarbonate
Ethanol

Quebrachitol Nitrate
(Secondary explosive)
May flash upon ignition

Quebrachitol
99% Nitric acid
98% Sulfuric acid
10% Baking soda

RDX (see figure next page)
(Cyclonite: hexahydro-1,3,5-trinitro-1,3,5-triazine)
(Secondary explosive)
Requires primer
Also see Blasting caps

1.
DAPT
Glacial acetic acid
Ammonium nitrate
70% Nitric acid
Acetic anhydride
2.
Propionitrile
99% Nitric acid
98% Sulfuric acid
Trioxane
3.
Hexamine
99% Nitric acid
Sodium nitrite
4.
Tripropionylhexahydrotriazine
99% Nitric acid

SATP
(Di-silver-aminotetrazole perchlorate)
(Primary explosive)
Detonates upon ignition

Silver perchlorate
5-Aminotetrazole
70% Perchloric acid

Silver Acetylide

Silver metal
Acetylene
70% Nitric acid
Alcohol

Silver Fulmanate
(Primary explosive)
Explodes when heated
UNSTABLE

Variation 1

70% Nitric acid
Silver nitrate
Ethanol

Variation 2

Silver
70% Nitric acid
Ethanol

Silver NENA
(Silver salt of (*N*-(2-nitroxyethyl)nitramine))
(Primary explosive)
Detonates upon ignition

Silver nitrate
NENA
Ethanol
Ether

Silver Nitride
(Primary explosive)
Detonates upon ignition
UNSTABLE

Ammonia gas
Silver (I) oxide

Silver Nitroform
(Primary explosive)
Detonates upon ignition

Nitroform
Ether
Silver iodide

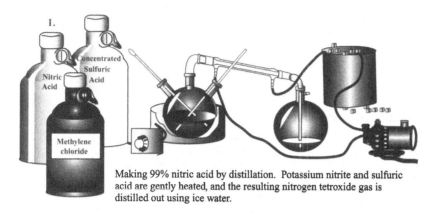

Making 99% nitric acid by distillation. Potassium nitrite and sulfuric acid are gently heated, and the resulting nitrogen tetroxide gas is distilled out using ice water.

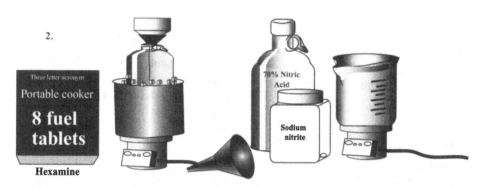

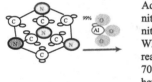

Add sodium nitrite to 70% nitric acid. When it reaches 70°C, add the hexamine/nitric acid mixture.

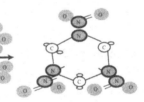

RDX a solid is formed

3.

The solid made above is filtered, then boiled in water. The pH should be basic.

The solid made is filtered again, dried, then plasticized by adding it to mineral oil, and mixing.

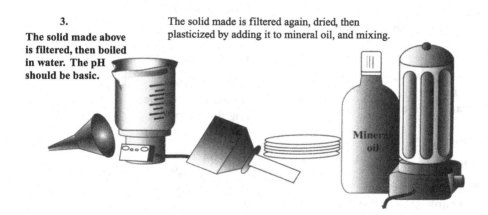

**Smokeless Powder
(Nitrocellulose blasting powder)**

Wet nitrocellulose
Glycerin
Oleic acid
Ammonium nitrate

Variation 5. Making Nitromethane

Sodium nitrite
Monochloroacetic acid
Sodium hydroxide

**Supplementary 1. Amines That Enhance
Nitromethane**

Diethylamine
Triethylamine
Ethanolamine
Ethylenediamine (best)
Morpholine

Supplementary 2. Other Enhancers

Microballoons
Polymeric foam
Steric acid
Aluminum

**Supplementary 3. Other Nitroparaffins
That Could Be Used**

Nitroethane
Nitropropane

**Sodium Azide
(Primary explosive)**

Variation 1

Sodium hydroxide
Ethyl alcohol
Hydrazine hydrate
Butyl nitrile
Dextrin

Variation 2

85% Hydrazine hydrate
Butyl nitrite or isopropyl nitrite
Ethanol
Sodium hydroxide

**Sodium Picramate
(Primary explosive)
Highly flammable; deflagrates**

Picric acid
Sodium hydroxide
7% Ammonium hydroxide
Carbon disulfide

**Solex
(1-(N)-Acetyl-3,5,7-trinitrocyclo-
tetramehylene-tetramine)
(Secondary explosive)
Requires primer**

TAT
99% Nitric acid
Phosphorus pentoxide

**Styphnic Acid
(Secondary explosive)
Flammable; deflagrates;
may detonate
TOXIC**

1.
Resorcinol
16% Nitric acid
Sodium nitrite
60% Nitric acid
2.
Resorcinol
98% Sulfuric acid
Sodium nitrite
40% Nitric acid
2% Nitric acid
3.
98% Sulfuric acid
68% Nitric acid
Resorcinol
Ethanol

**Sulfur Nitride
(Primary explosive)
Highly flammable**

1.
Toluene
Sulfur chloride
Ammonia gas
2.
Toluene
Sulfur

Ammonia gas
Chlorine gas

SZ
(Silver azide)
(Primary explosive)

Silver nitrate
Sodium azide

TA
(Trinitroanisole)
(Secondary explosive)
Burns with smoky flame

Picryl chloride
Methanol
Potassium hydroxide

TACC
See Tetraaminecopper(II) chlorate

TADA
(5,5-Bi-1*H*-tetrazole diaminosodium salt)
(Primary explosive)
Flammable; deflagrates

99% Hydrogen cyanide
Sodium azide
Copper(II) sulfate
36% Hydrogen peroxide
88% Formic acid
Ammonium chloride

TAEN
(Triazoethanolnitrate)
(Secondary explosive)

99% Nitric acid
98% Sulfuric acid
Triazoethanol
10% Baking soda
Sodium sulfate

TATB
(1,3,5-Triamino-2,4,6-trinitrobenzene)
(Secondary explosive)

1.
3,5-Dichloroanisole
90% Nitric acid
98% Sulfuric acid

Ammonia gas
Toluene
Acetone
2.
1,3,5-Trichloro-2,4,6-trinitrobenzene
Xylene
Ammonia gas
Hexane
3.
1,3,5-Trichloro-2,4,6-trinitrobenzene
Dioxane
Ammonia gas

TATP
(Triacetone triperoxide)
See Acetone peroxide

TBA
(4,4,4-Trinitrobutyraldehyde)
(Secondary explosive)
May burn upon ignition

Nitroform
Acrolein
Methylene chloride
Magnesium sulfate

TCTNB
(Trichlorotrinitrobenzene)
(Secondary explosive)
Burns with smoky flame

1,3,5-Trichlorobenzene
70% Nitric acid
98% Sulfuric acid

Tetracenetetrazene
(Primary explosive)
Flashes when ignited

Aminoguanadine·HCl
Acetic acid
Acetic anhydride
Sodium nitrite

Tetraminecopper(II) Chlorate
(Primary explosive; detonates C4)

Sodium chloate
Copper sulfate
Ammonium hydroxide
95% Ethanol

Wax or clay
Water

Tetraniline
(Primary explosive)

1.
98% Sulfuric acid
Sodium nitrite
M-Nitroaniline
70% Nitric acid
2.
98% Sulfuric acid
99% Nitric acid
M-Nitroaniline

Tetrazide
(Primary explosive)

Isocyanogen tetrabromide
Acetone
Sodium azide

Tetryl
(Secondary explosive)

99% Nitric acid
98% Sulfuric acid
N,N-Dimethylamine

TEX
(4,10-Dinitro-2,6,8,12-tetraoxa-4,10-
diazatetracyclododecane)
(Secondary explosive)
Requires blasting cap

98% Sulfuric acid
30% Fuming sulfuric acid
90% Nitric acid
THDFP
Urea
5% Baking soda

TNA
(1,3,5,7-Tetranitroadamantine)
(Secondary explosive)
Flammable; burns with smoky flame
May flash

Potassium permanganate
Adamamtane

Bromine
Sodium sulfite
18% Hydrochloric acid
Acetic acid
Aluminum foil
Toluene
Methylene iodide
Acetonitrile
Tetrahydrofuran
Sodium hydroxide
Acetone
Magnesium sulfate
Aluminum trichloride
Chloroform

TNAD
(1,4,5,8-Tetranitro-1,4,5,8-tetraazadecalin)
(Secondary explosive)
Requires primer

Ethylenediamine
Gloyoxal
Sodium niytrite
38% Hydrochloric acid
99% Nitric acid
Ethanol

TNB
(4,4,4-Trinitrobutanol)
(Secondary explosive)
May ignite

4,4,4-Trinitrobutyl aldehyde
Methanol
Sodium borohydride
18% Hydrochloric acid
Methylene chloride
Baking soda
Magnesium sulfate

TNBCl
(Trinitrobenzyl chloride)
(Secondary explosive)
Burns with smoky flame

TNT
5% Sodium hypochlorite (Chlorox®)
Tetrahydrofuran
Methanol
38% Hydrochloric acid

TND (see figure next page)
(1,4,6,9-Tetranitrodiamantane)
(Secondary explosive)
Burns with smoky flame
May flash

Aluminum foil
Iodine
Carbon disulfide
1,4,6,9-Tetrabromodiamantane
Sodium bisulfite
5% Hydrochloric acid
Methanol
Acetonitrile
38% Hydrochloric acid
Acetone
Sodium hydroxide
Magnesium sulfate
Potassium permanganate
Toluene
Ultraviolet light

TNEN
(2,2,2-Trinitroethyl-2-nitroxyethyl ether)
(Secondary explosive)
May burn upon ignition

1.
1,2-Dichloroethane
Hexamethyldisilane
Iodine
Cyclohexene
1,3-Dioxolane
Nitroform
Methylene chloride
Dimethylformamide
Sodium sulfate
38% Hydrochloric acid
Magnesium sulfate
90% Nitric acid
98% Sulfuric acid
2.
Methylenechloride
2-Bromoethanol
Trioxane
Aluminum trichloride
Magnesium sulfate
Potassium nitroform
Acetone

5% Baking soda
Hexanes
Silver nitrate
Acetonitrile

TNM
(Tetranitromethane)
(Secondary explosive)
May burn upon ignition

1.
90% Nitric acid
20% Fuming sulfuric acid
Melon amide
98% Sulfuric acid
2.
Sulfuryl chloride
Acetic anhydride
99% Nitric acid
10% Baking soda
Sodium sulfate

TNN
(Nitronaphthalene: 1,3,6,8-Tetranitro-
naphthalene)
(Secondary explosive)

70% Nitric acid
98% Sulfuric acid
Naphthalene
Potassium or sodium nitrate
99% Nitric acid
5% Sodium carbonate
Ethanol
Chloroform

TNP
(1,1,1,2-Tetranitropropane)
(Secondary explosive)
Flammable

Nitroform
Ether
1-Bromo-1-nitroethane
Sodium sulfate

TNPU
(*N,N'*-Bis(2,4,6-trinitrophenyl)urea)
(Secondary explosive)
Burns with NO ignition

TND

1,4,6,9-Tetranitrodimantane

1.

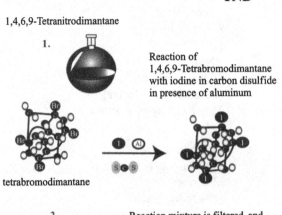

Reaction of
1,4,6,9-Tetrabromodimantane
with iodine in carbon disulfide
in presence of aluminum

tetrabromodimantane

2. Reaction mixture is filtered, and
washed with 5% hydrochloric acid
and collected and mixed with acetonitrile.

3.

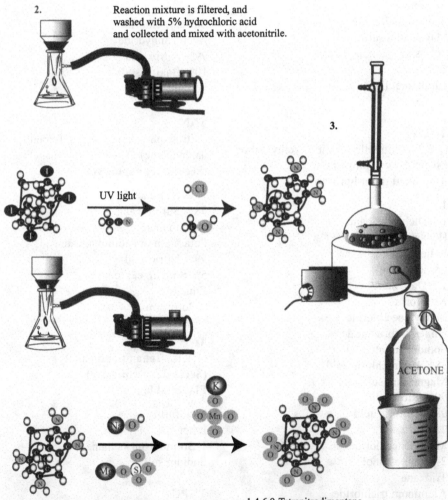

1,4,6,9-Tetranitrodimantane

98% Sulfuric acid
N,N-Diphenylurea
99% Nitric acid
5% Sodium bicarbonate

TNT
(Trinitrotoluene)
(Secondary explosive)
Requires nitramine for initiation
Burns with smoky flame

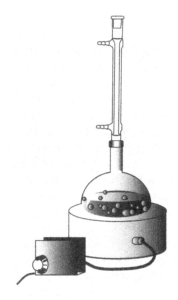

1.
99% Nitric acid
30% Fuming sulfuric acid
Toluene
70% Sulfuric acid
5% Sodium bicarbonate
2.
Toluene
99% Nitric acid
Gasoline
70% Sulfuric acid
5% Sodium bicarbonate
3.
Toluene
99% Nitric acid
Methylene chloride
70% Sulfuric acid
98% Sulfuric aicd
5% Sodium bicarbonate

4.
Toluene
Potassium nitrate
Methylene
70% Sulfuric acid
98% Sulfuric acid
5% Sodium bicarbonate
5.
Toluene
70% Nitric acid
98% Sulfuiric acid
70% Sulfuric acid
5% Sodium carbonate

TNTC
(2,4,6-Trinitro-2,4,6-triazacyclohexanone)
(Secondary explosive)
Requires blasting cap

Trifluoroacetic anhydride
Nitromethane
Ammonium nitrate
NIHT·HCl
Ethyl acetate

TNTPB (see figure)
(1,3,5-Trinitro-2,4,6-tripicrylbenzene)
(Secondary explosive)
Burns with smoky flame

Copper powder
Hydrochloric acid
Methanol
Ether

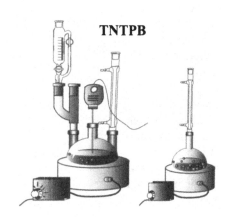

TNTPB

Mesitylene
Trichlorotrinitrobenzene
Picryl chloride
Diatomaceous earth
Charcoal
Acetone

Tovan
See Blasting Agents

TPG
(2,4,6-Trinitrophloroglucinol)
(Secondary explosive)
Burns with smoky flame

70% Nitric acid
98% Sulfuric acid
Phloroglucinol
5% Hydrochloric acid

Trimercury Chlorate Acetylide

Mercuric nitrate
Sodium chlorate
Acetylene

Trinitroanisole

Picryl chloride
Methanol
Potassium hydroxide

UNTNB
(5-Ureido-1,3-diamino-2,4,6-
trinitrobenzene)
(Secondary explosive)

30% Fuming sulfuric acid
Potassium nitrate
1,3,5-Trifluorobenzene
Methylene chloride
Hexanes
Charcoal
Sodium sulfate
2-Amino-2-methylpropane
Potassium hydrogen carbonate
1,2-Dichloroethane
Trifluoroacetic acid
Dimethylformamide
Urea

Yellow Powder
(Heated in contact with metal or metal
salt, detonates)

Potassium nitrate
Potassium carbonate

FIREWORKS

Fireworks is a difficult and precise chemical artform. Trying to put the various parts into formulations would have been like trying to explain how artists choose colors. Since this book is not designed to go into that much detail, we have listed chemicals in various processes, without giving the thousands of possible formulas. Note that depending on where the specific chemical is used, it can be a colorizer, an oxidizer, a fuel, or a retarder. The same chemicals are used over and over for many things. Most mixtures are made with a mortar and pestle.

Pyrotechnic mixtures are designed primarily to enhance burn rates, also often to retard or delay burn rates. Various oxidizers and fuels are mixed to provide mixtures that will ignite effectively and safely. They are designed to provide specific temperatures or conditions within a hot chemical soup to enhance color formation, spitzing strobes, flashing, and auditory reports. Pyrotechnics must also be designed to be propelled into the air, and to act as far away from their ignition source as possible.

Unlike other portions of this section, possible ingredients are provided rather than formulations. You will have to look through the various lists of ingredients for the chemicals that you have encountered. Many are given in more than one list. Here we list fireworks classified as UN 0336, UN 0337, UN 0431, or UN 0432 explosives by the U.S. Department of Transportation at CFR 172.101 and generally known as "consumer fireworks" or articles pyrotechnic."

Firecrackers

Flash powder

Variation 1

Potassium perchlorate
Barium nitrate
Aluminum
Sulfur

Variation 2

Potassium chlorate
Sodium chlorate
Aluminum powder
Black powder
Potassium nitrate
Charcoal
Sulfur

Firework rockets

Potassium nitrate
Fine charcoal
Charcoal (80 mesh)
Sulfur
Titanium

Snakes

Dinitrophenylhydrazine

Primes

Primes are usually an easily ignited covering
for polytechnics.

Rough black powder prime

Potassium nitrate
Fine charcoal
Sulfur
Silicon (325 mesh)

Potassium chlorate prime

Potassium chlorate
Accroides (red) gum
Fine charcoal
Potassium dichromate

Special primes may contain

Potassium nitrate
Potassium chlorate
Barium peroxide
Iron oxide (black)
Antimony sulfide

Binders
Hold prime composition together

Dextrin (methanol)
Nitrocellulose (lacquer, acetone)
Accroides (red) gum (alcohol)
Sulfur accroides (red) gum asphaltum

Fuel additives

Carbon
Aluminum
PVC
Dextrin

Burn-rate reducers

Zirconium powder
Lampblack

Catalyst

Manganese dioxide
Iron(III) oxide (red)
Potassium dichromate

Fuses

Black powder, paper wrap

Goldschmidt (thermite)

Silica
Pb_3O_4

Fuse-burn-rate adjusters

Tungsten
Barium chromate
Potassium perchlorate
Diatomaceous earth

Black match fuse

Cotton strings
Black powder
Cornstarch

Quick match

Black match fuse
Loose paper tubing

Colored flame

Fuels used in colored stares

Aluminum
Magnesium
Magnalium
Red gum
Shellac

Oxidizers used in colored stares

Ammonium perchlorate
Barium chlorate
Barium nitrate
Potassium chlorate
Potassium nitrate
Potassium perchlorate
Sodium nitrate
Strontium nitrate

Salts most used to generate colors

Red

Strontium chloride
Strontium hydroxide
Strontium oxide
Strontium nitrate
Strontium oxalate
Strontium sulfate

Green

Barium chloride
Barium hydroxide
Barium nitrate
Barium sulfate

Blue

Copper chloride
Copper carbonate
Copper oxide
Copper(II) oxychloride ($3CuO \cdot CuCl_2$)

Orange

Calcium hydroxide
Calcium chloride
Calcium carbonate

Calcium oxalate
Calcium sulfate

Yellow

Sodium metal
Cryolite (Na_3AlF)
$NaNO_3$
Sodium oxalate

Chlorine doners

Free chlorine in the hot chemical soup creates the chloride, which provides the truest color.

Dechlorane ($C_{10}Cl_{12}$)
Hexachlorobenzene (C_6Cl_6)
Saran resin ($C_3H_2Cl_2)_n$
Benzene hexachloride ($C_3H_2Cl_2)_n$
Parlon ($C_6H_6Cl_4)_n$
Chlorowax
Poly(vinyl chloride) ($C_2H_3Cl)_n$
Calomel (mercury(I) chloride (Hg_2Cl_2))

Coatings for magnesium in fireworks

Although magnesium is a good fuel, it can be eroded by oxidizers, lessening its value as a fuel.

Nitrocellulose and acetone
Parlon in acetone and methyl ethyl ketone
Red gum or shellac with alcohol

The coatings can be treated with

Potassium chromate
Ammonium vanadate
Ammonium molybdate
Ammonium phosphate
Boiled linseed oil

Flame-deoxidizing agents

Hexamine
Sulfur
Asphaltum
Steric acid
Paradichlorobenzene
Graphite

Sparks

Color is temperature dependent.

Titanium
Sulfur

Potassium nitrate
Meal powder
Steel filings
Cast iron borings
Charcoal
Coke grains
Porcelain grains
Zirconium
Magnesium

To protect sparking materials during storage

Linseed oil
Meal powder
Iron (course)
Aluminum (40 mesh)
Shellac

Spark retarder

Potassium nitrate
Potassium perchlorate
Titanium (10-2 mesh)
Dextrin
Sulfur
Charcoal
Zircalium (Al_2Zr) (100 mesh)
Accroides resin

Branching sparks

Iron
Ferroaluminum

Firefly

Carbon and aluminum

Glitter

Formulation 1
Meal powder
Sulfur
Dextrin

Formulation 2
Charcoal
Barium nitrate
Aluminum

Formulation 3
Meal powder
Aluminum

Antimony sulfide
Dextrin

Materials useful to delay flash reactions

Barium oxalate
Barium carbonate
Magnesium oxalate
Magnesium carbonate
Lithium oxalate
Lithium carbonate
Sodium oxalate
Sodium bicarbonate

Strobe formulations

Formulation 1
Barium nitrate
Sulfur
Magnalium

Formulation 2
Barium carbonate
Chlorowax
Guanadine nitrate

Formulation 3
Potassium perchlorate
Ammonium perchlorate
Barium sulfate

Formulation 4
Hexamine
Lithium perchlorate
Magnesium

Formulation 5
Magnesium copper
Parlon

Dark and flash effect

Ammonium perchlorate with

Magnesium
Magnalium
Zinc
Copper
Cyanoquanidine
Guanadine nitrate

Sulfur with

Magnesium
Magnalium

Titanium
Copper

Guanadine nitrate with

Ammonium perchlorate
Magnesium copper
Copper
Copper oxide

Smoke dye formulation

25% Potassium chlorate
20% Lactose
50% Smoke dye
Bicarbonate of soda

Smoke dyes

Red smoke

Rhodamine B
Oil orange
Aminoanthraquinone
Oil scarlet

Yellow smoke

Anramine

Green smoke

Diparatoluidinoanthroquinone

Blue smoke

Phthalocyanine blue
Methylene blue

Violet

Indigo

White smoke

Arsenic sulfide
Anthrocene

Whistle formula

Formulation 1
Potassium perchlorate
Sodium silicate
Sodium benzoate
Paraffin oils

Formulation 2
Potasium chlorate
Potassium perchlorate
Potassium dinitrophenate
Gallic acid
Red gum

Salutes

Potassium chlorate
Potassium perchlorate
Aluminum
Sulfur

Anticaking

Sawdust
Tricalcium phosphate
Petro-Ag® (sodium diethylnaphthalene sulfuric acid salt)

INCENDIARY DEVICES (Arson)

Arson is more technique than materials. A successful arsonist knows where to start a fire in a structure. Most of the time, arson is very straightforward. In real life, Situations 0.1 through 0.5 are the most likely way to start fires. There are very few or perhaps no professional arsonists. Situations 1 through 9 are published in underground sources and therefore are included here.

Situation 0.1.   Gas Accelerants

Gas is probably the best accelerant, as it leaves no trace after the fire, and it can be brought to the scene in small containers.

Acetylene
Bic® lighter
Butane
Natural gas
Propane
Propane match

Situation 0.2.   Liquid Accelerants

These can be sprayed under a door or splashed on surfaces. These do leave a trace after the fire. Creative persons have used such devices as fuel injectors to increase flammability simply by poking a hole in the hose below the fuel injector so that the liquid comes out vaporized.

Benzene
Benzine
Charcoal lighter
Coleman® white gasoline
Ether (sometimes)
Gasoline
Heptane
Hexane
Kerosene
Ligroin
Mineral spirits
Paint thinner
Pentane
Petroleum distillates
Petroleum ether
Petroleum naptha
(Other flammable liquids)

Situation 0.3.   Chemical Accelerants

Often discussed but probably not used often. More exotic accelerants are easier to identify after the fire. However, as a means to murder people, potassium chlorate can cause a fire to spread so rapidly that there is very little chance of escape.
Potassium chlorate
Painted on walls as accelerant

Situation 0.4.   Solid Accelerants

Duraflame® logs and dryer lint do not leave a trace after the fire that can be distinguished from the natural results of the fire in a building.
Duraflame logs
Clothes dryer lint
Newspaper

Situation 0.5.   BLEVE

Flame impingement on any pressurized container containing a flammable gas can BLEVE. Flame impingement on a propane cylinder could be very explosive. This method was employed at Walmart during an union action.

Sterno®
Aerosol cans

Situation 1.   Love Letter

Love is the name of a shampoo; any soap or shampoo rich in hydroxyl groups will do. It is placed in an envelope and topped off with solid hypo chlorite.
Calcium, sodium, or lithium hypochlorite
Prell® dishwashing detergent
Envelope

Situation 2.   Mothballs by Water Heater
Since the naphthalene in mothballs sublimes, it becomes a gas accelerant and may leave no trace.
Naphthalene
Water heater

Situation 3.   Daylight Exploding Bottle

The mixture of chlorine and turpentine is an explosive ignited by sunlight. *Also see* High Explosives, Process 5.

Turpentine
Chlorine gas

Situation 4.   Car Fire

The best methods use things that are already there. Although ethylene glycol is not flammable at ordinary temperatures, at the temperature of a catalytic converter, it is a flammable liquid. Two wires that have been affixed to a power source can be embedded in wax and the wax laid on the muffler, so that when the wax melts, the wires touch, producing a spark.

Radiator fluid
Muffler-attached container
Wax

Situation 5.   Thermite

Once ignited, thermite burns at 2200°F and is almost impossible to extinguish. This would be easily detected, so its use would be mostly by persons wanting to commit mischief, not just arson.

Iron filings
Aluminum powder
Magnesium powder

Situation 6. Spontaneous Combustion

Wet charcoal ignites spontaneously. The reaction is difficult to predict and unreliable, but when it works it is very effective and it is difficult to prove intent. A bag of wet charcoal laying against a house might not be noticeable.

Charcoal
Water

Situation 7. Potato Chip Bag in Circuit Box Ignites

Probably a McGyver trick; the oils on and in the potato chip bag will ignite when the electrical box has a short.

Potato chip bag
Electrical circuit box

Situation 8. Ignition Upon Evaporation

Once the quickly evaporating carbon disulfide is gone white phosphorus exposed to air ignites.

Carbon disulfide
White phosphorus

Situation 9. Oily Rags

Linseed oil will spontaneously ignite when left in bundled rags.

Linseed oil
Rags

Situation 10. Upgraded Molotov Cocktail:

Chemical Fire Bottle

Potassium chlorate
Sugar
Concentrated sulfuric acid
Gasoline

Situation 11. Slow Spontaneous Ignition When Wet

Potassium permanganate
Powdered sugar
Aluminum
Magnesium

Situation 12. Napalm

Napalm
Styrofoam® shavings
Gasoline

Napalm Bricks

Napalm
Newspaper bits
Corn flour

5

CHEMICAL WEAPONS

We are using a more extensive format for chemical weapons than we have for drugs or explosives since they present a more immediate public health concern.

NERVE AGENTS

Facility 1

The Abuzov reaction, the Michaelis reaction, and dexproportionation are utilized to make sarin. The primary clues that someone may be making sarin is the presence of compounds containing the elements phosphorus and fluoride, and the use of Teflon® plastic containers as reaction vessels. A defining chemical for sarin is isopropyl alcohol.

Historically, methylating phosphorus was difficult, requiring high temperatures and pressure. There are now several ways to methylate phosphorus compounds, which can be done at room temperature or 0°C, with no high-pressure requirement. This has made the difficult methods used prior to and through World War II obsolete and has brought the production of sarin, VX, and other nerve agent chemicals within the ability of less sophisticated chemists. The Abuzov and Michaelis reactions do not require high temperature or pressure, and can be carried out at room temperature or at a temperature of 0°C, requiring only ice for temperature control.

The *Abuzov reaction* to make most G-agents requires three steps. There are several additional intermediate activities required for chemical preparation, including filtering and desiccation. The apparatus displayed on the following page would be used to synthesize the precursor, methylphosphonic difluoride (Agent DF). Not shown in the diagram are the considerable requirements for ventilation and personal protection to prevent the death of the person making the compound. Hydrofluoric acid can be used in place of ammonium fluoride in a related method. It is unlikely that the full setup as diagrammed would be found set up in proper order, any more than a person making a cake would line up each step. Obviously, parts such as the vacuum pumps would be interchangeable from one step to another. None of the chemicals used in the Abuzov reaction are under the control of the Australian Group.*

*The Australian Group coordinates national export-control regulations to restrict the sale of key chemical weapon precursors to countries that might make chemical weapons.

Chemicals Used for Illegal Purposes: A Guide for First Responders to Identify Explosives, Recreational Drugs, and Poisons, By Robert Turkington
Copyright © 2010 John Wiley & Sons, Inc.

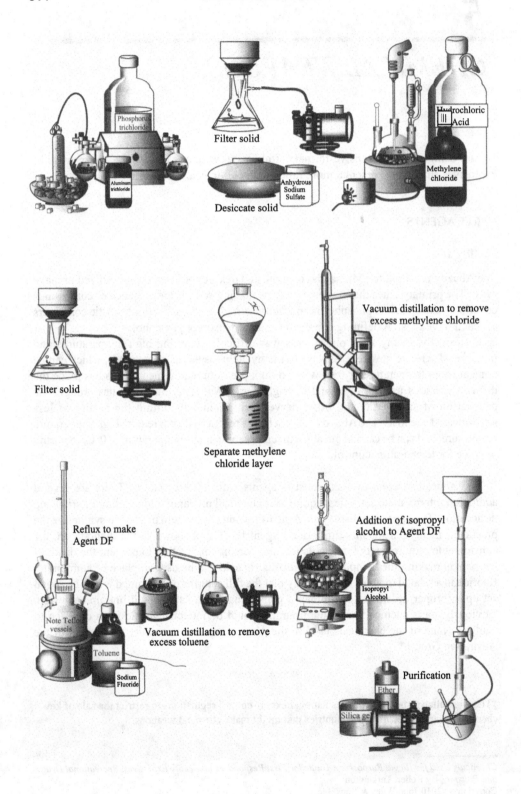

Phosphorus trichloride

Aluminum trichloride

Filter solid

Desiccate solid

Anhydrous Sodium Sulfate

Hydrochloric Acid

Methylene chloride

Filter solid

Separate methylene chloride layer

Vacuum distillation to remove excess methylene chloride

Reflux to make Agent DF

Note Teflon vessels

Toluene

Sodium Fluoride

Vacuum distillation to remove excess toluene

Addition of isopropyl alcohol to Agent DF

Isopropyl Alcohol

Purification

Ether

Silica gel

		Cost ($)
Aluminum trichloride		
Phosphorus trichloride	1 L	47.85
Methyl chloride	2.5 kg	198.45
Anhydrous sodium sulfate	500 g	18.15
Anhydrous sodium sulfate	500 g	18.15
Methylene chloride	1 L	20.80
Hydrochloric acid	500 mL	23.15
Toluene	1 L	17.45
Toluene	1 L	17.45
Sodium fluoride	500 g	35.30
Isopropyl alcohol		
Silica gel	250 g	48.80
Ethyl ether		

Estimated cost for the chemicals needed to make a gallon of sarin: $800. The cost of glassware and other equipment is $2000. This method was described in *Scientific American* (Dec. 2001).

The *Michaelis reaction* is another method to create a methyl phosphonate, using a halogenated phosponate without the high heat and pressure. This reaction requires an alkyl alkali metal. Methyl sodium is the most likely. This reaction uses many of the Australian Group precursor compounds (G-agents) listed below. Athough many of these chemicals do have other uses, which we have listed, their presence should be suspect if encountered. Phosphorus attached to methoxy, ethoxy, or related groups is convertible to methyl phosphonate. Methylated phosphonate halogens are the key component for preparing sarin precursors. Methoxy and ethoxy or related groups are used in the manufacture of some detergents or wash products. The presence of chemicals to add fluorine to organic phosphorus compounds and the presence of isopropyl alcohol provide all necessary ingredients for making sarin. If making soman, the isopropyl alcohol would be replaced by penacolyl alcohol. The two compounds, which at this point would be about the same to make, have different levels of toxicity as well as volatility. The volatility of the final product and the availability of penacolyl alcohol would determine the choice of final product. Formulas for other G-agents are also listed here.

Prealkylated compounds (look for methyl or ethyl group on phosphorus, along with a double-bonded oxygen)

Precursor	Where available	Cost		
Diethylethyl phosphate	Gasoline additive	$22.50	for	50 g
Diethylethyl phosphite	Lab chemical	$19.00	for	250 g
Diethylmethyl phosphonate	Lab chemical	$27.50	for	0.5 g

Diethyl phosphite	Paint solvent, lubricant additive	$19.00 for 250 g
Dimethylethyl phosphonate	Antifoam, gasoline additive	not found
Dimethylmethyl phosphonate	Hydraulic fluid, antifoam, antistatic, manuf. of flame retardants	not found
Dimethyl phosphite	Lubricant additive	not found
Dimethyl phosphite	Lubricant additive	not found
N,N-Dimethyl phosphosamidic dichloride	Lab chemical	not found
Ethylphosphonothioic dichloride	Manuf. of pesticide	not found
Methylphosphinyl dichloride	Lab chemical	not found
EthylPhosphonyl dichloride	Lab chemical	not found
Methylphosphinyl difluoride	Lab chemical	not found
Triethyl phosphite	Lubricant additive, plasticizer, manuf. of pesticide	$16.00 for 250g
Trimethyl phosphite	Gasoline additive, manuf. of paints, pesticide	not found
Methylphosphonic acid	Lab chemical	$8.00 for 1 g
Methylphosphonic dichloride	None	$30.00 for 1 g
Methylphosphonothioic dichloride	Lab chemical	not found
Ethylphosphinyl difluoride	Lab chemical, preservative	not found

Chemicals used to alkylate phosphorus

Phosphorus oxychloride	Manuf. of pesticide, pharmaceutical, dye, flame retardant
Phosphorus pentachloride	Manuf. of plastic, pesticide
Phosphorus trichloride	
Aluminum trichloride	
Alkyl chloride	

Chemicals used to fluoridate G-agents or make methylphophonic bifluoride

Ammonium bifluoride	Disinfectant, glass etching
Potassium bifluoride	Glass etching, welding fluxes, wool
Hydrofluoric acid	Industrial acid, glass etching, gasoline manufacturing
Potassium fluoride	Electroplating, manuf. of pesticide
Sodium bifluoride	Glass etching, rust removal, welding flux

Binary agents

Methylphosphonic dichloride	$112 for 25 g
Isopropyl alcohol	Solvent, cleaner, disinfectant
Isopropylamine (Scavenger®)	Manuf. of pharmaceutical, dye, pesticide, textile surfactant rubber accelerator

Methylphosphonic difluoride   None
Penacolyl alcohol   Lab chemical (difficult to find)

All these chemicals are manufactured in the United States. The chemicals with prices after them were all found available for sale by chemical supply houses. The easiest reaction to make Sarin used in this example would cost roughly $10,000 per gallon (a second estimate, by a far better chemist, was $12,000.

Dexproportionation is a process where by two molecules, in this case Agent DF and Agent DC, are mixed, resulting in a new compound which contains both Cl and F. If sarin is going to be stored for a period of time, the excess fluoride must be removed or the product will quickly degrade. This is done using the di-di method, in which, methylphosphonic dichloride is mixed with methylphonic difluoride to make di-di before storing and mixing with isopropyl or penacolyl alcohol.

The binary agent di-di can be stored indefinitely. Isopropyl alcohol containing isopropylamine can be added at any point to create sarin. Mixing the di-di with any alcohol, or even ethylene glycol, can make an effective junk nerve agent. The reaction, which could take place in a plastic barrel, is instantaneous. The two components can be stored side by side in plastic containers in munitions. Military munitions often contain these two compounds in two distinct plastic bags on top of an explosive charge. Only when the charge explodes are the two products mixed. Sarin is formed as the munitions explode. Agents DF and DC are water reactive as well as corrosive.

We have described only two of many ways to make sarin, soman, and other G-agents, starting with standard laboratory chemicals. The Japanese cult Aum Shinrikyo purportedly used a method using acetonitrile. We did not find that method.

Facility 1

Sarin/Soman

Variation 1

Step 1
Phosphorus trichloride
Cylinder of methyl chloride
Aluminum trichloride
Anhydrous sodium sulfate

Step 2
Methylene chloride
Hydrochloric acid

Step 3
Sodium fluoride
Toluene

Step 4
Isopropylamine
Isopropyl alcohol or penacoyl alcohol
Ethyl ether
Silica gel

Variation 2

Step 1.   Making Agent DC
Triethylphosphite
Methylene chloride

Step 2.   Making DF
Ammonium fluoride
Agent DC
Ethanol

Supplementary 1.   Abuzov Reaction
Aluminum trichloride
Phosphorus trichloride
Methyl chloride gas

Supplementary 2.   Michaelis Reaction
Methyl sodium
Halogenated compound from list below

Supplementary 3.   Making di-di
Agent DC
Agent DF

Chemicals that can be used to alkylate phosphorus

Aluminum trichloride	From chemical supsply house
Phosphorus oxychloride	Manuf. of pesticide, pharm, dye, flame retardant
Phosphorus pentachloride	Manuf. of plastic, pesticide
Phosphorus trichloride	
Thionyl hloride	

Prealkylated compounds (look for methyl or ethyl group on phosphorus, along with a double-bonded oxygen)

Diethylethyl phosphate	Gasoline additive
Diethylethyl phosphite	Lab chemical
Diethylmethyl phosphonate	Lab chemical
Diethyl phosphite	Paint solvent, lubricant additive
Dimethylethyl phosphonate	Antifoam, gasoline additive
Dimethylmethyl phosphonate	Hydraulic fluid, antifoam, antistatic, manuf. of flame retardant
Dimethyl phosphite	Lubricant additive
N,N-Dimethylphosphosamidic dichloride	Lab chemical
Ethylphosphonothioic dichloride	Manuf. pesticide
Methyl chloride	Lab lecture bottle
Methylphosphinyl dichloride	Lab chemical
Ethylphosphonyl dichloride	Lab chemical
Methylphosphinyl difluoride	Lab chemical
Triethyl phosphite	Lubricant additive, plasticizer, manuf. of pesticide
Trimethyl phosphite	Gasoline additive, manuf. of paint, pesticide

Methylphosphonic acid   Lab chemical
Methylphosphonic dichloride   Lab chemical
Methyl phosphonothioic dichloride   Lab chemical
Ethyl phosphinyl difluoride   Lab chemical

Other G-agents; the manufacture of these will be similar to making sarin

Cyclosarin

Variation 1

Phosphorus trichloride
Sodium fluoride
Anhydrous aluminum chloride
Anhydrous cyclohexanol
Methyl chloride
Dry silica gel
Methylene chloride
Isopropyl ether
37% Hydrochloric acid
Toluene

Variation 2

Phosphorus trichloride
48% Hydrofluoric acid
Anhydrous aluminum chloride
Anhydrous cyclohexanol
Methyl chloride
Dry silica gel
Methylene chloride
Iosopropyl ether
37% Hydrochloric acid

Thiosarin

Phosphorus trichloride
Toluene
Methyl disulfide
Anhydrous sodium fluoride
Methyl iodide
Isopropyl alcohol

Soman

Ammonium bifluoride
Isopropyl amine
Pinacoyl alcohol
Cylinder of methyl chloride
Methylene chloride
Aluminum trichloride
Phosphorus trichloride
Hydrochloric acid
Anhydrous sodium sulfate

Toluene
Sodium fluoride
Isopropyl ether

Chemicals used to fluoridate G-agents

Ammonium bifluoride
Ammonium fluoride
Potassium bifluoride
Hydrofluoric acid
Potassium fluoride
Sodium fluoride

Due to the extreme hazard of chemical weapons, we will provide further information that might become necessary if one of these facilities is encountered.

Preservative history

On March 20, 1995, Aum Shinrikyo, a cult group, released sarin in commuter trains on three different Tokyo subway lines. Sarin was concealed in lunch boxes and soft-drink containers and placed on subway train floors. It was released as terrorists punctured the containers with umbrellas before leaving the trains. The incident was timed to coincide with rush hour, when trains were packed with commuters. Eleven persons died and over 5500 were injured in the attack.

On the day of the event, 641 victims were seen at St. Lukes's International Hospital. Among them five displayed cardiopulmonary or respiratory arrest with marked miosis and extremely low serum cholinesterase values. Two of the five died and three recovered completely. Of the 641, 106 were hospitalized with mild to moderate symptoms.

In March 2004, Libya declared that thousands of tons of precursors that could be used to make sarin nerve gas were stored in a facility near Tripoli.

Chemical characteristics

Sarin is a odorless, colorless liquid. The vapor pressure is 2.1 mm Hg @ 20°C; the vapor densitiy is 4.8; the density is 1.1; the boiling point is ~316°F (~157°C). Sarin is miscible in water and has a volatility of 22,000 mg/m^3 @ 25°C. Sarin is 32 times more volatile than tabun.

OVERALL RATING (1–10)

Effectiveness	8	Field stability	8
Persistence, open	7	Storage stability	8
Persistence, closed	8	Toxicity	7.5
Total effectiveness	7.7		

Protection

V-5 ointment; totally encapsulated suit. The vapor pressure of sarin is right at the point where it can be used as a vapor or as an aerosol. It is therefore an inhalation as well as a skin-contact hazard. There is one recorded case of a person being stricken while wearing a completely encapsulated suit and mask because of a small crack in the facepiece.

Detection method

Draeger phosphoric ester acid tube 0.05 ppm (qualitative) FID (real time) oxime solution containing a silver electrode with a platinum reference electrode. Infrared. Dark blue M-8. Organic Phosphorus Test and Fluoride Test. The only test specific to sarin is a solution of o-dianisidine and sodium pyrophosphate peroxide.

HF gas, which is very corrosive and difficult to contain, is created in this process. In some of the processes, HCl gas is also produced. These tend to corrode piping, which may leak. There is a Draeger tube for both of these compounds. The tubes are not specific.

Decontamination

Even after personal decontamination, sarin already in the skin may continue to work, causing a delayed reaction. It is therefore important to wash any skin-contact area as soon as possible. Inhaled sarin can linger in still airways and be absorbed in the lungs much later, due to Brownian movement in the lungs.

The importance of early decontamination cannot be overemphasized. Decontamination of the skin must be done quickly if it is to be effective. Liquid agent can be removed by Fuller's earth or chemically inactivated by the use of reactive decontaminates. The following calcium hypochlorite solutions can all be used for decontamination: The military's special STB (Super Tropical Bleach), 93% calcium hypochlorite and 7% sodium hydroxide, which corrodes metal; DANC, DuPont® RH-195 and acetylene tetrachloride, eats paint; a solution of 70% diethylenetriamine, 28% Cellosolve®, and 2% sodium hydroxide leaves permanent stains; M258A1 diluted or a M291 kit. Sarin is persistent on surfaces and has a half-life in soil of from 2 to 24 hours at standard temperatures. Sandia National Laboratories has developed and commercialized an aqueous-based decontamination technology (DF-100 and MDF-200). This neutralization technology is attractive for civilian and military applications for several reasons, including: (1) a single neutralization solution can be used for both chemical and biological toxicants; (2) it can be deployed rapidly; (3) mitigation of agents can be accomplished in bulk, aerosol, and vapor phases; (4) it exhibits minimal health and collateral damage; (5) it requires minimal logistics support; (6) it has minimal runoff of fluids and no lasting environmental impact; and (7) it is relatively inexpensive. Information on the product can be obtained at www.deconsolutions.com. This was developed to decontaminate facilities or equipment to an acceptable level in a very short time so that casualties can be located and treated. However, complete decontamination is still a very important aspect of these compounds. Using Sandia products, the hydrolysis of sarin occurs rapidly under alkaline conditions and breaks down to o-alkyl-methylphosphonic acid. Sandia products can be used as foam or as an aerosol (fog). If the contaminant is known, the pH can be adjusted for minimum neutralization time. For sarin that is a pH of 8. The pH of the material being used by a first responder who is not certain of the contaminant is 9.2.

Decon Tech	1 Minute	15 Minutes	60 Minutes
DF-100 (pH 8)	53%	ND	ND
DF-100 (pH 9.2)	ND	ND	ND
DF-200	ND	ND	ND

Symptoms

Pinpoint pupils	Sweating
Runny nose	Nausea

Difficulty breathing	Vomiting
Loss of consciousness	Seizures
Convulsions	

Increased sweating at site, muscular fasciculations at site, nausea, vomiting, diarrhea, generalized weakness, loss of consciousness, convulsions, flaccid paralysis, apnea, generalized secretions, involuntary defecation. The eyes will go to a pinpoint even in complete darkness.

Antidote

Atropine, 2-PAM-C1. Pyridostigmime bromide tablets are a pretreatment. The carbamates pyridostigmine and physostigmine will tie up the colinesterase until the organophosphate has passed.

Storage

If the distillation step is not done properly, the product will contain significant amounts of HF, which will quickly degrade the sarin. In two years it deteriorates to between 1 and 10%. Refrigerating the product can slow down degradation. Such refrigeration would be an indication that the person intends to use the product fairly soon. 90% pure sarin has been stored for as long as three decades.

Dispersal

Explosive is the dominant mode. An explosion creates droplets ranging from ~100 μm to 1 mm. This is a fine rain to coat surfaces. Dispersal can also be achieved by piston action, a spray device, and dispersal by a vehicle such as an airplane or an automobile. Increased humidity will increase the particle size by hygroscopic effects. Increased particle size may decrease exposure because of precipitation prior to inhalation. The more soluble gases are deposited almost entirely in the upper respiratory system. The less soluble gases tend to produce effects in the peripheral airways or alveoli. Sarin is miscible (totally soluble). In peripheral airways the air motion is relatively slow, occurring by molecular diffusion. The substances may stay longer and induce a surprising degree of damage, due to their prolonged effects.

Other concerns

Teflon® Erlenmeyer flasks when burned produce the single most toxic gas ever created by human beings.

Facility 2

Tabun

Variation 1

Phosphorus trichloride
Diethylamine
Ethyl alcohol
Cyanogen iodide
Hydrochloric acid
Sodium hydroxide

Some Equipment and chemicals associated with Facility 2, Variation 1

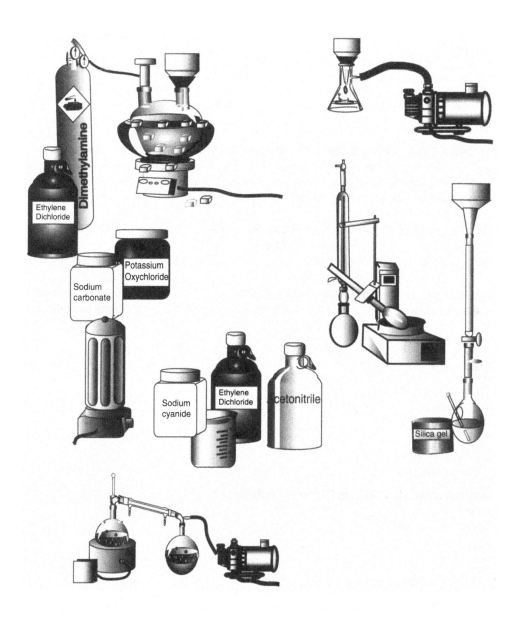

Variation 2

Step 1
Phosphorus oxychloride
Diethylamine

Step 2
Ethyl alcohol
Ethylene dichloride
Sodium carbonate
Pyridine
Acetonitrile
Sodium cyanide

Step 3
Silica gel

Variation 3

Variation 3 is a two-step process. The first reaction produces hydrochloric acid as a by-product; the second involves the generation of hydrogen cyanide gas. Although not necessary, distillation can provide an 80% product. Vacuum distillation will remove the carrier solvent. The cost of ingredients:

Phosphorus chloride	Used in pesticide manufacture, gasoline additive.	$23.70 for 250 mL
Diethylamine	Used in manufacture of soaps and detergents	7.50 for 500 mL
Ethyl alcohol	Drinking alcohol, solvent	50.00 for 100 mL
Sodium cyanide	Used in metal plating (can be made; see Facility 10)	27.00 for 500 g

Variation 3

Phosphorus oxychloride
Ethyl alcohol
Dimethylamine
Sodium cyanide
Ethylene dichloride
Sodium carbonate

Supplementary 1.   Making Cyanogen Iodide
Iodine
Sodium hydroxide
Sodium cyanide
Nitric acid

Summary of chemicals used to make tabun

Acetonitrile
Diethylamine   Used in manufacture of soaps and detergents
Ethyl alcohol   Drinking alcohol, solvent
Ethylene dichloride

Hydrochloric acid	Standard Industrial chemical
Iodine	
Phosphorus oxychloride	
Phosphorus trichloride	Difficult to obtain; pesticide manufacture, gasoline additive
Pyridine	
Sodium carbonate	
Sodium chloride	Table salt
Sodium cyanide	Used in metal plating
Sodium hydroxide	Standard industrial chemical

Due to the extreme hazard of chemical weapons, we provide further information that might become necessary if one of these facilities is encountered.

Chemical characterization

Colorless to brown liquid with a fairly fruity odor, Vapor pressure 0.037 mm Hg at 20°C. Vapor density 4.86; density 1.1. Volatility 22,000 mg/m^3 @ 25° C. 9.8 g/100 g in water at 25°C. Soluble in most organic solvents.

History

Tabun was made by Germany. During manufacture in World War II problems making tabun resulted in several deaths. Iraq has manufactured a 40% pure tabun. Variation 1 using cyanogen iodide makes the process much safer.

Protection

V-5 ointment; totally encapsulated suit. The vapor pressure of tabun is sufficiently low that it is most effective as an aerosol. Vapor pressure is 0.037 mmHg @ 20°C. It is therefore less of an inhalation hazard. It is an extreme skin-contact hazard.

Detection method

Draeger phosphoric ester acid tube 0.05 ppm (qualitative) FID (real time) Oxime solution containing a silver electrode with a platinum reference electrode. Infrared. Dark blue M-8. Organic Phosphorus Test and Fluoride Test. In this facility, some consideration should be given to testing the air for cyanide. Draeger hydrogen cyanide tubes are readily available.

Decontamination

It is important to wash any skin-contact area as soon as possible. The importance of early decontamination cannot be overemphasized. Decontamination of the skin must be done quickly if it is to be effective. Liquid agent can be removed by Fuller's earth or chemically inactivated by use of reactive decontaminates. Calcium hypochlorite solutions (the military has a special STB, Super Tropical Bleach) (which corrodes metal), DANC (DuPont® RH-195 and acetylene tetrachloride, which eats paint), a solution of 70% diethylenetriamine, 28% Cellosolve®, and 2% sodium hydroxide (which leaves permanent stains), M258A1 diluted, and M291 kit can all be used for decontamination. Tabun is persistent on surfaces and has a half-life in soil of from 1 to 1.5 days at standard temperatures. Sandia National

Laboratories has demonstrated and commercialized an aqueous-based decontamination technology (DF-100, and MDF-200).

Symptoms

Pinpoint pupils	Sweating
Runny nose	Nausea
Difficulty breathing	Vomiting
Loss of consciousness	Seizures
Convulsions	

Increased sweating at site, muscular fasciculations at site, nausea, vomiting, diarrhea, generalized weakness, loss of consciousness, convulsions, flaccid paralysis, apnea, generalized secretions, involuntary defecation.

Antidote

Atropine, 2-PAM-C1, The carbamates pyridostigmine and physostigmine will tie up the colesterase until the organophosphate has passed.

Storage

The storage area may have more than just tabun. The phosphorus oxychloride is water reactive, producing strong hydrochloric acid. If stored near the sodium cyanide, hydrochloric acid could pose a serious potential hazard.

Dispersal

Explosive is the dominant mode. An explosion creates droplets ranging from ~100 μm to 1 mm. This is a fine rain to coat surface. Dispersal can also be achieved by a piston action, a spray device, and dispersal by a vehicle such as an airplane or an automobile. Increased humidity will increase the particle size by hygroscopic effects. Increased particle size may decrease exposure because of precipitation prior to inhalation. The more soluble vapors are deposited almost entirely in the upper respiratory system. The less soluble vapors tend to produce effects in the peripheral airways or alveoli. Sarin is miscible (totally soluble). In peripheral airways the air motion is relatively slow, occurring by molecular diffusion. The substances may stay longer and induce a surprising degree of damage, due to their prolonged effects.

Facility 3. VX

We are listing three variations for making VX. Any variation can be stopped after making Agent QL, producing only a binary agent, which is safer to handle and transport. Other methods exist, it is important to note the list of chemicals that can also be used to make VX. If you encounter these materials, **be extremely careful; carelessness may result in death.**

Variation 1

Step 1. Making QL
Anhydrous ethyl ether
Dichloromethylphosphine

Equipment that might indicate the manufacture of VX

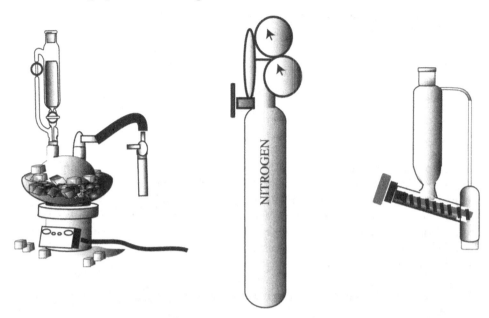

Ethyl alcohol
N,N-diethylanaline
Mercury
Nitrogen gas

Step 2
2-Diisopropylaminoethanol

Step 3. Making VX
Rhombic sulfur

Variation 2

O-Ethylmethylphosphonothioic acid
N,N-Diisopropyl 2-aminoethanol
Sulfur

Variation 3

Ether
Dichloromethylphosphine
Ethanol
N,N-Diethylaniline
2-Dimethylaminomethanol
Sulfur

Chemicals that could be used to make Agent QL

O-Ethylmethylphosphonothioic acid
O-Ethyl 2-diisopropylaminoethylmethyl phosphonite
O-Ethyl 2-diisopropylaminoethylmethyl phosphonate

More common amines that can be used to make Agent QL

2-Chloro-*N*-*N*-diisopropylethyl amine	Lab chemical
Diethylaminoethanol	Textile softener
Diisopropylamine	Antifoam agent, detergents
Diisopropylaminoethanothiol	Lab chemical
N,N-Diisopropyl-2-aminoethanol	Lab chemical
Dimethylamine	Gasoline additive, manuf. of pesticide, pharmaceutical

Agents NE and NM

Once the primary binary product QL has been made, there are two secondary binary products that can be used to make the final product. NE is sulfur, usually mixed with something to keep it from caking (Cab-O-Sil® is a common one). NM can be a mixture of dimethylpolysulfides. QL + NE(NM) = VX. This final step is so quick that there is a strong warning against getting any sulfur near agent QL. The primary task is to make agent QL. Like the G-agents sarin and soman, if this product is made from scratch, the most difficult task is to alkylate the phosphorus.

Chemicals that could be used as or to make Agents NE and NM

Phosphorus pentasulfide	Lubricating oil additive, pyrotechnics, manuf. of pesticide
Sulfur, Agent NE	Manuf. of pyrotechnics, manuf. of pesticide agriculture
Cab-O-Sil	Industrial fill, used for plastics
Dimethylpolysulfides, Agent NM	

V-Sub X

Ether
Dichloromethylphosphine
N,N-Diethylaniline
2-Ethylthioethanol
Due to the extreme hazard of chemical weapons, we will provide further information that might become necessary if one of these facilities is encountered.

Chemical characteristics

VX, colorless-to-straw-colored odorless liquid. Vapor pressure 0.0007 mmHg @ 20°C. Vapor density 9.2; density 1.008. Miscible in water. Soluble in all solvents. Volatility 10.5 mg/m³ @ 25°C.

Protection

Until the components are mixed, minimum protective clothing would be necessary. Agent QL is a mild cholinesterase inhibitor, so prolonged breathing may produce headaches and nausea. It is important to remember that any QL contact with sulfur or sulfur compounds can be very dangerous.

Detection method

Turns M8 paper to dark green, The M256A1 detector kit can detect VX.

Decontamination

Deluging with water is very effective in reducing the effects of VX. However, the sulfur in VX makes it very susceptible to destruction by oxidation by hypochlorite. It is best if the hypochlorite solution is alkaline. M291 resin, a black powder (mixture or resin and charcoal) that comes in packets, is very effective for skin decontamination.

Sandia National Laboratories has developed and commercialized an aqueous-based decontamination technology (DF-100 and MDF-200).

Decon Tech	1 Minute	15 Minutes	60 Minutes
DF-100 (pH 8)	45%	99%	ND
DF-100 (pH 9.2)	33%	71%	93%
DF-200	66%	99%	ND

Symptoms

Pinpoint pupils	Sweating
Runny nose	Nausea
Difficulty breathing	Vomiting
Loss of consciousness	Seizures
Convulsions	

Increased sweating at site, muscular fasciculations at site, nausea, vomiting, diarrhea, generalized weakness, loss of consciousness, convulsions, flaccid paralysis, apnea, generalized secretions, involuntary defecation.

Antidote

Atropine, 2-PAM-C1, pyridostigmime bromide tablets. The carbamates pyridostigmine and physostigmine will tie up the colesterase until the organophosphate has passed.

Storage

While neither of the binary components is particularly hazardous, once mixed they are extremely toxic. Considerable effort is important to keep the two components separate. QL should be treated as a flammable liquid, because it reacts with moisture to produce highly flammable diethyl-methylphosphonite (TR). Addition of water can cause QL to burst into flame. "The pure material is many times less toxic than VX but is by no means harmless. It reacts with moisture and other substances to produce highly toxic materials as well as flammable materials. It will ignite without application of spark or flame at 265°F. Hydrolysis products of QL ignites at a much lower temperature."

Dispersal

1. Explosive release; although some product is lost to decomposition, their simplicity makes explosives the dispersal of choice. 2. Bulk release from high-velocity projectiles, where the agent spills out into the airstream. 3. Base-ejection devices (pistons) are relatively uncommon, owing to their complexity. 4. Spray delivery can be used to achieve large-area coverage from various vehicles.

VESICANTS

Facility 4. Mustard

There are several ways of making mustard, utilizing various types of chemicals. The simple (almost a bucket) Variation 3 was used throughout World War I using only the chemicals below.

Sodium sulfide	250 g	$93.00
Ethanol (reagent)	1 L	$18.10
Calcium hypochlorite	1 kg	$40.00

Variation 1

2-Chlorohydrin
Hydrochloric acid
Sodium sulfide

Variation 2

Ethylene oxide
Hydrogen chloride
Hydrogen sulfide

Variation 3

This variation, which uses chemicals that are easily available, will always be possible for the home chemist. Most of the other variations are more likely to be used for large-scale production.

A better product is produced if it is washed with water before distillation. Even very pure product must be washed with water prior to distillation. If the product is not distilled, the product tends to polymerize upon storage; particles fall out and make distribution less effective. In a number of weeks, the active portion can be as low as 8%.

The wash water is a very dangerous. Waste product which should be located if possible. Fully 50% of the product can be expected to end up as waste. One cannot use current U.S. safety and environmental standards as the norm when judging these facilities.

This variation was used to make thousands of tons of mustard gas in Chapayevs, Russia during World War II, in an open operation. The plant was often idle due to the fact that all of the workers had died, and reinforcements could not be brought in quickly enough to replace them. Other indicators as to the hazard of this material are that both the person who made the initial product and the man who refined the product refused to continue working with it.

Sodium sulfide
Ethanol
Calcium hypochlorite

Variation 4

Thiodiglycol
Toluene
Thionyl chloride

Summary of Ways to Manufacture Mustard

Ethylene + Sulfur monochloride = Mustard

Victor Meyer-Clarke process

1. Starting from absolute scratch, ethylene would be reacted in oxygen to make ethylene oxide:

Ethylene +	2. Ethylene oxide +	3. Thiodiglycol and	=	Mustard	5. Mustard is
	can be reacted	hydrogen chloride gas		4. A far superior	is about 40%.
	with hydrogen	are reacted to make		mustard is made	To obtain a
	sulfide to make	mustard		if the product is	better and more
	thiodiglycol.			washed in water	stable product,
				before distillation.	mustard should
					be distilled.
					About an 80%
					product can be
					produced.

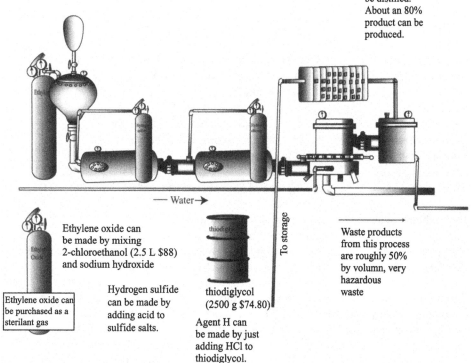

— Water→

Ethylene oxide can be made by mixing 2-chloroethanol (2.5 L $88) and sodium hydroxide

Ethylene oxide can be purchased as a sterilant gas

Hydrogen sulfide can be made by adding acid to sulfide salts.

thiodiglycol (2500 g $74.80)

Agent H can be made by just adding HCl to thiodiglycol.

To storage

Waste products from this process are roughly 50% by volumn, very hazardous waste

Variation 4 also requires standard laboratory chemicals, which can be obtained easily:

Thiodiglycol	500 g	$14.80
Thionyl chloride	1 L	$38.60
Toluene	1 L	$18.40

Note that this variation is closely related to Variation 5 below.

Variation 5
Levinstein

Chlorine gas
Sulfur
Ethylene
Activated charcoal
Nitrogen gas

Variation 6

Ethylene
Sulfur monochloride

Chemicals that can be used to make mustard

There are nine documented ways to make mustard using the compounds listed below.

Activated charcoal
Sodium sulfide   Standard laboratory chemical
Sodium sulfide   Dehairing, manuf. of paper, metallurgy
Chlorine gas
Nitrogen gas
Hydrogen chloride gas
Sulfur dichloride   Vulcanizing rubber, oils, manuf. of pesticide
Sulfur monochloride   Gold extraction, wool treatment, manuf. of pesticide,
                              pharmaceutical, dye
Sulfur
Thionyl chloride   Manuf. of plastic, pesticide, pharmaceutical
2-Chloroethanol   Solvent, manuf. of pesticide, pharmaceutical
Muriatic acid   Used to clean cement
Ethylene diglycol   Brake fluids
Thiodiglycol   2.5 kg for $74.80; solvent, antioxidant, lubricant, in ballpoint
                              ink pens
Hydrochloric acid   Standard industrial acid
Ethylene (ethene)   Petroleum manufacturing by-product
Hydrogen sulfide   Industrial gas; can be made from decaying organic materials
Ethylene oxide   Sterilizer used in medical applications

The most important of these compounds is thiodiglycol. Using thiodiglycol, a very pure mustard can easily be made in bulk. This compound has been controlled internationally by the Australian Group, which coordinates national export-control regulations to restrict its sale to potential manufacturers of chemical weapons. However, there are five manufacturers of thiodiglycol in the United States, and the manufacture of this product is fairly simple. The Victor Meyer-Clark process reaction below is for large-scale operations; however, it could be done on any scale. The United States used primarily the reaction between ethylene and sulfur monochloride to make mustard.

Chemicals for making bulk mustard from scratch

Ethylene (ethene) gas   Petroleum manufacturing by-product
Sulfur monochloride   Gold extraction, wool treatment, manuf. of pesticide,
                              pharmaceutical, dye
Ethylene oxide   Sterilizing agent for medical purposes
Hydrogen sulfide gas   Industrial gas
Hydrogen chloride gas   Industrial gas
Sulfur dichloride   Vulcanizing rubber, oils, manuf. of pesticide

Supplementary 1. Making Ethylene Oxide

2-Chloroethanol
Sodium hydroxide

History

Although there is no history of terrorists using mustard, this chemical weapon has been utilized in several military struggles, including World War I, by all sides, some in World War II, by the Japanese in China, by the Egyptians in Yemen during a civil war there, in Iran by the Iraqis, and in Iraq on the Kurds. In March 2004, Libya disclosed that it had 22 tons of mustard gas stored near Tripoli.

Chemical characteristics

Mustard, pale yellow-to-dark brown liquid with a garlic/roses odor. Vapor pressure 0.07 mmHg @ 20°C. Vapor density 5.4; density 1.2. 0.092 g/100 g water soluble. Very soluble in most organic solvents.

Protection

1–1.5 teaspoons = LD_{50}. Mustard is a liquid over 38°C. At 100°F becomes an inhalation hazard. A detector 3–6 feet above the ground may not register; however, at 6 to 12 inches, vapor may be 1–25 mg/m^3. Most injury is from the vapor, not the liquid. Mustard is considered to be a radio-mimic; like radiation, it is nucleophylic, causing DNA disruption. Many people who have worked with this product have been seriously hurt. Although it smells like mustard (garlic, garlic and roses), the odor is not strong and provides little or no warning. Fully encapsulated clothing and SCBAs are necessary when working in this facility. A facility like this is most certainly going to be booby-trapped. The most injury is from mustard vapor, not the liquid. However, in a bucket operation in Chapayevsk, Russia during World War II, the plant had to be closed down due to the fact that the Russians could not get replacement conscripts there quickly enough to keep it running.

Detection method

Mustard forms colored complexes with p-nitrobenzpyridine. Draeger phosphoric thioether tube, FID, and PID*. Gold chloride solution will flocculate mustard.

That mustard can be identified by PID or FID is stated in the Army literature. We have not found the IP for mustard in any of the literature.

Decontamination

Soap and water are the best decontaminate for skin. Mustard is persistent and can last for 30 years in concrete. Decontaminants include the M295 decontamination kit, which includes a pouch containing four wipe-down mitts: STB, fire, or DS2. Decontaminate liquid agent on the skin with the M258A1, M258, or M291 skin-decontaminating kit. Decontaminate individual equipment with the M280 individual equipment decontamination kit.

Sandia National Laboratories has developed and commercialized an aqueous-based decontamination technology (DF-100 and MDF-200)*.

Decon Tech	1 Minute	15 Minutes	60 Minutes
DF-100 (pH 8)	18%	42%	81%
DF-100 (pH 9.2)	16%	38%	83%
DF-200	94%	98%	ND

*There is some controversy regarding this product.

Symptoms

Reddening of skin   Sluggish
Blistering   Apathetic
Convulsions   Lethargic
Respiratory failure

The effects of exposure may not be apparent for as long as 2 hours. The reaction is concentration dependent, and the effects may not appear for even a considerable time after 2 hours. Blisters begin as a string of pearls around the affected area and grow into a single large blister. The eye is the organ most sensitive to mustard. Mustard produces dose-dependent damage to the mucosa of the respiratory tract, inflammation beginning with the upper airways and descending to the lower airways. Damage results from the chlorine bonding by electomagnetic attraction to the sulfur. The final effect is the same as if the portion of the skin were severely burned. The reaction takes longer than in the case of a burn.

Antidote

Immediate removal. After years of research there is NO antidote to mustard.

Storage

Mustard that has not been distilled properly will not last for a very long time. The breakdown creates corrosive gases, which will attack the container. Mustard freezes at 57°F.

Dispersal

1. Explosive release; while some product is lost to decomposition, their simplicity makes these the weapons of choice.
2. Bulk release from high-velocity projectiles, where the agent spills out into the airstream.
3. Base-ejection devices (pistons) are relatively uncommon, owing to their complexity.
4. Spray delivery can be used to achieve large-area coverage from various vehicles. Mustard is mixed with chlorobenzene or chloropicrin to lower its freezing point. It is mixed with lewisite in warm areas to increase persistence. High pressure from compressed air would work here.

Nitrogen Mustard N1-1

Diethanolethylamine
Acetone
Thionylchloride
Anhydrous sodium carebonate
Chloroform

Nitrogen Mustard N1-2

Diethanolethylamine
Sulfur
38% Hydrochloric acid
Chlorine gas
Chloroform
Anhydrous sodium carbonate
Acetone

Nitrogen Mustard N1-3

Ethyldiethanolamine
Methyldiethanol amine
Triethanolamine

Nitrogen Mustard N2-1

Diethanolmethylamine
Acetone
Thionylchloride
Anhydrous sodium carbonate
Chloroform

Nitrogen Mustard N2-2

Diethanolmethylamine
Sulfur
Hydrochloric acid
Chlorine gas
Chloroform
Anhydrous sodium carbonate

Nitrogen Mustard N3

Anhydrous triethanolamine
Acetone
Thionyl chloride
Anhydrous sodium carbonate
Chloroform

Arsenicals

Facility 5. Lewisite

Variation 1

The production of lewisite is a two-step method. The product of the first step is sufficient
to be a chemical weapon; however, by distilling the product into three subproducts increases
the potency considerably.
Acetone
Arsenic trichloride
Acetylene or calcium carbide (used to manufacture acetylene) or dry ice

Variation 2

Acetone
Arsenic trichloride
Acetylene or calcium carbide (used to manufacture acetylene) or dry ice
Hydrochloric acid
Mercuric chloride

The yellow liquid in the large main vessel is arsenic trichloride, a choking agent. This is
probably more dangerous than the product being produced, lewisite. The end product will
have to be distilled. (This requires a triple distillation apparatus.) Lewisite is a vesicant.

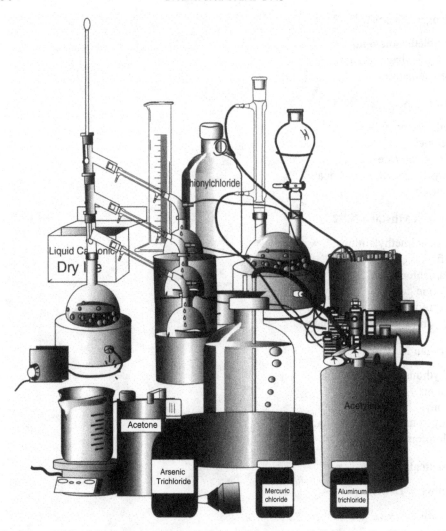

There is considerable distillation necessary to create high-grade lewisite. Lewisite can also be made using mercuric chloride, and hydrochloric acid Variation 2 to catalyze the mixture. This procedure requires less temperature control.

History

Lewisite works best at temperatures below 0°C. It does not do well in high humidity or in moist areas. Mustard works best in warmer climates and is not affected by moisture. The Soviets favored a mixture of lewisite and mustard, due to their climatic situation.

Chemical Characteristics

Pure lewisite is an oily colorless liquid; the agent is amber to dark brown, with the smell of geraniums. The vapor pressure is 0.39 mmHg @ 20°C. Vapor density 7.1; density 1.89. Volatility 4480 mg/m³ @ 20°C; slightly soluble in water; soluble in most organic solvents. The reaction above creates three types of lewisite: 1, 2, and 3. The number indicates the number of chlorine atoms in the final product. Lewisite 1 and 2 are considered binary agents. Lewisite 3 is a vesicant; a mixture of the three is effective as a chemical weapon.

Protection

The final product is dangerous by skin contact and can be lethal by inhalation. Lewisite has a low vapor pressure, but it is sufficient to get into the atmosphere up to 8.6 mg/m^3 at 30°C. The LD$_{50}$ is 30 mg/kg. The LCI$_{50}$ (respiratory) is 1400 mg-min/m^3. As an aerosol lewisite acts much like mustard, except that eye contact is likely to cause permanent blindness. Lewisite is destroyed by high humidity, and water vapor which could be employed to protect people entering the facility. Arsenic trichloride is a fuming liquid; the fumes are intensely poisonous. SCBA and total encapsulation would be required for entry into this facility.

Detection method

Lewisite is a clear-to-amber or even black oily liquid that has the odor of geraniums. Lewisite at dangerous levels will burn the eyes and skin immediately. Arsenic trichloride and some of its breakdown products smell like garlic. Draeger arsenic tube 0.1 ppm (qualitative) will detect lewisite. A PID can be used to measure atmospheric lewisite (real time). The ability of a PID to measure aerosol is questionable. A combustible gas indicator to determine that acetylene is not reaching its explosive level would be a good protective measure in this facility. Lewisite can be detected using a Ilosvay test paper. An Arsenic Test and a Chloride Test will both be positive.

Decontamination

Strong alkalis destroy the blister-forming properties. HTH, STB, household bleach, DS2, or caustic soda can be used to decontaminate surfaces. Decontaminate liquid agent on the skin with the M258A1 or M291 skin decontaminating kit. Decontaminate individual equipment with the M280 individual equipment decontamination kit. Lewisite has a short persistency in humid conditions. Depending on temperature and humidity, lewisite can last for days. Remember that even if it is broken down, it is still arsenic. Any residue remains a hazard. Products of hydrolysis products are hydrochloric acid (a vesicant) and chlorovinyl arsenous oxide, a nonvolatile solid.

Symptoms

Immediate eye and skin pain	Vomiting
Loss of sight	Hoarseness
Reddening of skin	Fever
Irritation of mucous membranes	Apathy
Nausea	Depression

Lewisite produces pain or irritation within seconds. The blisters begin as a small blister in the center of the erythematous area and expand to include the entire inflamed area. Eye injury is generally not severe, due to the pain that produces blepharospasm, effectively preventing further exposure. Direct contact will destroy the eye. Lewisite is extremely irritating to the nose and lower airways. Inhalation may be fatal in as short a time as 10 minutes. Lewisite shock is the result of extreme exposure, and all of the capillaries hemorrhage with subsequent hemolytic anemia.

Comparison of the three primary vesicants:

Lewisite:

A blister starts at the center of the red area and moves out; pain is immediate.

Mustard:

Blisters form around the red area and fill in; pain is delayed.

CX:

Area of contact becomes gray and necrotic; wheals form instead of blisters; pain is immediate and intense.

Antidote

BAL, British anti-lewisite, a cheating agent, can ameliorate the effects of lewisite if used soon after exposure.

Storage

Lewisite will break down if it becomes moist.

Dispersal

Lewisite works best in winter or in cool weather. Its low freezing temperature makes it attractive as a carrier for other chemical weapons. HD is a mixture of mustard and lewisite which can be used a very low temperatures.

Lewisite lends itself to the four modes of chemical agent release: 1. Explosive release; although some product is lost to decomposition, their simplicity makes the weapons of choice. 2. Bulk release from high-velocity projectiles, where the agent spills out into the airstream. 3. Base-ejection devices (pistons) are relatively uncommon, owing to their complexity. 4. Spray delivery can be used to achieve large-area coverage from various vehicles.

BLOOD AGENTS

Blood agents are chemical asphyxiates. The gas or vapors react with the blood more quickly than oxygen and tie up the oxygen receptors, causing death by asphyxiation. Hydrogen sulfide and carbon monoxide could legitimately be included here; however,

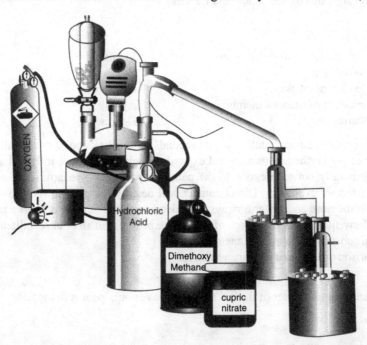

neither is considered to be a blood agent. Note all of the blood agents are INORGANIC chemicals, making them considerably different than most of the chemical weapons.

Facility 8

Variation 1. Hydrogen Cyanide (Liquid)

This would be a setup to make liquefied hydrogen cyanide. Sodium ferrocyanide and sodium or potassium cyanide could be used in this process. Refrigeration is very important, as hydrogen cyanide becomes a gas at 25.6°C. The liquid is pale blue with a bitter almond odor. Oral LD_{low}(man) less than 1 mg/kg.

Potassium ferrocynide
Concentrated sulfuric acid
Ice

Variation 2. Making Cyanogen

Cyanogen is a colorless gas with a pungent almond-like odor. The odor is NOT detected by everyone.

Hydrogen cyanide liquid
Cupric nitrate
Dimethyoxymethane
Oxygen gas
Ice

Variation 3. Making Cyanogens Chloride

Cyanogen chloride, colorless liquid or gas with a pungent, biting odor. The odor will cause discomfort. Boiling point 55°F; vapor pressure 1010 mmHg @ 20°C. Vapor density 2.1. Cyanogen chloride, which is heavier and less volatile than either hydrogen cyanide or cyanogen, is the more effective weapon at low concentrations.

Sodium cyanide
Acetic acid
Carbon tetrachloride
Ice
Chlorine gas

Variation 4. Making Cyanogen Bromide

Bromine
Sodium hydroxide
Sodium cyanide
Sulfuric acid

Supplementary 1. Making Sodium Cyanide from Scratch

Platinum can catalyze ammonia, sodium hydroxide, and methane to form hydrogen cyanide/sodium cyanide. Sodium cyanide is a salt which has the appearance of table salt. A grain of this salt on the tongue is the LD_{50} for a 180-lb man. However, the low vapor pressure, negligible at 20° C, 1 mmHg @ 817°C (1503°F), makes sodium cyanide fairly easy to handle. Despite this low vapor pressure, the presence of sodium cyanide can be detected by odor by 1 out of 10 people. This reaction should be done in alcohol. Sodium cyanide is easily available. Cost at present is 500 g @ $22.50. Cyanide is naturally occurring in chokecherries, almonds, apricot pits, lima beans, casava beans and roots.

Ammonia
Chloroform
Sodium hydroxide
Alcohol
Platinum

NOTE: None of the chemicals used in this process would attract attention when purchased.

Supplementary 2. Arsine Gas

Arsenic salts
Zinc metal
Hydrochloric acid

Chemicals used to make cyanide blood agents

Acetic acid	Standard laboratory chemical
Ammonia	Household chemical
Bromine	Laboratory chemical
Calcium chloride	Drying agent
Calcium hypochlorite	Household chemical
Carbon tetrachloride	Cleaning solvent
Chlorine gas	Swimming pool disinfectant
Cupric nitrate, trihydrate	Standard laboratory salt
Dimethoxyethane	Unusual laboratory solvent
Ethyl alcohol	Standard laboratory solvent
Methane	Natural gas
Oxygen gas	Medical
Potassium cyanide	
Potassium ferrocyanide	Standard lab chemical
Sodium cyanide	Plating chemical
Sodium ferrocyanide	
98% Sulfuric acid	Standard lab chemical

Due to the extreme hazard of chemical weapons, we provide further information that might become necessary if one of these facilities is encountered.

History

Sodium and potassium cyanide are easily obtained and very toxic. It has been used to poison foods in various schemes. In January 2004, a couple in Texas were found making cyanide bombs. Cyanide is one of the chemical weapons that has actually been used in warfare.

Protection

Despite the fact that cyanide is very toxic, it is among the least toxic of the WMD compounds. It can be lethal when absorbed through the skin at very high levels (@ 20,000 ppm); however, the primary hazard is by inhalation. An SCBA must be worn.

Detection method

Hydrogen cyanide indicator tubes (qualitative), FID (real time, not specific).

Decontamination

Remove exposed person to fresh air. Remove clothes and any liquid on the skin using copious amounts of water. Wash water must be collected. Cyanide can be neutralized with hypochlorite. Acidification of any liquid can produce large amounts of hydrogen cyanide gas.

Symptoms

Increased rate and depth of breath	Headache
Great difficulty breathing	Convulsions
Dizziness	Cardiac symptoms
Nausea	Odor of burnt almonds or peach kernels
Vomiting	

Although these are generally considered to be very toxic substances compared to the others discussed here, they are among the least toxic. Exposure is accompanied by decreased awareness, followed by loss of consciousness and convulsions. The body may take on a red or blue tinge. There may be salivation, urination, and defecation.

Antidotes

Sodium nitrite and sodium thiosulfate(IV) (30 mg/mL, within 5 to 15 minutes, never over 0.33 mL of the 10% solution per kilogram of body weight). 4-Dimethylaminophenol and sodium thiosulfate (3 mg/kg intravenous injection), dicobalt edetate. (The severe toxicity from cobalt can be seen even after initial recovery from acute cyanide poisoning.) Hydroxocobalamin (vitamin B_{12}) was not very effective in World War I.

Storage

Cyanogen chloride is capable of polymerizing during storage, and could explode.

Dispersal

Cyanogen and hydrogen cyanide are gases, whereas cyanogen chloride can be a liquid. Although these have been used in munitions, they were very ineffective. They do, however, lend themselves more to urban warfare. These volatile gases would be most effective if released in enclosed spaces.

CHOKING AGENTS

Facility 9. Nickel Carbonyl

Nickel carbonyl is not considered to be a metal fume choking agent. These reactions include finely divided nickel over which carbon monoxide is being force-drafted. Carbon monoxide could be manufactured on site. It would be as easy to obtain carbon monoxide gas in gas cylinders. Carbon monoxide can also be generated by mixing equal amounts of formic acid (any organic acid will work) with concentrated sulfuric acid. Nickel carbonyl explodes if heated, and temperature control would be an important part of this facility. Nickel carbonyl is generally stored as a liquid under low pressure (about 5 lb) in cylinders, so that it can be transferred without opening.

Variation 1

Nickel powder
Charcoal

Variation 2

Nickel powder
Carbon monoxide gas cylinders

Due to the extreme hazard of chemical weapons, we will provide further information that might become necessary if one of these facilities is encountered.

Chemical characterization

Nickel carbonyl is a colorless-to-yellow volatile liquid with a musty or sooty odor, which can explode at 60°C. Vapor pressure 315 mmHg @ 30°C (acetone is 215 mmHg @ 30°C). Nickel carbonyl is 0.05% soluble in water. Flash point <–4°F.

Protection

Inhalation is the primary hazard. An SCBA would be minimum protection.

Detection method

A Draeger colorimetric tube has been developed to test for nickel carbonyl in the air. However, the sensitivity of the calorimetric tube is not low enough to indicate the presence at levels required to provide personal protection. Nickel can be determined by dimethylglyoxime. Using field methods to identify this compound would be very dangerous. A PID can detect relatively large amounts of nickel carbonyl in the air; the IP is 8.3 eV.

Decontamination

Vapors, by ventilation. Liquids require immediate removal of clothing and washing exposed skin with large amounts of water. Puddles may be contained by covering with vermiculite, diatomaceous earth, clay, fine sand, sponges, paper towels, or cloth towels. Absorption of gas by porous materials can extend the period of time that the compound will be found as a vapor in the air. Remove all material and place in a container. Wash impregnated solid materials with large amounts of water.

Symptoms

Not all symptoms appear in all persons exposed. Most experience a frontal headache and/ or giddiness, tightness in the chest, nausea, weakness of the limbs, perspiring, cough, vomiting, and a few exhibit cold and clammy skin or shortness of breath. High exposures where there is survival cause a reversible fibration of the lungs, and lung cancer.

Antidote

Use of dithiocarb orally relieved the symptoms of poisoning, and the delayed reactions were minimal and convalescence uneventful.

Storage

Nickel carbonyl has been stored in glass bottles, but this is very hazardous. Transferring the material to and from glass bottles can be difficult because of the high volatility.

Generally, nickel carbonyl is stored in low-pressure metal cylinders. The primary clue is the presence of a gauge on a compressed gas cylinder with 5 lb of total pressure.

Dispersal

This volatile liquid oxidizes quickly in the air; therefore, it might be impregnated in a solid for dispersal.

Facility 10

Chloropicrin is available commercially, as is picric acid. Making all of these from the chemicals below might be done to avoid detection. It would also be possible to continue on and make CX using these same chemicals. Chloropicrin is used as a carrier for many other chemical weapons. It is almost necessary for the dispersion of CX.

Chloropicrin

Variation 1

Step 1
The addition of phenol to sulfuric and nitric acids produces picric acid. Picric acid and calcium hypochlorite produce chloropicrin. *See* Explosives, picric acid. Distillation will be required.

Nitric acid
Sulfuric acid
Phenol

Step 2
Calcium hypochorite powder
Picric acid made in Step 1

Variation 2

Picric acid
Calcium hypochlorite

Variation 3

Sodium hydroxide
Chlorine gas
Nitromethane

Variation 4

Sodium hypochloride
Nitromethane
Calcium chloride

Supplementary 1. Making Picric Acid from Aspirin

In San Diego County, two teenage boys set up a small chemical workshop a shed behind one of their houses. They possessed a significant library. The boys were known as very smart, slightly odd loners, who espoused an odd political philosophy mixed with a kind of mysticism. After they were killed in a car explosion, investigators found many 1-gallon bottles of concentrated nitric acid and concentrated Sulfuric acid. There are several glass jars of sodium nitrate. The trash cans were filled with many empty aspirin containers. They also had a great deal of anhydrous ethanol.

Aspirin
Nitric acid
Sulfuric acid
Potassium nitrate

Due to the extreme hazard of chemical weapons, we provide further information that might become necessary if one of these facilities is encountered.

Chemical characteristics

Chloropicrin is a colorless, slightly oily liquid with a very intense odor and definite lachrymatory effects. Chloropicrin is very toxic by inhalation. Specific gravity 1.69; boils at 112°C.

Protection

Besides the final product, chloropicrin, which is a very nasty inhalation hazard, there is nitric acid, which can produce large amounts of nitrogen oxides. Phenol is a skin-contact poison. Picric acid, especially as a liquid, is far more stable than most people think (being a secondary explosive); however, if contaminated, or if it is in contact with metal, it can form picarates (primary explosives), which can easily set off the picric acid. Preliminary monitoring should be done prior to entering this facility. The level of protection would be based on the results of the monitoring. Although chloropicrin is an extreme inhalation hazard, most of these are not only inhalation hazards but skin-contact hazards.

Detection method

A halogen candle, which consists of a propane torch and a copper wire, Draeger tubes, or a Freon® gun would detect the presence of chloropicrin in the air. There are also available phenol Draeger tubes and nitrogen dioxide Draeger tubes. Picric acid is a distinctive bright yellow water-soluble crystal.

Decontamination

Decontamination is by ventilation. Chloropicrin is used to fumigate houses, where it is quickly dissipated by opening up the doors and windows. Picric acid can be diluted and removed with water, 1 g in 78 cm³ of water, and will not explode as long as it remains wet.

Symptoms

Tearing and a peculiar form of frontal headache is characteristic of persons exposed to chloropicrin. Chloropicrin is a lacrimator and a lung irritant. Exposure to 4 ppm for a few seconds renders a person unable to perform even the slightest work, and 15 ppm for the same amount of time results in respiratory tract injury. Chloropicrin is also a potent skin irritant.

Antidote

Picric acid can go through intact skin. Overexposure to both picric acid and chloropicrin would have to be treated symptomatically.

Storage

Storage of homemade explosives should terrify any responder. Nitric acid in contact with most metals will release a red cloud, which is greater than the IDLH for nitrogen dioxide. Chloropicrin is very deadly at high concentrations.

Dispersal

This volatile liquid has been used as a fumigant for years. For dispersal it was simply poured into a dish and left in the tented house.

Facility 11. Arsenic Trichloride

Besides being an off-the-shelf chocking agent, arsenic trichloride is the main ingredient in many arsenical vesicants and vomiting agents.

Chloroform
Arsenic sponge*
Chlorine gas

Due to the extreme hazard of chemical weapons, we will provide further information that might become necessary if one of these facilities is encountered.

Chemical characteristics

Arsenic trichloride is a colorless-to-yellow slightly oily liquid with a garlic odor. Arsenic trichloride fumes in the air. This is an unusual salt, as it is a liquid.

Protection

A fully encapsulated suit would be necessary.

Detection method

Arsine indicator tubes can identify this material in the air.

Decontamination

Arsenic cannot be denatured. This product must be removed completely.

Dispersal

Any method to aerosolize arsenic trichloride would cause a lethal cloud.

Facility 12. Metal Fumes

Stannic chloride
Cadmium metal
Selenium

Elements of Death

Organic chemical weapons tend to be very toxic, a level of toxicity that would be difficult to achieve using inorganic compounds. However, there are inorganic compounds that are sufficiently hazardous to be considered as possible chemical weapons. *Hazardous* can be defined as high toxicity (low LD_{50}) and high vapor pressure. Although there are many elements and inorganic compounds with low LD_{50}'s, there are other factors that may still keep them from being considered highly toxic: (1) Most inorganic elements and compounds will have lower vapor pressures, making them less hazardous but adding to their persistence; and (2) since they are elemental, they may be more difficult to decontaminate. Low vapor pressure can be compensated for by atomizing a material; however, this gives rise to another challenge—that of dispersing and keeping the material in the air. Although some inorganic compounds can be transported through the skin, they are not as efficient as organic compounds going through the skin. This means that generally, the targeted route of entry is inhalation.

The blood agents are made up of inorganic compounds. Except for arsine and hydrogen sulfide, the blood agents rely on cyanogens as the knockdown agent. Arsine is one of three gaseous hydrides (salts?). The other two, stibine and phosphine, are so similar to arsine that although they are not listed as chemical weapons, they would be good candidates. Most of these inorganic compounds are either gases or liquids with high vapor pressures. (Cyanogen is considered organic in many classification systems, but its actions are inorganic, and we will treat it as such here.) Carbon monoxide also has characteristics that would make it a blood agent if it were considered as a chemical weapon. Most of the other inorganic compounds used or considered as candidates for use are choking agents.

An elemental weapon that has been used in the field is chlorine. This was among the first chemical agents used in World War I. The first use, against the French, was very successful, as they had no warning or protection. Other attempts were not as successful, as the gas was caught up in the wind and returned to the senders. Bromine and chlorine trifluoride have been considered for use because they have qualities similar to those of chlorine. Other halogen-containing compounds either suggested or used as chemical weapons include phosgene, diphosgene, disulfur decafluoride and arsenic trichloride.

Some metals (metalloids) fit the following criteria: (1) they can have devastating effects; (2) the onset of symptoms may take several months; (3) these metal poisons are difficult to diagnose; and for the most part; (4) when the effects manifest themselves, it is too late to reverse the effects. The following are candidates: beryllium, very expensive; mercury widely available but generally very slowly effective; thallium, inexpensive and widely available; lead, inexpensive and widely available; cadmium; arsenic; and selenium. Arsenic compounds, selenium oxide, and cadmium oxide have been designated as choking agents. For most of these elements there is no easy method for direct real-time analysis. Mercury, the one element for which there is an effective detection method, is so rare that testing for mercury is rarely done. These substances lend themselves to smaller "poisoning" events. However, for certain terrorist groups, availability does make these materials attractive. Although no real-time monitoring instrumentation is available, most of these toxins lend themselves easily to field chemical analysis.

INCAPACITATING AGENTS

Facility 13

In this facility we cover two reactions: making benzilic acid and making 3-auinuclidrinyl benzilate (Agent BZ) or buzz. Agent BZ is a drug, and made in a similar manner using typical reaction vessels, and the final product is salted out. BZ and related compounds are used as designer drugs. *Also see* Designer drugs.

The manufacture of benzilic acid from benzaldehyde is an important step, as benzilic acid is both rare and expensive. Benzaldehyde is far less expensive and easier to obtain. Benzaldehyde is a scheduled DEA methamphetamine precursor.

There are other ways to carry out this rearrangement reaction. Although some are more complex, they can be done in fewer steps. One is a condensation reaction using a cyanide salt to make benzoin into benzyl. Another uses a reflux condenser, copper sulfate, pyridine, hydrochloric acid, carbon tetrachloride, and water to make benzyl from benzoin. Look for dark green reaction mixtures, and filtering.

Buzz

Benzilic acid
3-Hydroxy-1-methylpiperidine
Propylene glycol
Sodium hydroxide
Hydrogen chloride gas

Making benzilic acid

Benzaldehyde
Potassium cyanide
Lead oxide
Ethanol
Potassium hydroxide
Hydrochloric acid

Other chemicals used to make buzz

Methyl benzilate
3-Quinuclidinol

Making benzyl from benzoin
Look for dark green reactions mixtures, and filtering.

Copper sulfate
Pyridine
Hydrochloric acid
Carbon tetrachloride
Water

Due to the extreme hazard of chemical weapons, we provide further information that might become necessary if one of these facilities is encountered.

History

Although buzz has never been used in the field, the Russian incapacitating agent fentanyl has been. The Russian military used fentanyl against terrorists in a hostage situation at a theater in Moscow in October 2002. Although it is designed to incapacitate, there were over 100 fatalities attributed to the agent in the incident.

Chemical characterization

White crystalline solid with no odor. Vapor pressure is negligible. Buzz has an ID_{50} that is approximately 40-fold lower than the LD_{50}. For the white crystalline solid, ID_{50} is 6.2 g/kg free base MED_{50} (minimal effective dose) 2.5 g/kg.

Protection

The primary route of entry into the body is by inhalation. The secondary route is through the digestive tract. Skin absorption is possible with proper solvents. Depending on the situation, an air-purifying respirator would be adequate; however, if mixed with certain solvents and aerosolized, an encapsulating suit could be required.

Detection method

No device is available at present for detecting Agent BZ.

Decontamination

For area contamination, wash with a bleach solution. The bleach solution should be no less than one part household bleach in nine parts water. Personnel should be washed with soap and water. Rinse with copious amounts of water. Delayed effects can occur as much as 24 hours after exposure. Buzz is very persistent in soil water and on surfaces. The half-life is 95 hours at a pH of 7.4. In moist air, the half-life can be 3 to 4 weeks.

Symptoms

Delusions.

Antidote

Phenothiazines, benzodiazepines, physostigmine, neostigmine, pyridostigmine, and tetrahydroacridine will immediately reduce the effects of BZ.

Dispersal

Buzz is dispersed as an aerosol, in smoke-producing munitions, or just as an aerosol, solid, or in solvents. Propylene glycol allows 10 to 15% of Agent BZ to pass through intact skin. The effects may take 2 to 3 hours.

VOMITING AGENTS

Facility 14

Vomiting agents were initially designed to be sneezing agents. Most will cause sneezing initially in most people.

Adamsite

Adamsite is a vomiting agent. It is made by adding diphenylamine to arsenic trichloride, which is a choking agent that can be made in Facility 11.

Arsenic trichloride

Diphenylamine
Benzene

Diphenylcyanoarsine

Sodium arsenide
Zinc metal
Chlorobenzene
Hydrochloric acid

Diphenylcyanoarsine

Arsenic salt
Zinc metal
Hydrochloric acid
Potassium cyanide

Due to the extreme hazard of chemical weapons, we provide further information that might become necessary if one of these facilities is encountered.

Chemical characteristics

Agent DM, adamsite, CAS 578-94-9 is made up of light to yellow crystals. The vapor pressure is negligible at 2×10^{-13} mmHg @ 20°C. Melting point 195°C; boiling point 410°C. Breaks down at 400°C. Adamsite is not flammable under standard conditions. Adamsite is not soluable in water or in any of the other chemical warfare agents. Acetone is the most likely solvent for adamsite. LCT_{50} @ 11,000 mg-min/m^3, ICT_{50} 150 mg-min/m^3.

Protection

An SCBA should be worn. Adamsite has no pronounced odor, but it can be irritating. Requires only about 1 minute to incapacitate temporarily at a concentration of 22 mg/m^3. Exposure will cause strong pronounced sneezing. Although adamsite is irritating to the skin and eyes, it is relatively nontoxic to these organs.

Detection method

None. There are tests for arsenic, chlorine, and the benzene ring. A combination of these three elements could indicate a number of CW agents, including all of the vomiting agents.

Decontamination

This compound has very short persistency. No decontamination is required outside. Caustic soda and chlorine may be used as remedial action in the case of gross contamination in enclosed areas. Remember that the residue will contain arsenic, which is a carcinogen. Adamsite breaks down to hydrochloric acid and diphenylarsenous oxide. Taken internally the oxide is very poisonous.

Symptoms

Regurgitation is the primary symptom. This does not occur for all persons exposed. Adamsite causes coughing, sneezing, pain in the nose and throat, nasal discharge, and/or tears. Headaches often follow exposure to vomiting agents. Vomiting agents often cause dermatitis on exposed skin. In enclosed or confined spaces, vomiting agents can cause serious illness or death. The arsenic can cause many secondary problems.

Storage

Should be stored in brown glass jars. As such, can remain indefinitely.

Dispersal

Agent DM is disseminated as an aerosol.

SMOKE AGENTS

Facility 15

Burning a mixture of aluminum zinc and hexachloroethane produces a nasty very cloudy, toxic smoke. Although this agent was initially proposed as an obscuring smoke screen, it proved to be very irritating and lethal. Exposure causes a cough, which often produces bloody secretions. Exposure may lead to death within hours.

Zinc chloride

Zinc
Hexachloroethane
Aluminum
Titanium tetrachloride

Due to the extreme hazard of chemical weapons, we provide further information that might become necessary if one of these facilities is encountered.

Hexachloroethane, in the presence of aluminum as an accelerator, burns with the zinc to produce white smoke, which also contains carbon tetrachloride, ethyl tetrachloride, and hydrochloric acid. This is a very different from corrosive smokes (below), which are made from an anhydrous acid.

Corrosive Smokes

The corrosive smokes actually consist of the fumes of concentrated acid formed when anhydrous acid or a strong hydrolyzing agent is spattered into the air by munitions. Corrosive smokes were not terribly successful in the field, but lend themselves to urban terrorism because these chemicals are easily obtained, and it is the nature of the urban setting that people are gathered together in large crowds, often where escape in mass would be difficult. Candidates for corrosive smokes, anhydrous acids or strong hydrolyzing agents, which could be placed on explosives include:

Thionyl chloride
Boron trichloride
Phosphorus trichloride
Sulfur trioxide
Hydrogen chloride
Acetic anhydride

Protection

Acid smokes, if released, would require encapsulation and a SCBA.

Detection method

The presence of zinc could be proved with dithizone. Aluminon could determine the presence of aluminum. A PID could determine the presence of hexachloroethane nonspecifically in the air. The chlorinated nature of the agent could be determined using the PCB Clor-n-oil kit. The presence of the HCl in the smoke could be determined by an acetic acid Draeger tube.

Decontamination

A minimum of decontamination would be required unless the product had been released.

Storage

No specific problems in storing. The storage time would be indefinite, with little deterioration.

Dispersal

This agent is designed to be used in munitions. Any explosive device would suffice to spread the agent.

TEAR AGENTS

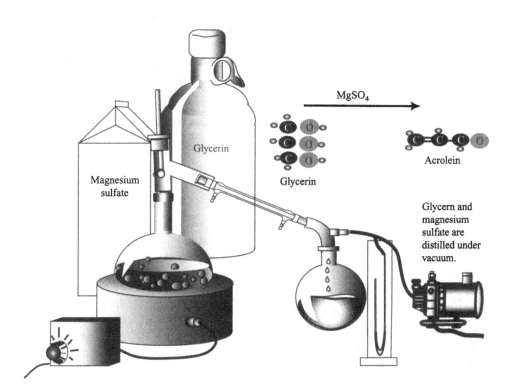

MgSO$_4$

Glycerin

Glycerin

Acrolein

Magnesium sulfate

Glycern and magnesium sulfate are distilled under vacuum.

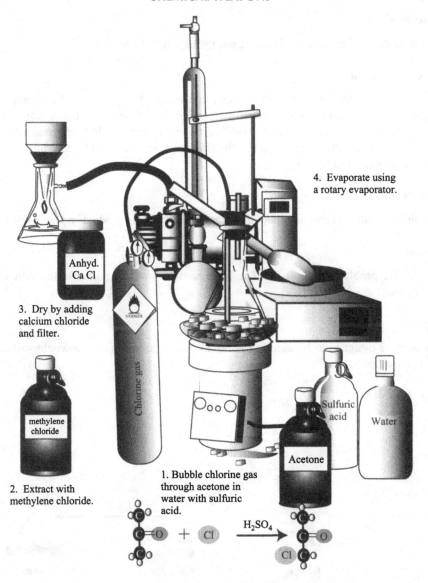

3. Dry by adding
calcium chloride
and filter.

Anhyd.
Ca Cl

methylene
chloride

2. Extract with
methylene chloride.

Chlorine gas

OXIDIZER

4. Evaporate using
a rotary evaporator.

Sulfuric
acid

Water

Acetone

1. Bubble chlorine gas
through acetone in
water with sulfuric
acid.

$$C\text{---}O + Cl \xrightarrow{H_2SO_4} C\text{---}O$$
$$Cl$$

Acrolein

Anhydrous MgSO$_4$
Glycerin

Chloroacetone

Acetone
Methylene chloride or chloroform
98% Sulfuric acid
Anhydrous calcium chloride
Chlorine gas

Tear gas (unspecified, from *The Terrorist's Encyclopedia*)

Glycerin
Sodium bisulfate
Alcohol lamp

o-Chlorobenzylidenemalonitrile
Knoevenagel condensation

2-Chlorobenzaldehyde
Malanonitrile
Piperidine or pyridine

Pepper spray chemicals, tear agents and riot control chemicals

Bromobenzyl cyanide
Capsaicin
Chloroacetopherone
o-Chlorobenzylidene
Dibenz[b,f]-1,4-oxazephine
Malononitrile
o-Chlorostyrene
Phenylacyl chloride
Trans-8-methyl-N-vanilyl-6 nonenamide

TOXINS

Toxin is a very broad category. We have tried as best we could to stay with homespun poisons and poisons that have been seen in underground literature, on the Web, or have been tried either historically or recently. We have avoided regurgitating the many lists of such compounds that have been published.

Certain things, such as exotic animals, are not included. We find the possession of a ringed octopus, cobra, or a black widow farm as not being very likely.

Mushrooms are a readily available source of very effective poisons. However, identification of mushrooms is well beyond the scope of this book. Mushroom identification is difficult even for mushroom experts. If mushrooms were encountered, you should find an expert. Death cap mushrooms are determined in the field by amateurs by holding them over black paper and tapping them to knock out the spores, which are obviously white on the black paper. If you do not see white spores, the test is NOT negative; many other mushrooms are poisonous. We have included toxins derived from bacteria (such as botulism) only at the urging of our reviewers.

Some of the substances below are considered chemical weapons, and others, such as 1080 are on a short list to become classified as chemical weapons. We have separated toxins from chemical weapons, as these all seem more personal murder-type chemicals. We feel that these are more appropriate to murdering one's mother-in-law or an entire Odd Fellows' meeting than really being a method of mass murder.

Ricin

There are many recipes on the Web and in the literature, even short videos showing how to make ricin. Not all of these will work. It is certain that there are recipes not listed below.

Variation 1

The DMSO is supposed to transfer the ricin through the intact skin. There is some controversy if this really works. There is an urban legend describing a sheriff deputy who was killed by putting DMSO ricin on the steering wheel of his car.

Castor beans
Methanol
Sulfuric acid
Dimethyl sulfoxide (DMSO)

Variation 2

The solvents are used first to extract and then to isolate the ricin from the caster beans so it can be salted out using magnesium sulfate or sodium disulfite. The real trick is the preparation of the beans, which usually requires sodium hydroxide.

Castor beans
Table salt
Water
Magnesium sulfate

Variation 3

Considered to be a faster method, as the acetone evaporates more quickly than water. Acetone shortens the process.

Castor beans
Sodium hydroxide
Acetone

Due to the extreme hazard of chemical weapons, we provide further information that might become necessary if a person making one of these compounds is encountered.

Ricin

9009-86-3

Appearance: May be an aqueous suspension or in the form of a fine white powder.

Odor: None found.

Hazards: Extremely toxic, may be fatal if swallowed, inhaled, or absorbed through the skin. LD_{50}(human) 0.2 mg/kg. Although recovery from lower doses often occurs, there is NO antidote.

Regulations: A person caught possessing or making ricin is liable for up to 30 years of prison.

History

In London in January 2003, five people associated with Al Qaeda were caught making ricin in an apartment. Making ricin was part of the curriculum in training camps in northern Iraq. On February 2, 2004 a powdery substance found in the mailroom of U.S. Senate Majority Leader Bill Frist indicated the presence of ricin. On finding the ricin, it was noted that a similar letter had been found earlier at a postal handing facility in Greenville, South Carolina. Ricin was also indicated indirectly on July 28, 2004, when ground-up castor beans were found in two jars of baby food sold in Irvine, California. Some of these ricin events indicate a low level of sophistication on the part of the would-be terrorists. Mail is not the most efficient way to distribute ricin.

The manufacture of ricin is discussed in Saxon's *Poor Man's James Bond*. It has been considered a chemical weapon since an event in London, England, when a Bulgarian dissident, Georgi Markov, was killed in 1978 by a poison (umbrella) dart filled with ricin.

Incomplete reactions

Ricin is an albumin (agglutinin), a ribbon-like structure, with an A and a B part that are linked to each other. Ricin A is toxic to the cell by interfering with the RNA function, and thus protein synthesis, whereas ricin B is important for assisting ricin A into a cell by binding with a cell surface component. Many plants, such as barley, have the A chain but not the B chain. Since people do not get sick from eating large amounts of such products, ricin A is of extremely low toxicity if and only if the B chain is not present.

Characteristics

White powder, destroyed by 20 minutes of 140°F heat. Ricin does not survive oxidation.

Protection

Although ricin can poison when inhaled or ingested, the primary hazard is injection through the skin. The vapor pressure is low, and except were it is adsorbed in a dust (which is not a usual method of delivery for this toxin), the primary concern is skin contact, especially where there is a break in the skin, such as a cut. An air-purifying respirator with a HEPA filter, gloves, and impervious coveralls would be a minimum.

Detection method

Scott's reagent, cobalt thiocyanate, will effervesce vigorously in the presence of ricin. Small amounts of ricin could be detected using ninhydrin, the Amino Acid Test. Exposure could be determined by antibody production. There is also a ticket detection system which is very specific to ricin.

Decontamination

Remove contaminated clothing. Wash skin with soap and water. Wash all surfaces with undiluted household bleach, ensuring a minimum contact time of 10 minutes. Wash the area with soap and water followed by rinsing with copious amounts of water. Ricin is persistent in the environment.

Symptoms

Inhalation: Cough, tightness of chest, difficulty breathing, nausea, muscle aches, cyanosis, death.

Ingestion: Nausea; vomiting; bleeding in the gastrointestinal tract; failure of liver, spleen, and kidneys; death.

Injection: Marked death of muscles and lymph nodes near site of injection; failure of major organs; death. Symptoms depend on the route of entry. Inhalation: respiratory distress and airway and pulmonary lesions; Ingestion: gastrointestinal signs and gastrointestinal hemorrhage, with necrosis of liver, spleen, and kidneys; intramuscular intoxication: causes severe localized pain, a feeling of weakness in about 5 hours; in 15 to 24 hours—temperature, nausea, and vomiting. After 24 hours, death from vascular collapse and shock will occur. The white blood count can be as high as $33,200/mm^3$.

Antidote

None. Oxygen therapy, with or without ventilation, may relieve symptoms.

Storage

Ricin is very stable and would not be difficult to store indefinitely just about anywhere.

Dispersal

Ricin is toxic by all three routs of entry. Traditionally, it is injected; however, mixing it into *DMSO and spreading it on touch surfaces has been suggested in the *Poor Man's James Bond*, and apparently tried. It has been suggested that this does not work very well. Also suggested is a cool explosion such as the use of dry ice and plastic bottles, possibly with acetone peroxide, using flechettes. Flechettes are grooved metal spikes. The grooves allow the spike to carry the toxin and easily disperse the toxic in an explosive. Aerosolizing the compound could be used for inhalation.

Methyl mercury

Variation 1

Chloromethane gas
Magnesium shavings
Mercuric chloride
Ether
Calcium chloride

Variation 2

Iodomethane gas
Magnesium shavings
Mercuric iodide
Ether
Calcium chloride

Due to the extreme hazard of chemical weapons, we provide further information that might become necessary if a person making one of these compounds is encountered.

Methyl mercury (Dimethyl mercury)
$(CH_3)_2Hg$
593-74-8
UN 1992

Appearance: Colorless liquid.

Odor: A faint sweet smell.

Hazards: Lethal. Absorbed through the skin rapidly, 1 or 2 drops can be fatal. LD_{50} less than 1 mg/kg. Reported that 1 or 2 drops on the gloved skin can cause levels of mercury in the body that are 80 times lethal. Causes metal fume fever by inhalation or ingestion as well as irreversible central newous system damage (one of the most poisonous neurotoxins known). Methyl mercury is also extremely flammable, flash point 5°C.

Incompatibilities: Strong oxidizing agents.

Regulations: Hazard Class 3; PG I.

Noncatagorized poison

Methyl mercury is NOT listed as a chemical weapon, although it would certainly be effective as such.

History

In 1971, at Minimata Bay in Japan, 700 people were poisoned with methyl mercury created naturally by life in the bay from discharged mercury waste. In 1971, in Iraq, 500 people died and more than 6000 were admitted to the hospital from eating wheat treated with methyl mercury as a pesticide. The seeds were to be used for planting only. In 2000 a researcher working with methyl mercury was killed when a drop went through her gloves.

Detection method

The Lumex® mercury detector, which is made by a Russian company, is available off the shelf. This device with a minor modification from the manufacturer can detect the presence of methyl mercury in commercial fish.

Decontamination

Organic mercury can remain in soil for 1000 years. Contaminated soils and objects must be removed.

Symptoms

The target organ is the central nervous system. The symptoms may appear weeks to months after exposure to toxic doses. There will be a progression of symptoms. Numbness and tingling of lips, mouth, hands, or feet; constriction of visual field, blurred vision, blindness; deafness; and death.

Antidote

None. Immediate removal.

Storage

Methyl mercury is fairly stable and can be stored indefinitely in glass.

Dispersal

There is no history of any weapon of mass destruction like methyl mercury. Deaths would be slow and painful and would definitely tie up local health services for a considerable time. Methyl mercury could be distributed in food. Symptoms would not become apparent for months, and a significant population could be poisoned prior to discovery of the distribution method.

Nicotine Extraction

Variation 1. Death Dealer

Chewing tobacco
Water
Bandanna or handkerchief

Variation 2. 95% Recovery

Tobacco leaves or chewing tobacco or pipe tobacco or
Many cigarettes
Short-chain organic acid
Acetic acid or formic acid
Ammonium hydroxide

Variation 3. CTRI

Tobacco leaves or chewing tobacco or water
Resin column
Ammoniacal alcohol
Sulfuric acid

Due to the extreme hazard of chemical weapons, we provide further information that might become necessary if a person making one of these compounds is encountered.

Nicotine
$C_5H_4NC_4H_7NCH_3$
54-11-5
UN 1654

Appearance: Oily (viscosity 4 or 5) liquid, which can be colorless or pale yellow to brown, depending on how old it is and how it was made; miscible in water. Variation 1 will give a sticky brown liquid. Nicotine is available commercially in the pesticide Black Leaf 40® as a 95% solution of nicotine sulfate, which is used to kill plant-sucking insects such as aphids, lice, and mites.

Odor: Pyridine, or slight fishy odor.

Hazards: Quickly lethal if absorbed through the skin. May be fatal if inhaled, PEL 0.08 ppm, Vapor pressure 0.04 mmHg @ 20°C or swallowed, LD_{50}(rat) 50 mg/kg. (There are some claims that the LD_{50} for human is 0.1 mg/kg, which is the equivalent of 60 mg/

150-lb person.) Nicotine ingested will cause vomiting nausea, headaches, and difficulty breathing, possibly seizures, due to excessive stimulation of the cholinergic neurons (similar to the effects of cholinesterase) By comparison, Strychnine is 75 mg and arsenic is 200 mg. A cigar contains sufficient nicotine to kill a man; a cigarette can be very toxic to children; one cigarette contains about 1 mg of nicotine.

Incompatibilities: Strong oxidizers and strong acids.

Regulations: Hazard Class 6.1.

Toxic Plants

Toxic plants are a readily available source of very effective poisons. Therefore, this book could not be complete without some discussion of toxic plants. Setting up a classification proved to be difficult, as there are so many criteria for toxic plants that no two published lists of toxic plants agree. Many other references on this subject exist and should be consulted if you are interested in a complete list of toxic plants.

Some of these plants are also listed under hobbies. Having a large amount of one of those plants can in no way be an indication of a potential threat. Only clear intent would have any significance. However, all of the plants below have been used with a criminal intent.

This is a short list of toxic plants. The criteria that we used is as follows:

1. The plant must occur in the United States.
2. The effects of the poison are fairly quick and are difficult to counteract.
3. Historically, poisonings have occurred.
   a. The plant is either palatable or toxic at levels not detected by taste or smell.
   b. The plant can be extracted to concentrate the poison, or added to food without notice.

Many of these plants are listed with full descriptions in Section 2 under the common name. Where possible, a picture of the plant is included in that list. Below is a list of all the poisonous plants considered in this book by scientific name, with the common name.

Vegetable poisons tend to decay with the body, and are usually undetectable after time.

Aconitum napellus
Monkshood, friar's, garden wolfsbane

Atropa belladonna
Belladonna, deadly nightshade

Conium maculatum
Hemlock

Convallaria majalis
Lilly of the valley, May lilly

Datura stramonium
Jamestown weed, gymson weed, apple of Peru, stinkweed, locoweed

Juniperus sabina
Juniper

Nerium oleander
Oleander

Nicotiana tobacum
See the extraction of nicotine above; tobacco

Taxus baccata
Yew

Toxic chemicals

This is also a short list. I have not included chemicals that are no longer available. I have tried to list chemicals that are difficult to detect by taste or in an autopsy. (Autopsies in smaller counties or towns may not be as thorough as those in larger counties or cities, and if there is no particular reason to suspect poison, there may be no attempt to look for it.) Chemicals such as cyanide, strychnine, and arsenic have been used so often to poison people that they are included even though their detection is easy.

Abrin

Abrin is reputed to be 75 times more toxic than ricin. Abrin bonds to transport proteins on cell membranes, which transport it into the cell, where it deactivates ribosomes. One molecule of abrin will inactivate up to 1500 ribosomes per second. Abrin is extracted from the jequirity bean (probably the same process used to extract ricin, which abrin is very much like) to make a yellowish-white powder. The powder has been made into pellets.

Although the jaquirity bean is a poisonous plant, because it is not found in the United States, we have not included it in the section on poisonous plants. The Jequirity bean is found in Brazil. Ricin is the only plant toxin that occurs naturally in large quantities. A large cash of jequirity beans would definitely be an indicator of terrorism.

These beans must be imported illegally. This often happens by accident, as tourists in Brazil will purchase the beans in handmade jewelry on the street and unwittingly bring them home. It is certain that a knowledgeable person could obtain a number of these beans by doing the same. It is only with reluctance that we have included abrin here, as it seems a lot of trouble to go through to obtain a toxic material. However, the CRS Report for Congress listed this as feasible.

The beans can be swallowed whole as long as they have not been punctured, or chewed. However there are reported cases of people making necklaces from the beans pricking their fingers after pushing a needle through a bean to make a hole in order to string the bead, and dying. This would not be true of castor beans.

Alkali cyanide salts

Cyanide is easy to obtain, and can be made using the formula under "Chemical Weapons, Facility 8." Cyanide is very easy to detect in ingestion and inhalation, and in the postmortem. The primary attraction for alkali cyanide salts is the quick time to death.

Amatoxin

Mushrooms are beyond the scope of this book; however, since this commodity is fairly easily obtained and extremely toxic, we have included it here for information only. The identification of mushrooms is very difficult and requires an expert. The toxin in the death cap mushroom is amatoxic and must be ingested. Amatoxin destroys the liver and kidneys over several days while the victim remains conscious in excruciating pain. The spores of this mushroom are white, which is an identifying mark, and can be seen by holding the mushroom cap over a piece of black paper and thumping it.

Antimony and its salts

Antimony and many of its compounds are poisonous. Antimony poisoning is very similar to arsenic poisoning. Small doses have been administered in fruit drinks, where the acidity of the drink is sufficient to dissolve the antimony or salt.

Arsenic and its salts

Arsenic has been the poison of choice for years, as it is not detectable by taste. It is a nonmetallic element. It is a silver-gray brittle crystalline metallic appearing solid, which darkens in moist air. Arsenic and its salts are highly toxic by ingestion and inhalation.

Compound 1080

One teaspoon of Compound 1080 (sodium monofluoroacetate) could kill 100 people. Death is slow; symptoms occur an hour or two after ingestion. Symptoms include vomiting, involuntary hyperextension of the limbs, convulsions, and collapse. There is no antidote. It is colorless, odorless, and tasteless. It dissolves easily in water.

Mercury and its salts

Mercury metal is highly toxic by skin absorption and inhalation. All soluble or slightly soluble inorganic compounds of mercury are highly toxic by ingestion, inhalation, and skin absorption.
Most organic compounds of mercury are also highly toxic.

Paraquat

Paraquat is a poison that can easily be put into any salty food. It could also be put into cigarettes. It has been used in this manner to poison people. It would, for example, be unnoticeable in a bowl of pea soup. This poison will initially make a person very sick, a condition from which he or she will recover, but which will eventually cause death by fibrosis of the lung. The speed with which the fibrosis kills is increased by oxygen, so as a person loses lung capacity, there is no way to help. This is fatal.

Polonium

This is a very toxic, slow-acting poison that is difficult to diagnose and has no antidote. It is included here only because of the death of Russian anti-spy Alexander Litvinenko in November 2006, when the compound was sprayed on his sushi. The way that the newspapers reported it, everyone could purchase a static-free brush and poison their mother-in-law. Within days there were recipes on the Web to retrieve polonium from the brushes for the use of poisoning. The brushes contain only microcuries of polonium, and since the half-life is 138 days, it is likely that the brushes may already contain as much lead. The lethal dose is reputed to be 300 mg, which would require thousands of the brushes, which at this time are about $40 each. Against making use of this element just to kill your mother-in-law is the fact that once the capsule inside the brush is opened, about one-third of the polonium will evaporate off in a day. There are reported accounts of large amounts of polonium having been stolen, but with a half-life of 138 days, they would quickly disappear. Polonium is very soluble in acids; polonium nitrate is water soluble. Although there is no mention of taste, since polonium is considered a metalloid, it is likely that there is no taste. As in the case of ricin, the Russians appear to be the leaders in new and unusual toxins.

Polonium
Nitric acid

Potassium chloride

Potassium chloride has an LD_{50} of 2500 mg/kg, which for a 190-lb person would be about 6.5 oz, which is about the same as table salt. It is consumed in such foods as Gatorade® where it is the electrolyte. In this case it is the route of entry that determines the real hazard. Injected potassium chloride has an LD_{50} of 100 mg/kg, which is sufficiently low that potassium chloride is used for lethal injections. This poison would be a little tricky to find on autopsies.

Selenium and its salts

Selenium is not always detected or even looked for postmortem. Selenium is a black powder, which can be added to mascara or added to pepper (both methods have been used). The salts are usually white and look like table salt. This is not a fast poison, but the symptoms are difficult to diagnose, being in some cases similar to congestive heart failure. Sometimes if produces blind staggers. Some selenium compounds will off-gasdimethyl selenide after ingestion. This gives a strong smell of garlic on the breath of a sick person or from a body.

Strychnine

Strychnine is one of the bitterest compounds in the world. Its taste is detectable at concentrations as low as 1 ppm. Despite this, it has been used as a poison for years. Strychnine seems to be the poison of choice for people who kill dogs. The approximate lethal dose for a dog is 0.75 mg. It is an urban legend that strychnine is used to cut herion.

Thallium and its salts

Along with arsenic, thallium sulfide has been called "inheritance powder." It takes about 1 g to be lethal, but this can be given over a period of time. Almost an ideal poison, its salts are colorless, odorless, and tasteless, as well as being freely soluble in water. In the body the salts are converted to thallium chloride, which is somewhat insoluble and therefore slow-acting. General symptoms of nausea and intestinal problems are rarely noticeable before 12 hours, and may be delayed as long as 48 hours. It is 2 to 5 days before specific symptoms occur. A characteristic symptom of thallium poisoning is significant loss of hair, which comes out in tufts. A characteristic sign of thallium poisoning is the *signe de sourcil*, a loss of hair on two-thirds of the eyebrow, leaving only the inner third intact. Thallium can easily be detected in the urine.

Toxins found in common household products

This area was difficult to explore, as no one advertises the fact that their product contains compounds that have LD_{50} values that are close to or below cyanide. There is, however, some information on these products circulating in the underground or on the Web. It is difficult to say what would indicate that an illegal use of these compounds was being contemplated. Vicks® Sinex, Afrin, Nasin, Afrin Extra Moisturizing, Afrin Sinus, Afrin w/ Menthol, Muramist Plus, 4-Way Long Lasting, Genasal, NRS-Nasal Relief, Sacon, Neo-Synephrin 12 hour, Nostrilla, Duration, and Good Neighbor Nasal Spray pump all contain oxymetazoline·HCl at a content near 0.05%.

Oxymetrazoline·HCl
$C_{16}H_{24}N_2O·HCl$
2315-02-8
UN 2811

Appearance: Solid white crystalline powder.

Odor: None.

Hazards: A classic example of how important dose and route of entry are. As a 0.05% solution this is used as a nasal decongestant. Looking at the LD_{50}(rat) value of 0.63 mg/kg and the NFPA Health rating of 4, this is obviously very toxic material. Although oxymetrazoline is not considered to be a great hazard for skin contact (strong irritant), ingestion of the compound can easily be fatal. Like all the zolines, this compound activates the adrenaline system, creating a hazard to the heart.

Regulations: Hazard Class 6.1; PG I; toxic solid, organic n.o.s.

Visine®
Tetrahydrozoline·HCl
$C_{13}H_{16}N_2·HCl$
522-48-5

Appearance: Called "Mickey red eyes." Although this can be used safely in the eyes, one or two drops in an alcoholic beverage theoretically can be lethal. (We could find no instance where anyone was actually killed using this product.) Probably included as a toxic product because of an urban legend that Visine will cause immediate diarrhea, which is not true, but like all zolines, it does cause the adrenaline system to be activated, which can be deadly.

Odor: Not found.

Hazards: Very hazardous in case of ingestion. LD_{50}(mouse, rat not found) 345 mg/kg; NFPA Health 2.

Toxins derived from bacteria

These include some of the most toxic compounds known. In selecting which toxins to include, we have followed the *CRS Report for Congress* criteria for "ease of acquisition" for inclusion of toxins. However, we have not included bacteria, which are found only on foreign soil such as Brazil or Arabia. For a complete list, consult any WMD document.

Clostridium botulinum

Clostridium butlinum is an anerobic, gram-positive, spore-forming rod that produces a potent neurotoxin. The LD_{50} of the neurotoxin by ingestion is estimated to be 2 ng/kg, and by inhalation, 3 ng/kg. This is considered to be the most toxic compound known. This very low level of toxicity is probably due to the fact that the toxin acts a little like a hormone in the body, which amplifies its activity. It works by inhibiting the motor nerves to release acetylchloline.

There are on average 30 outbreaks a year in the United States. There is no indication that this toxin has ever been used with criminal intent. If it was used, they got away with it. There is some speculation (unsupported) that botulinum toxin was used as a biological weapon in World War II in the Heydrich assassination.

However, since the most toxic compound known can be made accidentally, it seems likely that someone might consider trying to make it purposefully. If they were trying to make this toxin, the following information might be found. Animals most commonly affected are wild fowl, poultry, cattle, horses, and some species of fish. Dirt collected around these animals may contain the bacteria. Botulism has been seen with the preservation of meat, fish, peppers, green bears, beets, asparagus, mushrooms, ripe olives, spinach, tunafish, chicken livers, luncheon meats, ham, sausage, stuffed eggplant, lobster, and smoked and salted fish. The food product can never be heated above 80°C for a period of 10 minutes or longer. Other than that, it need only be sealed in some manner to prevent contact with the air. In some cases, plastic bags were sufficient.

Genetic research on botulinium is against the law in most Western nations, including the United States; however, there is no international law against experimenting with weaponizing botulinum toxin, according to work published during World War II. The first step to weaponize botulinium would be crystalization.

Acid precipitation, no acid specified.

Dissolve in buffer solution, not specified.

Ammonium sulfate to produce crystalline form.

Although the crystalline form is more than adequate for weapons, the crystal can be further purified using sodium dodecyl sulfate polyacrylamide gel in electrophoresis.

Shigatoxin

Shigellosis is an infectious disease caused by a group of bacteria called *Shigella*. Most people who are infected with *Shigella* develop diarrhea, fever, and stomach cramps, which start a day or two after exposure to the bacterium.

OTHER ACTIVITIES INVOLVING HAZARDOUS CHEMICALS

Circuit board reclamation

Sodium cyanide
Mercury

DIRTY BOMBS

A dirty bomb or radiological dispersion device (RDD) uses conventional explosives to disperse radioactive elements. The radioactive materials needed to build a dirty bomb can be found in almost any country in the world. In eastern Europe there are many "orphan" sources. Most countries have little or no regulation, and even in the United States, where there is significant regulation, 1500 sources have disappeared since 1996, and almost half of those have still not been found. At this time there has only been one attempt to deploy a dirty bomb: in 1996 at Izmailove Park in eastern Moscow.

There is considerable difference of opinion as to whether a dirty bomb would be particularly effective. Those that feel that they are not particularly effective point to Chernobyl, where cancer rates spiked but quickly dropped to normal as time passed. They also point to the limited spread of radioactive materials. Those that feel it would be very effective point to events in Goiania, Brazil, where the discovery of discarded cesium-137 affected hundreds, killing four. There is some consensus that the most serious effects would be social disruption and associated economic costs.

The highest-risk materials are gamma ray–emitting elements, the most common of which in industrial use are cesium-137, cobalt-60, iridium-192, and strontium-90. Cesium-137 is easily considered to be the highest risk element, as cesium is one of the most reactive elements; if it were spread into a civic area, it would react with such things as the roofing and plants, making it very difficult to decontaminate.

Elements of concern

Americium-124
Californium-252
Cesium-137
Cobalt-60
Iridium-192
Plutonium-238
Radium-226
Strontium-90
Uranium-238 (depleted uranium)

PATHOGENS

Pathogens are beyond the scope of this book. We have included some information here, but prefer to stay with chemical aspects. Growing pathogens is mostly a matter of technique, and it is very unlikely that there will be labeled containers or original stock. Check the agar type, which may help in determination of what is being grown.

Chemicals and equipment that might be used when working with pathogens

Glove box
Dewars of nitrogen liquid
Cylinders of Freon®
Bentonite
Agar
Carbon dioxide gas
Bicarbonate of soda

Agars used to grow various pathogens

Blood agar	Anthrax
Bile esculin agar	Steptococci
Chocolate agar	Anthrax
Enriched media	*E. coli*

Hektoen enteric	*Salmonella* and *Shigella*
Onoz	*Salmonella* and *Shigella*
Peptone	*E. coli*
Phenylethyl alcohol agar	*Staphylococcus*
Salmonella agar	Salmonella
Soy agar	General purpose
Trytptic	General purpose
Thayer–Martin agar	Gonorrhea
Yeast extract agar	*E. coli*

Pathogens considered as possible chemical weapons

Anthrax (*Bacillus anthracis*)

This is easy to acquire and easy to disseminate; however, it is very difficult to weaponize.

Brucellosis (*Brucella abortus*)

This is easy to acquire and easy to disseminate.

Cholera (*Vibrio cholerae*)

Escherichia coli

This is easy to acquire and difficult to disseminate. This is the leading cause of foodborne illness. It is found on most cattle farms and often in petting zoos.

Glanders (*Burkholderia mallei*)

This is easy to acquire and easy to disseminate.

Pneumonic plague (*Yersinia pestis*)

This is easy to acquire and easy to disseminate.

Meliodosis (*Burkholderia pseudomallei*)

Yellow fever

This is easy to acquire and easy to disseminate.

Q fever (*Coxiella burnetti*)

This is easy to acquire and easy to disseminate.

Salmonella typhimurium

This is easy to acquire, and difficult to disseminate. *Salmonella* is one of the only pathogens that have actually been employed in the field. The incident in Oregon with the Rashneesh is usually quoted when discussing pathogens. The attack made hundreds of people sick, but the event was not successful as a terrorist event in that no one knew that it had occurred. *Salmonella* is associated with chicken, birds, and eggs.

Typhus (*Rickettsia prowazekii*)

This is easy to acquire and difficult to disseminate.

HOBBIES THAT USE SUSPICIOUS OR HAZARDOUS CHEMICALS

Beer (general)

Basic formula: can be made in a coffee pot
Malted barley
Water
Yeast

Variation 1. Beer

May contain one or more of the following grains:
Rye
Barley
Wheat
Corn
Rice

Other materials used to make beer

β-Glucanase
Yeast
Hops

Filtering materials

Diatomaceous earth
Kieselguhr paper filters

Variation 2. Ale

May contain one or more of the following grains:
Rye
Barley
Wheat
Corn
Rice
Product may be dark yellow to black.

Supplement

β-Glucanase
Yeast
Magic
Carbonated soda drinks
Menthol

Entomology

Sodium cyanide
Chloroform

Gardening

This section contains only plants that are contained in other sections of this book but are mainly thought of as garden plants. For instance, as toxic as rhododendrons and junipers are, there is little mention of such in the literature.

Belladonna: Belladonna is not common as a garden plant and is considered a weed in some areas. Best growth is under conditions not normally considered for gardens.

Castor bean: The plant is grown commercially in California; it is often cultivated as an ornamental annual in gardens.

Junipers: A very popular low-spreading ground cover as well as an upright conifer.

Lily of the valley: This is a popular garden plant grown for the scented flowers.

Oleander: The bush is cultivated as a decorative plant in many countries all over the world. It has been introduced as an ornamental shrub in the southern United States. In the northern United States it is grown as a houseplant. It is very prevalent in California.

Peyote cactus: The millions of cacti that are harvested each year go into the nursery industry. Despite being a Schedule 1 drug, peyote is found in many gardens. It is a spineless, low-growing cactus and very useful in landscaping.

Poppies (opium): Poppies are sold over the Net and are very popular for their many bright colors. Occasionally, they are even encountered growing wild in vacant lots and fields. Opium poppies are grown commercially for their seeds, which are used in baking.

Rhododendron: This is a very popular plant. It is the state flower of West Virginia, and rhododendron clubs are found in every state. This is a garden plant that embellishes many parks.

Gold (Silver) mining

Variation 1

Sodium cyanide

Variation 2

Mercury metal

Variation 3

Sodium bromate

Photography

One or all of these chemicals may be found:

Acetanilide	Beeswax (white)
Aluminum sulfate	Benzotriazole
Amidol	Borax
Ammonium bromide	Boric acid
Ammonium chloride	Cadmium bromide
Ammonium citrate	Cadmium iodide
Ammonium dichromate	Calgon
Ammonium hexachloropalladate(IV)	Canadium balsam
Ammonium iodide	Catechol
Ammonium persulfate	Chloroauric(III) acid
Ammonium sulfate	Chloroform
Ammonium tetrachloroplatinate(II)	Citric acid
Ammonium thiocyanate	Collodion
Ammonium thiosulfate	Copper(II) chloride
Ascorbic acid	Copper sulfate

Ethylenediamine tetraacetic acid
Ferric ammonium oxalate
Ferric oxalate
Ferrous sulfate
Gelatin
Glycerin
Glyoxal
Gold chloride
Gum arabic
Gum sandarac
Gum sundarac
Hexachloroplatinicd(IV) acid
Hydroquinone
Hydroxylamine sulfate
Iridium(IV) chloride
Lampblack
Lanolin
Lavender oil
Lithium chloride
Metol
Nitric acid
Oxalic acid
Palladium on barium sulfate (5%)
Palladium on charcoal (5%)
Palladium(II) chloride
Palladium chloride
Phenazone
Potassium aluminum
Potassium bromide
Potassium carbonate
Potassium chloride
Potassium citrate
Potassium dichromate
Potassium ferricynide
Potassium formate
Potassium hexachloropalladate
Potassium hydroxide

Potassium iodide
Potassium metabisulfite
Potassium nitrate
Potassium oxalate
Potassium tetrachloropalladate(II)
Propylene glycol
Pyrocatechol
Pyrogallic acid
Pyrogallol
Resorcinol
Rodinal
Rhodium(III) chloride
Ruthenium(III) chloride
Selenium dioxide
Silver chromate
Silver nitrate
Silver oxide
Silver(I) selenide
Sodium acetate
Sodium ascorbate
Sodium betanborate
Sodium bicarbonate
Sodium bisulfite
Sodium bromide
Sodium carbonate
Sodium hydroxide
Sodium metabisulfite
Sodium phosphate
Sodium sulfide
Sodium thiocyanate
Sodium thiosulfate
Sodium tungstate
Sulfamic acid
Tannic acid
Thiourea
Thymol crystals
Triethanolamine

Photography on glass

Sodium cyanide

ROCKETS

Relates to the importation, distribution, and storage of fireworks classified as UN 0336, UN 0337, UN 0431, or UN 0432 explosives by the U.S. Department of Transportation at CFR 172.101 and generally known as "consumer fireworks" or articles pyrotechnic." Rocket fuels are hard-classified UN 0431 (1.4G Articles Pyrotechnic for Technical Purposes) (former Class C and thus exempt under solid as well).

Variation 1. Commercially Available Rocket Fuels

Ammonium perchlorate
Nitrogen oxide

Variation 2. RDX, rocket propellent

Hydroxyl-terminated polybutadiene (HTPB, R-45 M, Alco Co.)
Dioctyl adipate
Triethanolamine
Polybutadiene (Nippon Zeon Co. Ltd)
Isophorone diisocyanate (VEBA Chemical Co.)

Variation 3. High-Performance Rocket Fuel

Kraton 1107 (plastic elastimer)
Toluene
Aziridine BA 114
Aluminum powder
Ammonium perchlorate

Variation 4. High-Performance Rocket Fuel

RB-810
Dioctyl adipate
Lecithin
Ammonium perchlorate

Variation 5. High-Performance Rocket Fuel

Polyethylene glycol
NENA
Ammonium perchlorate
Ammonium nitrate

Variation 6. DIANP Propellant

Nitrocellulose
Ethyl cellulose
2-Nitrodiphenylamine
RDX
DIANP

Variation 7. Fireworks Rocket

Potassium nitrate
Fine charcoal
Charcoal (80 mesh)
Sulfur
Titanium

Variation 8. Sport Rockets

Ammonium perchlorate
Aluminum (200 mesh)
Hydroxy-terminated polybutadiene

Variation 9. Liquid Rocket Fuels

Ammonium thiosulfate

Variation 10. Rocket Fuel Used on Space Shuttle

Ammonium perchlorate
Iron oxide
Magnesium

ROOT BEER

Root beer is included here only because it can use sefrole, or safrole. These are used to
make root beer, also these are the main ingredients in the manufacture of Ecstasy.

General Formula

Sassafrax
Extract
Any of the following sugars:
Sucrose
Corn syrup
Brown sugar
Fructose
Honey
Molasses
Malt extract
Alcoholic root beer (2%)
Yeast

Additives

Must have
Vanilla
Wintergreen (leaves)

Could have
Ginger
Licorice
Sarsaparilla

Other possible additives
Birch bark
Dog grass
Ginger
Juniper berries
Chirreta
Hops
Yerba maté
Water

TAXIDERMY

Like so many other hobbies, taxidermy is an art. Like fireworks and root beer, there are
basic formulas, but everyone will find what works best for them and then jealously guard

their "secrets." There are different methods suggested for different animals, and different formulas that work better, for certain animals. A clue that animal preservation is occurring is that at the location there will be many examples of the work. A remarkable collection of odd knives and scrapers will also indicate this particular activity. However, at dump sites there may be no such clues.

Smaller animals are now more commonly freeze-dried, which might be sufficient for biologists making collections, but proper mounting may improve the quality of the specimen. Some people collecting specimens may also be collecting insects; *see* Entomology. Making trophies from fish, or preserving fish, often includes paint and urethane plastics.

A lot of this waste may look as though it came from a restaurant. Whey grains are leached of their organic acids to effect the tanning process.

Cleaning/softening

Borax
Formic acid
Sawdust
Soap
Soap bark

Fleshing

Look for unusual scrapers and knives
Technical butter
Bran
Sodium hydroxide
Chalk
Calcium chloride
Lactic acid

Converting to leather

Alcohol
Potassium carbonate
Steric acid
Gelatin
Tannin

Pickling method

Sodium chloride
Sulfuric acid
Tanning liquor

Schrot-beize

Wheat
Barley
Bran
Rye
Whey
Sodium chloride

Mineral tans

Alum or aluminum sulfate (chromium salt)
Alkali carbonate

Alum tan

Aluminum sulfate
Potassium sulfate
Sodium chloride

Chrom tan

Chrom alum

Chamois tan

Animal fats or mineral or vegtable oils
Linseed oil
Vasoline®
Butter
Egg yoke

Formaldehyde tan

Neradol D (sulfonated phenol formaldehyde)
Tanning liquor

RECON–RENDER SAFE OPS–SAMPLING–TAKE DOWN/DISMANTLING*

All small-scale-production illicit operations have elements in common that will enable a hazmat response team to perform "Recon, render safe ops, sampling, and take down/ dismantling" in a systematic manner that ensures the best chance of identifying the purpose of the operation, does not destroy evidence, and secures scene safety.

Typical objectives for a response team to an operation could be to identify and quantify all hazardous materials within the site, to describe any processes that are involved, and to ascertain what the perpetrators are attempting to produce.

Depending on the background intelligence known prior to a response team's first entry, the focus of recon efforts will ideally be the room containing the operation. Recon priorities may be stated as follows:

1. Monitor any places where air exchange with the operations room could occur.
2. Enter the operations area and attempt quickly to determine the type of product being synthesized.
3. Based on the determination above, monitor the operation environment for a release.
4. Document and "render safe" the processes.
5. Upon exit, if a WMD (weapons of mass destruction) agent release is deemed probable, mitigate to the extent possible by sealing the room, ceiling vents, windows, etc.: any places where air exchange might occur.

If the location of the operation within a building is known; decisions such as the manner in which to conduct the initial recon (left- or right-hand search, etc.) should be considered

*Material used with the permission of Dan Keenan, Oakland City Fire Department.

carefully so as to guarantee that the high-priority location in the building is the first room entered. These approach recon efforts should be concentrated on the place in the building where the largest hazard and potential for a release exists. Time on-air places considerable restrictions on hot-zone ops. If SCBA is used instead of rebreathers or umbilicals, time on-air will probably be the most limiting factor and should be used wisely.

Let's examine the recon Priorities set forth above in greater detail:

1. Monitor any places where air exchange with the operations room could occur. During a 360° recon made of the building's perimeter, special attention should be paid to windows, doors, and vents connected with the operations room. Recon objectives should follow a checklist of potential air-exchange locations that prompts the response team to monitor and record results in a logical manner.

2. Enter the operations area and attempt quickly to determine the type of product being synthesized.

    Upon entering operation-flammable atmospheres, IDLH atmospheres and life-threatening radiation sources should be ruled out. With this hurdle passed, the entry team gains access to the operations room. Radiation detectors, combustible gas detectors, etc., get you into the room and allow you to work knowing that these particular life hazards are not a concern.

    The key to determining the type of chemical being synthesized is to hunt for suspicious elements, specific precursors, and the end product itself. Systematic sampling and screening will not occur until the second entry, so the recon entry team will have to rely on quick screening of "samples of opportunity," discovery of labeled precursor reagent bottles, and documents revealing synthetic routes.

    Hunting for suspicious elements means looking for precursors that contain any of the following elements: arsenic, Fluorine, chlorine, phosphorus, nitrogen, sulfur, and/or the cyanide functional group for chemical weapons; iodine and phosphorus for methamphetamine; nitric acid and sulfuric acid for explosives; ammonium perchlorate and sulfur for pyrotechnics. Certain combinations will point toward a specific type of chemical being synthesized.

3. Based on the determination above, monitor the operation environment for a release. Instruments such as the ICAM, Chem Agent Alarm, APD 2000, SAW Mimicad, and detection papers M8 could be used to provide vague hints as to the presence of the type of chemical operation. Given the large number of interferences from organic solvents that could be present in the operation, any positive readings on these monitors must be viewed with suspicion. Tools better suited to narrow down the search include the M256 kit or the Draeger CDS kit, which has specific tubes for phosphoric acid esters. Monitors such as the ICAM were not designed to be used in the organic synthetic laboratory environment.

4. Document and "render safe" the processes.

    As part of recon the operation should be studied and divided into processes that are separate from each other. All temperatures and pressures that can be, must be documented prior to shutting off heating and vacuum systems (i.e., "render safe" ops). The importance of documenting temperatures and pressures is derived from the fact that these numbers can identify distillates and constitute important evidence of the purpose of the operation. Ideally, the render safe ops should also be conducted by the recon team (on first entry) so that the glassware is given time to cool off before samples are drawn.

    A simple "left-to-right" shutdown of additions, heating, vacuum, and stirring accompanied by cooling maintenance (i.e., add more ice, keep water pump operating) should complete the render safe ops. Once again this is a job for the recon team (first entry), not the screening/sampling team. Much has been written regarding how to deal safely with an operation that is currentlty operating. Some of these documents suggest turning off all the electricity to the building. Although this might help in some cases, it could also cause a runaway reaction. If addition funnels were not shut off, exothermic reactions would potentially continue to occur without the benefit of temperature control, cooling water pumps, etc. Also, if electricity is shut down, the chance of documenting the operating temperatures and pressures is lost, and entry will have to be performed using portable lighting. The other approach often touted, which is equally bad, is to leave the operation running as is. This could be very dangerous. If there is any chance of an operation being in a critical stage, it is all that more important to get in as quicly as possible to render it safe. Delaying render safe ops may result in one of the reaction vessels undergoing distillation to boil dry, overheating, catching on fire, even exploding. Any number of not-so-good outcomes could lead to the need for a hasty retreat from the building. The best approach is to conduct a 360° recon of the operations's building perimeter and then, upon entering, go straight to the operation, examine it, document temperatures and pressures, and render it safe.

5. Upon exit, if a WMD agent release is deemed probable, mitigate to the extent possible by sealing the room, ceiling vents, windows, etc: any places where air exchange might occur. If the end product was an explosive, prepare for the possibility of detonation. At a minimum, by the end of recon the entry team should be able to say definitively if a gas or vapor had been released in the area of the operation, based on their monitoring results. Is it necessary to seal the room in order to mitigate and shrink the hot zone?

A comprehensive detailed recon entry, complete with pictures and diagrams, if properly debriefed, sets the stage for a more effective second entry, wherein screening/sampling ops should be performed. Based on recon photos and diagrams, a systematic screening/sampling plan should be developed with input from a subject-matter expert who has a thorough familiarity with basic and advanced synthetic organic chemistry laboratory techniques. From the photos taken of the lab, a list of vapor, liquid, and solid samples should be developed.

Comprehensive sampling ensures that smoking guns are not missed. There is no guarantee that reagent bottles and household containers are holding what their labels declare. A criminal might attempt to disguise an operational setup to appear to be an essence-oil extraction process or other innocuous hobbyist venture. An all-inclusive sampling approach allows for a complete characterization of the operation's process.

The contents of cooling/heating baths are important because they are evidence of the temperatures the operator was trying to maintain; for example:

- Potassium chloride/ice baths are used for temperatures ranging from° −10 to 0°C.
- Calcium chloride/ice baths are used for temperatures ranging from −55 to 0°C.
- Dry ice/acetone baths are used for temperatures close to −75°C.
- Mineral oil heating baths are used for temperatures ranging from 100 to 200°C.

The purpose of sampling at this stage should be analytical, not evidentiary. Fairly small sample sizes are sufficient for GCMS analysis, hazcatting®, etc. Typically, the samples taken from operational apparatus are of highly concentrated pure chemicals or reaction mixtures derived from highly concentrated pure chemicals. Sampling should immediately follow screening so that the same sampling tool can be used. If the same tool is used for screening and sampling, the glassware will only have to be opened once, increasing efficiency and reducing the entire amount of time required to accomplish the objective.

ODORS

At no time should anyone intentionally smell anything. Nevertheless, odors are often sensed by those responding to and working at sites, chemical storage facilities, or waste disposal locations. Conspicuous odors are frequently reported to law enforcement personnel or emergency responders by witnesses or informants and by complainants or reporting parties. Like other elements of sensory information, odors should be considered in determining the type of activity that has been encountered and the chemicals likely to be present, with their associated hazards. The following is a list of odors that may be present and commented on by people in the vicinity. These summaries are provided as a means of interpreting odors and to assist in forming opinions as to the processes, chemicals, and wastes found at the scene. The general descriptions of likely odors are listed alphabetically. Remember that describing an odor is a very subjective process. Do not restrict yourself to only one possibility. We report the odors as they are reported, some since antiquity; when we disagree, we report what we smell. Sometimes we do not agree with the reported smell, but cannot give a better description. We have made NO effort to smell all of these compounds, nor would we ever do so intentionally. NEVER VERIFY WHAT SOMEONE ELSE REPORTS BY SMELLING THE AREA YOURSELF!

1. Airplane glue
Toluene

2. Alcohol
Ethanol, isopropanol

Most people can recognize ethanol and isopropyl alcohol. Methanol is not easy to recognize. Alcohols are easily acquired organic solvents and work well in the extraction of ephedrine and pseudoephedrine from pharmaceutical tablets.

3. Almonds
Benzaldehyde, benzoic acid, cyanide

An unpleasant odor similar to that of almonds should be a warning of the possible presence of hydrogen cyanide gas. All personnel should be evacuated until appropriate indicator tubes either indicate the absence of cyanide or respirators are provided. At levels greater than 20,000 ppm, cyanide can be lethal by skin absorption, even to people in SCBAs.

Cyanide indicator tubes should be used to establish the presence or absence of atmospheric hydrogen cyanide gas. Hydrogen cyanide is more associated with the production of chemical weapons. The presence of this gas should be an indication that something has gone wrong. The intentional generation of cyanide gas has been suggested by certain underground publications as a counter-intrusion device. Sharp chlorine-to-bitter almond odor is hydrogen cyanide, the gas released when sodium cyanide is in contact with acid.

Benzaldehyde is more associated with drug manufacture. This odor is most likely to be associated with a P-2-P process or the production of Agent BZ. A strong, attractive almond (some people recognize it as cherry) candy odor may indicate benzaldehyde. This is a very strong odor that most people can detect at low concentrations. Currently, both of these syntheses are rare.

4. Ammonia

Ammonia gas, ammonium hydroxide, ammonium carbonate, aziridine, ethylenediamine, diethanolamine, diethylamine, diisopropylamine, dimethylforamide, isopropylamine, morpholine

The odor of ammonia and amines are often confused. *Also see* Dirty diapers, Cat urine, *and* Dead animals or dead fish.

Indicator tubes are available to establish the presence of ammonia; however, all have significant interferences associated with the presence of the even more hazardous inhalation risk of amines. SCBAs should be worn in ammonia-leak situations. If the levels of ammonia have been quantified below their PELs, air-purifying respirators with NIOSH-approved green cartridges can be worn.

5. Animals (alive with unpleasant odors)

Organic acids

6. Ants and Bees

Formic acid

Most people cannot smell ants and bees; however, those who can will know the odor of formic acid, which is what makes up an ant's string.

7. Aromatic

Allylbenzene, anisol, benzene, Canadian balsam, dioctyl adipate, catechol, hydroquinone, mesitylene, resorcinol, sassafras oil, tymol crystals

Aromatic is a concept that any person working with these chemicals will understand; however, it would be impossible to explain it to someone who has never smelled an aromatic chemical.

8. Bar-B-que to very irritating

Acrolein

At the ppb level, acrolein is one of the odors that comes off a bar-B-que. As the atmospheric level rises, it becomes a very irritating odor.

9. Biology laboratory

Formaldehyde, formaline, paraformaldehyde

Most people will recall the odor as reminiscent of pickled frogs from their high school biology class. An associated hazard is the potential accidental production of bis(chloromethyl) ether by the combination of hydrogen chloride gas in the air with formaldehyde. BCME is one of the most potent and lethal carcinogens known. Formaldehyde can be detected using specific indicator tubes.

A certain percentage of the population cannot tolerate formaldehyde at any level. For most people, formaldehyde has sufficient warning properties to alert people as it approaches its PEL.

10. Biting, harsh

Acids, benzoic trichloride, bis(2-chloroethyl) ether, boron trichloride, boron trifluoride, bromine, bromylbenzyl cyanide, 2-bromoethanol, butyl alcohol, *N*-butyl chloride, 2-chloroacetyl chloride, chloroethanol, chlorosulfuric acid, hydrochloric acid, hydrogen chloride, iodine, mercuric nitrate, methamphethamine base, nitric acid, oleum, phosphorus tribromide, phosphorus trichloride, sodium metabisulfite, stannic chloride, styrene

Respiratory protection should be utilized in the presence of a harsh or biting atmosphere. It may be an indicator of an acid gas at levels greater than its OSHA PEL (for acids the PELs tend to be between 2 and 10 ppm. Most people experience a burning sensation in their upper respiratory system as atmospheric acids reach their PELs. These gases can cause severe lung damage.

Nitric acid will form NO_x (this is sometimes described as a green grass type of odor) when it is in contact with most metals. Hydrochloric acid, which fumes considerably in the standard laboratory reagent grade, is also a likely candidate. Pressurized hydrogen chloride gas is used to salt many of the chemicals discussed in this book. *Also see* Lachrymators.

Styrene, which is organic and has a boat resin odor, can be detected using a combustible gas indicator.

11. Boat resin
Styrene

12. Breath of an alcoholic
Acetaldehyde

Here we are talking about the type of alcoholic that becomes very red in the face. This is a very strong non-alcohol odor. Acetaldehyde evaporates quickly and has a wide flammable range. Consider this odor as a warning of a severe fire and explosion hazard. The presence of this odor indicates the need for monitoring with a combustible gas indicator prior to entry.

13. Camphor
Cyclohexanol, hexachloroethane

14. Cat urine
Phenylacetic acid, amines

This odor is quite subjective (even for those persons who have a cat box in their kitchen). *Also see* Ammonia, Dirty diapers, Dead animals or dead fish, *and* Honey.

For persons with considerable experience in investigating clandestine methamphetamine laboratories, this appears to be the most common street description of phenylacetic acid. Note that honey is also a common way to describe the odor of phenylacetic acid. The odor is permeating and lasts for days.

15. Characteristic (unspecified)
Amyl alcohol, agar, 5-aminotetrazole, amyl nitrite, aniline, γ-butryrolactone, diisopropyl ether, glyoxal, lidocaine, tannic acid

Many chemicals have characteristic easily recognized, yet difficult to describe odors. Generally, once you have smelled one of these chemicals, you will be able to recognize it when you encounter it again.

16. Chlorinated
Benzoyl peroxide, carbon tetrachloride, chloroform, 1,2-dichloro- 2-propanol, 1,2-dichloroethane, methylene chloride, nitroethane, trioxane

Chlorinated is a concept that any person working with these chemicals will understand; however, it would be impossible to explain it to someone who has never smelled a chlorinated chemical before. Note: NOT all of these are chlorinated compounds.

17. Chlorine
Chloramine chlorine, calcium hypochlorite, sodium hypochlorite, sulfur dichloride

Chlorine can do significant damage to the lungs in a very short time. The effects may not appear until several hours later. In some cases the person exposed felt no effects until a later time, when edema nearly caused them to drown in their sleep. A sharp chlorine-to-bitter almond odor is hydrogen cyanide. Indicator tubes are available to distinguish between these two compounds. Appropriate personal protective clothing and SCBAs should be worn in a suspected cyanide or chlorine atmosphere.

18. Chloroseptic®
Phenol

19. Chocolate (burnt)
Pentaborane

If this is at a very high concentration, it is a lachrymator. As such, any person smelling this compound should be on the way to the hospital immediately.

20. Cloves
Eugenol

21. Cinnamon
Cinnamic aldehyde

22. Cyanide
Copper cyanide, cyanide salts, cyanide solutions

1 out of 10 people can smell hydrogen cyanide at the ppb level. For those people who have smelled cyanide previously, cyanide salts and solutions are easily recognizable by smell.

23. Dead animals or Dead fish
n-Butylamine, tert-butylamine phosphine

Phosphine has been reported as having the odor of fish, dead fish, and dead cats by various persons who have encountered it in the research laboratory and in clandestine methamphetamine laboratories. Phosphine is an extremely toxic gas that is not easily monitored in the field. (A quick qualitative method is to put mercuric chloride on a piece of paper and look for a yellow coloration that occurs in the presence of phosphine.) Phosphine causes a brown color change on stripe 6 of the GasTec Polytec IV indicator tube. Death by inhalation of phosphine is very quick. Phosphine has caused a number of deaths to cooks in clandestine methamphetamine laboratories. *Also see* Garlic.

Phosphorus is the catalyst in the reduction of ephedrine using hydriodic acid (Synthesis 1). This method is found in cold cook laboratories, California laboratories, and Mexican national laboratories. Heating red phosphorus in a basic solution and the improper handling of red phosphorus can lead to formation of this toxic gas. Self-contained breathing apparatus should be required if there is any indication that this gas is present.

Amines like chlorine can do significant damage to the lungs in a very short time. The effects may not appear until several hours later. In some cases the person exposed feels no effects until a later time, when edema nearly caused them to drown in their sleep.

24. Dirty diapers
This may simply be attributable to the accumulation of disposable diapers at the site by tweakers. *Also see* Cat urine *and* Dead animals or dead fish.

25. Disagreeable
Isoamyl alcohol, nitromethane

To our way of thinking, almost everything described here is disagreeable. *Also see* Characteristic *and* Unpleasant.

26. Ether
Collodion, etherates, ethyl ether, Grignard reagents, methyl ether

Ethereal is a concept that any person working with these chemicals will understand; however, it would be impossible to explain it to someone who has never smelled an ethereal chemical. Ethyl ether is one of the most effective extraction solvents. Ether is used for extraction and purification in many methods. It is usually the most likely source of the odors reminiscent of medical facilities. *Also see* Hospitals *and* Phenol.

Ethyl ether evaporates quickly and has a high vapor density. Consider the high fire and explosive hazard associated with ether. This area should be tested using a combustible gas indicator prior to entry. Neighbors to the site will often report the odor of ether. A certain percentage of the population will have an immediate headache when exposed to ether. This is not a great marker, as there is a certain percentage of the population who maintain that they get a headache when they smell any "chemical". The latter may be psychosomatic. *Also see* Ethereal.

27. Ethereal
2-Chloroethanol, dimethyl ether, dioxane, 1,3-dioxolane, ethyl chloride, ethylene chlorohydrin, methyl chloride, methyl iodide, tetrahydrofuran
Also see Ether.

28. Fecal material
Indole, used in designer drugs.

29. Fingernail polish
Acetone, methyl ethyl ketone, toluene, methylene chloride, isoamyl acetate

Many evaporate quickly. Consider their fire and explosive hazard. Prior to entry, areas should be monitored with a combustible gas indicator. Flammable liquids are used to purify many precursors, intermediates, and finished product.

30. Fishy
Dimethylamines, ethyldiethanolamine hexamine, methyldiethanolamine

31. Floral
Benzyl chloride, benzyl cyanide, diphenylamine, phenylethyl alcohol

32. Fruity
Ethyl acetate, isoamylacetate, tabun

The artificial fruit flavors and odors are acetates. Grape tartrates, orange and lemon citrate, or limonene. *Also see* Juicy fruit.

33. Garlic
Acetylene gas, arsine, phosphine, arsenic, arsenic trichloride, dimethyl sulfoxide

A garlic-like odor is noted when acetylene gas is encountered at very high airborne concentrations. Look for a squat cylinder with a flat top. Acetylene presents an extremely high explosion hazard. A combustible gas indicator should be used to monitor areas prior to entering.

Arsine and phosphine can also have a garlic odor. There are arsine indicator tubes that detect arsine at its PEL. Phosphine can be detected with a Polytec IV tube.

Arsine can be associated with the charging of large banks of batteries. Arsine can also be created by acid attacking any metal that might contain arsenic. Lead that is found in lead acid batteries is almost always contaminated with arsenic or antimony (*see* Onions and mustard).

34. Garlic metallic
Arsenic, arsenic trichloride, bangsite, calcium carbide

Use arsine indicator tube.

35. Garlic/roses
Mustard gas

36. Gasoline
Prior to entry, combustible gas indicators should be used to monitor the area for organic vapors. These odors may be the result of incidental associations, such as in garages and farm/ranch outbuildings, simply because that is where petroleum products are likely to be stored. It may result from the use of such fuels for illumination at locations lacking readily available electricity. *Also see* Petroleum.

37. Geraniums
Lewisite

Use arsine indicator tubes, or ADP 2000.

38. Green grass
Dinitrogen tetroxide, nitrodioxide, nitrogen oxide, ozone, phosgene

A fresh clean odor. Phosgene is one of the several oxidizing gases that produce this odor. This odor may be noted where high energy comes in contact with chlorinated solvents. Silver soldering of copper pipes that contain residual Freon® or welding around Freon or methylene chloride are possible sources. *Also see* Biting, harsh.

39. Homeless person
Rich, pungent, unpleasant animal odors are associated with organic acids. Although formic and acetic acids are the most commonly encountered organic acids, other organic acids

might find use in a clandestine methamphetamine laboratory. You might also consider the cook as a potential source of such unpleasant odors.

40. Honey
Phenylacetic acid

Described as sweet floral or strong honey-like odor. Phenylacetic acid is used in the production of P-2-P. *Also see* Cat urine.

41. Hospitals
See Ether.

42. Incense
Gum sandarac

43. Irritating (pungent)
Acetyl chloride, acrolein, allyl bromide, allyl chloride, aluminum chloride, bischloromethyl ether, boron tribromide, boron trichloride, boron trifluoride, bromine, bromine trifluoride, 2-bromoethanol, n-butyl chloride, diethyl oxalate, ethyl chlorocarbonate, hydroiodic acid, hydrobromic acid, methyl disulfide, methylene iodide, methyl styrene, styrene, perchloric acid, phosphorus oxychloride, phosphorus pentachloride, resorcinol diacetate

Also see Pungent *and* Lachrymators. Entering into this space, you would expect to see something positive on an acid indicator tube, or a positive reading on a combustible gas indicator. (Most of these will destroy a CGI.)

44. Kerosene
The odor of petroleum products such as kerosene may be expected in association with many clandestine methamphetamine laboratories and waste abandonment sites. Prior to entry, combustible gas indicators should be used to monitor the area for organic vapors. The odor of kerosene and other petroleum products may simply be the result of incidental associations such as in garages and farm/ranch outbuildings, simply because that is where they are likely to be stored. Odors may result from the use of fuels for illumination at locations lacking readily available electricity. However, the most probable association results from Coleman® fuel and other petroleum distillates as an organic solvent used in the extraction of methamphetamine base.

45. Lacramators
Benzyl chloride, chloroacetone, chloroacetopherone, chloropicrin, o-chlorostyrene

Also see Irritating *and* Pungent.

46. Leather
Benzoyl chloride (a lacramator)

47. Leather dye
Chlorobenzene

48. Lemons
Citral, lemonine

These compounds are associated with illegal drugs.

49. Minty
4-Methyl-2-pentanone

50. Mothballs
Naphthalene, *p*-dichlorobenzene

51. Musty, sooty
Nickel carbonyl

Nickel carbonyl is a colorless-to-yellow volatile liquid with a musty or sooty odor. Although there is an indicator tube for this compound, it is not sensitive enough to protect you and you probably do not have this tube, as it is a special order.

52. Nailpolish remover
Acetone, methyl ethyl ketone

53. Nauseating
Diethylaminoethanol

54. Onions
Dimethyl sulfate

55. Onions and mustard
Stibine

There is no stibine indicator tube to detect this gas at its PEL. Stibine is not a likely find in clandestine laboratories. It can be associated with the charging of large banks of batteries. Stibine can also be created by acid attacking any metal that might contain antimony. Lead, which is found in lead acid batteries, is almost always contaminated with arsenic and/or antimony (*see* Garlic).

56. Paint (oil-based)
Linseed oil, turpentine

57. Peculiar
Ethyl nitrite, phosphorus pentachloride

58. Penetrating
Benzoic trichloide, cyanogen bromide, cyanogen chloride, sulfur monochloride, thiophosphorus trichloride

Also see Irritating *and* Lachrymators.

59. Peppermint
Cyclohexanone

60. Peppery
Diethyldiphenylurea, piperidine

61. Petroleum
Asphaltum, cyclohexane, diesel fuel, gasoline hydrocarbons

62. Phenol
Phenol

63. Phenyl-2-propanone (P-2-P)
This odor is unique. Until you have smelled it, there is no way to describe it. Once you have encountered it, you won't forget it. The odor of phenyl-2-propanone is, for many clandestine methamphetamine laboratory responders, the same odor that comes off the hydriodic acid/red phosphorus reaction. In an effort to eliminate this odor, many clandestine methamphetamine laboratories will filter the gases coming off the reaction through containers of kitty litter. Mexican national laboratories have been found using kitty litter in 5-gallon pails for filtering the odors. Apparently, the filtration method is not totally successful, as this odor has provided notice to neighbors that a clandestine methamphetamine laboratory has been established.

64. Pungent
Benzoyl chloride, 2-chloroacetyl chloride, crotonaldehyde, cyanogen iodide, dimethoxymethane, 2-ethylthioethanol, hydrogen selenide

65. Root beer
Anise, sassafras, safrole

Safrole is the beginning product in making Ecstasy (Synthesis 18). Safrole can be used in place of P-2-P in many reactions and has the characteristic root beer odor. The odor passes through the various reactions and may even be detected in the final product.

66. Rotten eggs
Ammonium sulfide, hydrogen sulfide, iron sulfide, sodium sulfide

Hydrogen sulfide is not a standard clandestine methamphetamine laboratory odor. Thio sulfates and sulfites are used in some laboratories. Under certain conditions these could release hydrogen sulfide gas. Hydrogen sulfide can be detected easily using an indicator tube or a four-gas detector fitted with an H_2S cell. This gas is on the list of methamphetamine chemicals not because it is used in any given methamphetamine method of production, but because it has been purchased by "cooks" thinking it was an effective substitute for hydrogen chloride gas. This has been a fatal mistake. Hydrogen sulfide can be bubbled through a suspension of iodine in water to make hydriodic acid. Usually, this gas is generated by adding acid to iron sulfide.

Lethal levels of hydrogen sulfide gas can be generated in any enclosed space by the action of organic materials rotting in water. Cardboard in contact with water has generated sufficient levels of hydrogen sulfide to knock people unconscious. This odor should be taken as a warning. Although not quite as toxic as hydrogen cyanide, hydrogen sulfide come in a close second. Beware: Higher levels of this gas cannot be detected by odor. The point at which a person stops smelling this gas is the point where lethal airborne levels have been reached.

67. Rotten soy sauce
Pyridine

Pyridine is used to neutralize red methamphetamine oil.

68. Sneezing caused by odor immediately
See Chemical Weapons, Vomiting Agents

69. Spearmint
Salicylates

70. Suffocating
Ethylene

71. Sulfurous (foul)
Thiodiglycol

72. Sweet
Ethylene glycol, Freon®, isobutanol

73. Tarry
Cresol, creosote

74. Unpleasant
Trimethylphosphite

Also see Disagreeable.

75. Urinal cake
p-Dichlorobenzene

76. Vanilla
Vanillin

This is a very strong odor. Vanillin (an aldehyde) provides a good precursor for many chemicals that contain a benzene ring.

77. Vinegar
Acetic acid, acetic anhydride, ammonium acetate, ammonium carbonate, ammonium citrate, sodium acetate

Acetic acid and acetic anhydride both smell like vinegar (which is actually a 5% solution of acetic acid in water). Indicator tubes are available that read acetic acid at and below the PEL. Be aware that hydrofluoric acid, which is a serious inhalation hazard, also smells like vinegar to many people. Acetic anhydride is by far the most likely source of this odor. Until the exact level and a definite identification of this compound are made, SCBAs are required. Air-purifying respirators with the yellow cartridge may be worn upon the quantification and qualification of this substance.

78. Vomit
Butyric acid

79. Waxy
Ozokerite

ARMY CODES FOR CHEMICAL WEAPONS

Agent	Chemical Name	Common Name	Type
AC	Hydrogen cyanide		Blood agent
BZ	3-Quinuclidinyl benzilate (QNB)	Buzz	Incapacitating
CA	Bromobenzylcyanide		Tear agent
CG	Carbonyl chloride	Phosgene	Choking agent
CH	1-Methoxy-1,3,5-cycloheptatriene		Tear agent
CK	Cyanogen chloride		Blood agent
Cl	Chlorine		Choking agent
CN	Chloroacetophenone	Mace	Tear agent
CR	Dibenz[b,f]-1, 4-oxazepine		Tear agent
CS	2-Chlorobenzalmalononitrile		Tear agent
CX	Phosgene oxime		Urticant
DA	Diphenylchloroarsine		Vomiting agent
DC	Methylphosphonic dichloride	MPOD	Binary nerve agent
DC	Diphenylcyanoarsine		Vomiting agent
DF	Methylphosphonic difluoride		Binary nerve agent
DM	Diphenylaminochloroarsine		Tear agent
DM	10-Chloro-5,10-dihydrophenbarsazine	Adamsite	Vomiting agent
DP	Trichloromethyl chloroformate	Diphosgene	Choking agent
DP			Nerve agent
EA	$(CH_3)_2NP(O)(F)OCH_2CH_2N(CH_3)_2$	DMAPDMAPP	Nerve agent
ED	Ethylene dichloroarsine	Arsenicals	Vesicant
FM	Titanium tetrachloride		Smoke
FS	Sulfur trioxide-chlorosulfonic acid		Smoke
GA	Ethyl N,N-dimethylphosphoramidocyanide	Tabun	Nerve agent
GB	Isopropylmethylphosphonofluoridate	Sarin	Nerve agent
GD	1,2,2-Trimethylpropylmethylphosphonofluoridate	Soman	Nerve agent
GF	Cyclohexylmethylphosphonofluoridate	Cyclosarin	Nerve agent
GP	$(CH_3)NP(O)(F)OCH_2CH_2N(CH_3)_2$	DMAEDMAFP	Nerve agent
GV			Nerve agent
H	Impure sulfur mustard	Mustard	Vesicant
HC	Zinc oxide, hexachloroethane, 7 % grained aluminum		Smoke
HD	Bis-2-chloroethyl sulfide	Mustard	Vesicant
HE	Hexachloroethane		Smoke
HL	Mixture of lewisite and mustard		Vesicant
HS		Mustard agent	Vesicant
HT	Mixture of HD and T		Vesicant
L	2-Chlorovinyl dichloroarsine	Lewisite	Vesicant
MC	Methyldifluoroarsine		
MD	Methyldichloroarsine	Arsenicals	Vesicant
NE	Sulfur in Cab-O-Sil	Agent VX	Binary nerve agent
NH	Three compounds, tertiary amines	Nitrogen Must	Vesicant
NM	Mixture of dimethylpolysulfides, containing sulfur	Agent VX	Binary nerve agent

Agent	Chemical Name	Common Name	Type
OC		Pepper spray	Tear agent
PD	Phenyl dichlorobenzene	Arsenicals	Vesicant
PS	Chloropicrin		Choking agent
Q	$ClCH_2CH_2SCH_2CH_2SCH_2CH_2Cl$	Sesqui Must	Vesicant
QL	O,O'-Ethyl (2-diisopropylaminoethyl) methylphosphonite	EDMP	Binary Nerve agent
SA	AsH_3 (arsine)	Arsine	Binary nerve
T	$(ClCH_2CH_2SCH_2SH_2)_2O$	O-Mustard	Vesicant
TR	Residue of mustard		
VM	$CH_3P(O)(OCH_2CH_3)SCH_2CH_2N$-$(CH_2CH_3)_2$		Nerve agent
VX	O-Ethyl S-[2-(diisopropylamino)ethyl] methylphosphonothiolate		Nerve agent
Vx	$CH_3P(O)(OCH_2CH_2)$	V-gas	Nerve agent
WP	White Phosphorus		Smoke

6

CONFIRMATION TESTS

CONFIRMATION TESTS

A confirmation test is only an indicator that a product is labeled correctly. If the product is what it is supposed to be, the test will be positive. However, there are many products that could also be positive. For instance, the Alcohol Test will pick all alcohols, glycols, polyols, and glycerides. This means that although the product is probably labeled correctly, there is the possibility that the product is something else. If the product is labeled, appears as it is supposed to appear, and the confirmation test is positive, the responder has three positive indications. In some cases we may give two confirmation tests. For instance, for barium sulfate we may give a Barium Test and a Sulfate Test; if both are positive, that is a positive identification.

It is best to think of the confirmation tests only as confidence boosters.

Acetic Acid Indicator Tube

Reagents needed: GasTec Acid Gases 80 indicator tube or Draeger Acetic Acid 5/a indicator tube.
   1. Break the ends off the indicator tube and push into the pump with the arrow pointing toward the pump.
   2. Collect a sample of the unknown by sampling the headspace of a test tube of the compound.
   3. Results:
      a. Indicator tube changes from purple to yellow: acetates are indicated.

Adamsite Test

Interferences: Some ketones will interfere with the first part of this test by turning sulfuric acid red, but not with the second part.
Reagents needed: concentrated sulfuric acid, sodium nitrite.
   1. Carefully add a few drops of the unknown liquid, or a pinch of the solid, to ¼ inch of concentrated sulfuric acid in a test tube.
   2. Agitate the tube gently, or mix using a pipette.
   3. The solution often will turn red.
   4. Add a few grains of sodium nitrite.
   5. Results:
      a. Solution becomes blue: indicates presence of secondary amine.
      b. Solution does not become blue: negative.

Chemicals Used for Illegal Purposes: A Guide for First Responders to Identify Explosives, Recreational Drugs, and Poisons, By Robert Turkington
Copyright © 2010 John Wiley & Sons, Inc.

Alcohol Test (Fearon–Mitchell)

Reagents needed: nitric acid, potassium dichromate solution.
1. Add several drops of nitric acid to a test tube containing ½ inch of the unknown liquid.
2. Add at least several drops of potassium dichromate solution until the entire solution turns orange.
3. **Results:**
   a. Slow color change from orange to blue: hydroxyl group present, alcohol indicated.
   b. Emerald green solution: DMSO indicated.

Aldehyde Indicator Tube

Reagents needed: GasTec formaldehyde tube 91L.
Interferences: Most low-carbon aldehydes.
1. Break the glass tips off a GasTec Formaldehyde Tube 91L and insert the indicator tube into the pump with the arrow pointing toward the pump.
2. Add ¼ inch of the unknown liquid to a test tube.
3. Line up the arrows and pull the handle on the pump all the way out until it catches.
4. Put the end of the indicator tube into the headspace of the test tube for 1 minute.
5. **Results:**
   a. Yellow-to-red/brown indicator tube change: indicates an aldehyde.

Aldehyde Test (Fehling)

Reagents needed: cupric sulfate solution, alkaline tartrate solution, 3 N hydrochloric acid.
1. Add 1 drop of 3 N hydrochloric acid to ½ inch of the unknown liquid in a test tube.
2. In a second test tube add ¼ inch of cupric sulfate solution to ¼ inch of alkaline tartrate solution. Shake the mixture until it turns a rich blue. If the mixture does not turn a rich blue, add a little more alkaline tartrate.
3. Add several drops of the mixture prepared in step 2 to the unknown solution.
4. If there is no reaction, heat the mixture a little gently over a torch flame.
5. **Results:**
   a. Yellow-orange-to-reddish-brown precipitate: aldehyde indicated.

Aluminum Metal Test

Reagents needed: sodium hydroxide, methanol, distilled water.
1. Add 2 pellets of sodium hydroxide to a test tube containing ½ inch of water. Place the tube in the test tube rack and allow the pellets to dissolve.
Warning! This solution will become very hot!
2. Place ½-pea-sized amount of the unknown metal on a watch glass.
3. Add several drops of the solution made in step 1 to the unknown on the watch glass.
4. If there is no effervescence, add a drop of methanol to break the surface tension.
5. **Results:**
   a. Effervesces: aluminum indicated.

Aluminum Test (Hammett–Sottery)

Reagents needed: aluminon, ammonium acetate/acetic acid solution, 3*N* hydrochloric acid, ammonium hydroxide, concentrated hydrochloric acid, pH test strips, distilled water.

1. a. **Liquid:** Add ½-inch of the unknown to a test tube. Add 3 N hydrochloric acid a drop at a time until the pH is less than 3.

   b. **Solid:** Add ½-pea-sized amount of the unknown to a test tube containing an inch of distilled water. Add 1/8 inch of concentrated hydrochloric acid, and allow the solution to stand for 2 minutes.

2. Add ¼ inch of aluminon to the tube containing the unknown solution.

3. Slowly add ¼ inch of an acetate/acetic acid solution, allowing the reagent to drip down the side of the test tube. Do not shake the contents.

4. **Results:**

   a. A red flocculating precipitate that forms slowly, floats, coagulates, and then sinks indicates the presence of aluminum.

Ammonia Indicator Tube

Test 1. Looking for Ammonia
Reagents needed: pH test strips, distilled water.

1. Add 1/2 inch of the unknown liquid to a test tube.

2. Wet a pH test strip with distilled water.

3. Insert the pH test strip into the headspace of the test tube.

Warning! Keep fingers away from the mouth of the test tube!

4. **Results:**

   a. pH near 11 from an inorganic liquid: Ammonia

Test 2. Looking for Ammonium Salts
Reagents needed: sodium hydroxide, pH test strips, distilled water.

1. a. **Liquid:** Add ½ inch of the unknown to a test tube.

   b. **Solid:** Add ½ inch of a pea-sized amount of the unknown to ½ inch of distilled water in a test tube. **Note:** If the unknown doesn't dissolve, it's not an ammonium salt.

2. Feel the test tube. If the solution has become very cold, indicating a nitrate, let it stand to reach room temperature before continuing to step 3.

3. Add 3 pellets of sodium hydroxide to the solution of the unknown.

Warning! The solution will become very hot, and may boil.

   Allow to stand for 1 minute.

4. Wet a pH test strip with distilled water.

5. Insert the pH test strip into the headspace of the test tube.

Warning! Keep fingers away from the mouth of the test tube.

6. **Results:**

   a. A pH near 11 from gas in the tube headspace from a pH 7 inorganic compound: ammonium cation indicated.

Test 3. Looking for Amines and Secondary Amines
Reagents needed: pH test strips, distilled water.

1. Wet a pH test strip with disilled water.

2. Insert the pH test strip into the headspace of a test tube containing ½ inch of the unknown solution.

Warning! Keep fingers away from the mouth of the test tube.
   3. Results:
       a. pH of 11: on an organic compound, indicates an amine.

Aniline Test (Letheby)

Reagents needed: manganese dioxide, concentrated Sulfuric acid.
   1. Add ½ inch of the unknown liquid to a test tube.
   2. Slowly add ¼ inch of concentrated Sulfuric acid.
Warning! Sulfuric acid reacts violently with most hydrocarbons.
   3. Add a pinch of manganese dioxide and gently heat the solution.
   4. Results:
       a. Blue liquid: aniline indicated.

Antimony Test

Reagents needed: ammonium hydroxide, 6 N hydrochloric acid, sodium thiosulfate.
   1. Add ammonium hydroxide to a test tube containing ½ inch of the unknown solution, until the solution is basic (pH > 7).
   2. Add 6 N hydrochloric acid until the solution is acidic (pH < 7). If a white precipitate does not form at this point stop, the unknown is NOT antimony.
   3. Add sodium thiosulfate and heat gently.
   4. Results:
       a. Orange-red precipitate indicates antimony.

Arsenic Test (Gutzeit)

Reagents needed: mercuric chloride in methanol, stannous chloride, potassium iodide, 2 N sulfuric acid, zinc metal, Whatman® filter paper strips, distilled water.
   1. Soak Whatman® paper strips with a drop of the mercuric chloride solution.
   2. To a solution of the unknown, add a pea-sized amount of stannous chloride, then potassium iodide, and then ¼ inch 2 N sulfuric acid.
   3. Sprinkle enough zinc metal into a test tube to cover the surface of the solution.
Warning! When arsenic is present, this test generates extremely toxic arsine gas. Mercury, phosphides, and antimony also produce toxic gases in this test.
       If effervescence forces the zinc metal up the side of the test tube, rinse it down with a few drops of 2 N sulfuric acid in a pipette.
   4. Immediately insert the activated wattman paper strip into the test tube.
   5. Allow to stand a few seconds before reading results.
   6. Results:
       a. The paper will turn yellow, orange, and finally brown, depending on the concentration of Arsenic. Yellowish-orange-brown spots may be from splashed material from the liquid in the bottom of the tube. This is NOT significant. Look for a general slow color change.

Barium Test (Feigl)

Reagents needed: 3 N hydrochloric acid, sodium rhodizonate, ammonium hydroxide, distilled water.
   1. a. **Liquid:** Add ½ inch of the unknown to a test tube.
      b. **Solid:** Add ½ of a pea-sized amount of the unknown to ½ inch of distilled water.

2. Add to a test tube sodium rhodizonate on a small spatula. Add distilled water to the test tube until it is mostly filled. Shake well.
3. Add 4 drops of the newly made barium test solution to the unknown solution.
4. Many initial colors may evolve.
5. Add ¼ inch of 3 N hydrochloric acid. Allow time for the solution to clear. If the unknown does not contain barium, the color should vanish completely rather than lighten.
6. **Results:**
   a. Cherry red: barium indicated.

Benzoyl Peroxide Test

Reagents needed: methanol, EM Quant Peroxide test strip, distilled water.
1. Place a pea-sized amount of the unknown solid on a watch glass.
2. Add a few drops of methanol.
3. Wet a peroxide test strip with a drop of Water Solubility Test.
4. Touch the peroxide test strip to the unknown.
5. **Results:**
   a. Blue color: benzoyl peroxide indicated.

Bismuth

See Nickel Test.

Boron Test (Rosenblatt)

Reagents needed: methanol.
1. Place a pea-sized amount of the unknown on a watch glass.
2. Add methanol until the pool is the size of a quarter.
3. Ignite the methanol with a match.
4. **Results:**
   a. Long-lasting green alcohol-type flame: boron indicated.

Bromide Test (Vitali)

Reagents needed: cupric sulfate solution:
1. a. **Liquid:** Add ¼ inch of the unknown to a test tube. Heat gently until it evaporates and leaves a dried residue.
   b. **Solid:** Add ½-pea-sized amount of the unknown to a test tube.
2. Add several drops of cupric sulfate solution, and heat gently.
3. **Results:**
   a. Reagent changes from blue to burgundy red: bromide indicated.

Cadmium Test (Feigl)

Reagents needed: α,α'-dipyridyl ferrous sulfate solution:
1. Add 2 drops of α,α'-dipyridyl ferrous sulfate solution to ¼ inch of the unknown solution in a test tube.
2. **Results:**
   a. Pink precipitate: cadmium. The precipitate is so fine that it may appear to be a pink solution. If allowed to stand, the precipitate eventually sinks to the bottom, making it easier to see.

Calcium Test

Reagents needed: 3 N hydrochloric acid, ammonium oxalate solution, distilled water.
   1. Add ½ inch of a pea-sized amount of the unknown solid to ½ inch of distilled water in a test tube.
   2. If the unknown sinks or forms a suspension, add 5 drops of 3 N hydrochloric acid.
      a. If a white or colorless unknown effervesces, it's probably calcium carbonate.
      b. If the suspension remains, filter using a 25-mm membrane filter in a Swynex filter holder. Collect the filtered solution in a new test tube and use this solution for step 3.
   3. Add ¼ inch of ammonium oxalate solution.
   4. **Results:**
      a. White precipitate: calcium indicated.
      b. Amber precipitate: cobalt or chromium indicated; continue testing.

Carbon Dioxide Indicator Tube

Reagents needed: 3 N hydrochloric acid, GasTec Polytec IV indicator tube or a Draeger carbon dioxide 0.01%/a indicator tube, distilled water.
   1. Break the tips off the indicator tube and insert with the arrow pointing toward the pump.
   2. If the unknown effervesces in water, add a pea-sized amount of the unknown to ½ inch of distilled water in a test tube. Go to step 4.
   3. If the unknown does not effervesce in water, add a pea-sized amount of the unknown to ½ inch of 3 N-hydrochloric acid.
   4. Put the tip of the indicator tube into the headspace of the test tube and draw a sample through the indicator tube.
   5. **Results:**
      a. GasTec tube (band 7 only): changes from purple to brown:
         i. Effervescing solid in water: bicarbonate of soda
         ii. Effervescing solid in 3 N hydrochloric acid: carbonate indicated.
      b. Draeger tube: changes from white to purple:
         i. Effervescing solid in water: bicarbonate of soda
         ii. Effervescing solid in Acid Test: carbonate indicated.

Carbon Monoxide Indicator Tube

Reagents needed: GasTec Polytec IV 27 indicator tube or Draeger Carbon Monoxide 8/a indicator tube.
   1. Break the tips off the indicator tube and insert with the arrow pointing toward the pump.
   2. Add ½ inch of the unknown solid to a test tube. Heat the test tube until the unknown glows orange.
   3. Put the tip of the indicator tube into the headspace of the test tube for a few seconds and draw a sample through the indicator tube.
   4. **Results:**
      a. GasTec tube (band 6 only) changes from yellow to dark brown: carbon or graphite indicated.
      b. Draeger tube changes from white to brownish green: carbon or graphite indicated.

Carbonyl Test

Reagents needed: pH test strips, distilled water, cobalt thiocyanate.
Interferences: Liquids with a pH of 8 or greater are always positive on this test and should not be tested.

1. Do a pH test on the liquid. If the pH is 17 or less, continue.
2. Add a pinch of cobalt thiocyanate to a test tube $3/4$ full of distilled water; it should dissolve to a pink solution.
3. Add several drops of the pink solution to a test tube containing $1/2$ inch of the unknown.
4. If it does not mix or change color, add 3 drops of methanol.
5. **Results:**
   a. Dark blue: indicates the presence of the carbonyl group.
   b. Solvent remains clear, water becomes dark blue: ethyl ether.

Chloride Test

Reagents needed: silver nitrate, ammonium hydroxide, 2 N sulfuric acid, pH test strips.

1. Add a few drops of silver nitrate to $1/2$-inch of the unknown solution.
2. **Results:**
   a. White, heavy, clotted precipitate: Chloride anion indicated. To confirm, add $1/4$ inch of ammonium hydroxide. The white precipitate should disappear as the solution becomes basic.

Chromium Test (Leitmeier–Feigl)

Reagents needed: diphenylcarbazide in ethanol, 3 N hydrochloric acid, pH test strip.

1. Dip a pH test strip in the unknown solution in a test tube Adjust the pH to between 1 and 3 by adding drops of 3 N hydrochloric acid as needed.
2. Add $1/2$ inch of diphenylcarbazide.
3. **Results:**
   a. Purple liquid: chromates indicated. In concentrated chromate solutions, the purple quickly fades to green; weaker solutions retain the purple color.

Copper Test (Sarzeau)

Reagent needed: ammonium hydroxide.

1. Add a dime-sized puddle of the unknown liquid, or place a $1/2$-pea-sized amount of the unknown solid on a watch glass.
2. Add several drops of ammonium hydroxide on the watch glass.
3. **Results:**
   a. Rich blue: copper

Beilstein Test

Warning! Prior to doing this test, add $1/2$ drop of the unknown to a test tube, and heat to see what it will do. If there is a violent reaction, do not do this test.
Reagents needed: methanol, copper wire.

1. Heat the copper wire until it glows red and no residual green flame is visible.
2. Dip the hot wire into $1/2$ inch of methanol in a test tube. If the liquid ignites, blow the flame out.

3. Allow the wire to cool in the methanol.
4. Remove the wire from the methanol and reheat it until it glows red.
5. Gently heat a test tube containing the unknown solution or solid until vapors reach the top of the test tube. Hold the tube near the air-intake holes in the propane torch head.
6. Place the copper wire in the torch flame.
7. **Results:**
   a. Intense blue flame that fades to green: Chlorinated or other homogenated organic indicated.

Hydrogen Cyanide Indicator Tube (ASTM)

Test 1. Looking for Liquid Nitriles

Reagents needed: GasTec Hydrogen Cyanide 12L indicator tube or Draeger Hydrocyanic Acid 2/a indicator tube.

1. Break the tips off the indicator tube and insert it into the pump with the arrow pointing toward the pump.
2. Add ¼ inch of the unknown liquid to a test tube.
3. Stopper and shake the test tube.
4. Remove the stopper and ignite the vapor with a match.
5. Put the tip of the indicator tube into the headspace of the test tube and draw a sample through the indicator tube.
6. **Results:**
   a. Indicator tube changes from yellow to red: nitrile indicated.
   b. Dark red ring only at the tip of the indicator tube: probably just water vapor.

Test 2. Looking for Insoluble Metal Cyanides

Reagents needed: GasTec hydrogen Cyanide 12L indicator tube or Draeger Hydrocyanic Acid 2/a indicator tubes.

1. Break the tips off of the indicator tube and insert with the arrow pointing toward the pump.
2. Add ½ a pea-sized amount of the unknown solid to a test tube.
3. Heat the unknown over a torch flame until it begins to change color.
4. Put the tip of the indicator tube into the headspace of the test tube and draw a sample through the indicator tube.
5. **Results:**
   a. Indicator tube changes from yellow to red: cyanide anion indicated.
   b. Dark red ring only at the tip of the indicator tube: probably just water vapor.

Cyanide Test (Kotney)

Reagents needed: ferrous ammonium citrate solution, ferrous ammonium sulfate, 3 N hydrochloric acid

1. Make a water solution of the unknown. If the unknown does not dissolve, go to the Cyanide Indicator Tube Test 2 for insoluble cyanides.
2. In a separate test tube, add a pinch of ferrous ammonium sulfate to ¼ inch of ferrous ammonium citrate solution and shake gently.
3. Add ¼ inch of the solution prepared in step 2 to the test tube containing the unknown from step 1, and shake gently.
4. Remove the stopper and add 3 N hydrochloric acid a drop at a time until a reaction occurs.

5. Results:

    a. Blue: cyanide indicated.

    b. Light blue: metal cyanides indicated.

Ephedrine Test

Reagents needed: Test Q is for ephedrine. Becton Dickinson Public Safety, 1720 Hurd Dr., Irving, Texas 75062 (214) 659-9595.

1. Remove the clip and insert into the test pack an amount of powdered suspect material that would fit inside this square; reseal with clip and tap gently to assure that material falls to the bottom of the pack.
2. With the printed side of the test pack facing you, break the ampoules from left to right. Break the ampoules by squeezing the center of the ampoule with the tips of thumb and forefinger.
3. Break the left ampoule. Agitate gently for 30 seconds. No color development is expected at this stage.
4. Break the right ampoule and agitate gently. If an immediate purple/violet color forms, presume that ephedrine is present.

If any other color develops, proceed to Test A (Narcotics Identification System) and continue polytesting as instructed by the Polytesting Manual.

Flame Test (Bunsen)

Reagents needed: platinum or stainless steel wire. A hairpin will work.

1. Place a small amount of the unknown solid or liquid on a watch glass.
2. Use a propane torch to heat the wire.
3. Dip the wire into the unknown.
4. Place the wire pack into the torch flame and look for the flame color.
5. **Results:**

    a. Very red: lithium

    b. Orange-red: short duration, calcium, strontium

    c. Yellow-orange: long duration, sodium

    d. Green: barium, boron, copper, thallium

    e. Blue: lead, selenium

    f. Lavender: potassium

Flour Test (Lugol)

Reagent needed: potassium iodide and iodine in an aqueous solution.

1. Place ½ a pea-sized amount of the unknown solid on a watch glass. Add 2 to 3 drops of the potassium iodide and iodine solution.
2. **Results:**

    a. Unknown turns blue or black: starch or flour indicated.

Fluoride Test (DeBoer)

Reagents needed: alizarin zirconyl chloride solution, 3 N hydrochloric acid, pH test strips.

1. Do a pH test on the unknown. If the pH is less than 7, go to step 2. If the pH is greater than 7 add a few drops of 3 N hydrochloric acid to bring the pH below 7 (use a new pH test strip for each measurement).

2. Add ¼ inch of alizarin zirconyl chloride solution. (Shake the bottle prior to use.)
3. **Results:**
   a. Reagent changes from purple to yellow: fluoride indicated. If the unknown is an acid, it is dangerous hydrofluoric acid.

Gold Test (Bayer)

Reagents needed: rubidium chloride, silver nitrate.
1. Add a ½ inch solution of rubidium chloride and 0.1% silver nitrate to a test tube.
2. To the solution add 1 drop of the unknown liquid or a pinch of the unknown solid.
3. **Observations:**
   a. Blood red indicates gold.

Hairpin Test

Reagents needed: standard hairpins.
1. Bend a hairpin until it is straight.
2. Use a torch to burn off the plastic tips on the hairpin.
3. Heat the hairpin until the end is glowing red.
4. Put the red-hot hairpin into the unknown.
5. **Results:**
   a. Purple fume: Iodine.
   b. Red unknown bursts into slow steady flame: red phosphorus.
   c. White unknown violent organic flame, copious white smoke: benzoyl peroxide.

Ignition of Off-Gas

Reagents needed: paper matches.
1. Place ¼ inch of unknown solid into test tube.
2. Add ¼ inch of water or 2 N hydrochloric acid until it effervesces.
3. Strike a match and hold the flame over the headspace of the test tube.
4. **Results:**
   a. Ignites with a pop: probably a IIA or IIB metal releasing hydrogen or a hydride.
   b. Burns very smoky (spider webs in smoke): most likely an alkali or alkaline earth carbide.
   c. Flame goes out: most likely a carbonate.

Indol Test

Reagents needed: metol.
Note: Indol smells intensely like fecal matter.
1. Add metol to the unknown.
2. **Results:**
   a. A blue color indicates the presence of indole.

Iodide Test

Reagents needed: concentrated sulfuric acid, 2 N hydrochloric acid, potassium iodide starch paper.
1. Add a few drops of the unknown liquid or ½ a pea-sized amount of the unknown solid to a test tube.

2. Add a few drops of concentrated sulfuric acid.
3. Gently heat the unknown.
4. Wet a potassium iodide starch paper strip with 3 N hydrochloric acid and wave it gently above the unknown.
5. **Results:**
   a. Purple oxidizing fume: the anion is iodine.

Determination with Iodine

Reagents needed: sublimed iodine.
1. Add ½ inch of the unknown liquid to a test tube.
2. Add a small iodine crystal. The smaller the crystal, the lighter the color change will be. If the color is too dark, the results may be confusing. Shake the test tube gently and look at the color on the side of the test tube.
3. **Results:**
   a. Purple: single bonds.
   b. Red: double bonds.
   c. Reacts with yellow smokes: amines, hydrazine.

Iron Test (Kim)

Reagents needed: 1,10-phenanthroline.
1. Add a few drops of 1,10-phenanthroline to ½ inch of the unknown in a test tube.
2. **Results:**
   a. Red: ferrous iron indicated.
   b. Yellow: ferric iron indicated.

Lead Test (Kotney)

Reagents needed: acetic acid/ammonium acetate, chloranilic acid.
1. a . **Liquid:** Add ¼ inch of the unknown to a test tube.
   b. **Solid:** Add ½ inch of acetic acid/ammonium acetate solution to a test tube, then add ½ a pea-sized amount of the solid unknown.
2. Add ¼ inch of chloranilic acid to the test tube.
3. **Results:**
   a. Purple-to-brown color change: lead.

Magnesium Test (Kim)

Reagents needed: 4-*p*-nitrophenylazo-1-naphthol, distilled water.
1. Add ¼ inch of the unknown liquid or ½ a pea-sized amount of the unknown solid to ½ inch of distilled water. Add a few drops of 4-*p*-nitrophenylazo-1-naphthol solution.
2. **Results:**
   a. Baby blue flocculate: magnesium.

Magnet Test (Sweetzer)

Reagents needed: magnet.
1. Hold the magnet alongside the unknown in a test tube.
2. Move the magnet up and down along the side of the test tube.
3. **Observations:**
   a. The unknown moves along with the magnet inside the test tube: iron.

Manganese Test

Reagents needed: ammonium oxalate solution, 3% hydrogen peroxide.
1. Add ½ inch of 3% hydrogen peroxide to the unknown solution. Add ½ inch of ammonium oxalate. At this point there may be strong effervescence.
2. **Results:**
   a. Black or very dark brown: manganese.

Mercury Test

Test 1 (Kotney)
Reagents needed: copper coin, 3 N hydrochloric acid.
1. Add 1 drop of 3 N-hydrochloric acid to the copper coin.
2. Allow to stand until the copper metal shines under the drop of acid.
3. Add 1 drop of the unknown solution to the drop of 3 N hydrochloric acid on the coin. Let the test stand another minute.
4. Wipe the coin with a paper towel.
5. **Results:**
   a. Silver shine or plating on the copper coin: mercury.

Nickel Test (Tschugaev)

Reagents needed: dimethylglyoxime in methanol, ammonium hydroxide.
1. Add 3 or 4 drops of dimethylglyoxime in methanol to the unknown solution.
2. Very slowly add ammonium hydroxide to the unknown solution, 1 drop at a time.
3. **Results:**
   a. Thick, distinct reddish-pink precipitate: nickel.
   b. Yellow: palladium or bismuth indicated.

Nitric Acid Tests

Reagents needed: EM Quant peroxide test strip; distilled water.
1. Add 1 drop of distilled water to the EM Quant peroxide test strip.
2. Add 1 or 2 drops of the unknown solution to the pad on the peroxide test strip.
3. **Results:**
   a. Bright yellow: nitric acid.

Organic/Inorganic Acid Gas Test

Reagents needed: Cresol Violet acetate.
1. To a test tube of the unknown solution, add several drops of Cresol Violet acetate.
2. **Results:**
   a. A blue color indicates phenol.
   b. A fluorescent red color indicates an organic acid.
   c. A light purple color indicates an acid with a pH near 2.
   d. A yellow color with precipitate indicates a pH near 0.

Organophosphate Test

Reagents needed: silver nitrate.
1. Divide the emulsified unknown into two test tubes, with equal amounts of the emulsion in each tube.

2. Add several drops of silver nitrate to one of the test tubes. Use the other test tube as a blank.
3. Compare the silver nitrate/unknown emulsion with the blank.
4. **Results:**
   a. Cream to yellowish brown: organophosphate indicated.

Oxalate Test

Reagents needed: barium chloride solution.
1. Add a few drops of barium chloride to ½ inch of the clear or filtered solution of the unknown in a test tube.
2. **Observations:**
   a. White precipitate: sulfate indicated.

Oxidizer Test (Runge)

Reagents needed: 3 N hydrochloric acid, potassium iodide, starch paper.
1. a. **Liquid:** Add the unknown to a watch glass to form a pool the size of a dime.
   b. **Solid:** Add ½ a pea-sized amount of the unknown to a watch glass.
2. Wet the potassium iodide starch paper with 1 or 2 drops of 3 N hydrochloric acid.
3. Touch the potassium iodide starch paper to the unknown on the watch glass.
4. **Results:**
   a. Test strip turns black, purple, or blue: oxidizer.

Perchlorate Test (Schilt)

Reagents needed: methylene blue in methanol.
1. Add 1 or 2 drops of methylene blue in methanol to ½ inch of the unknown solution in a test tube.
2. **Results:**
   a. Purple: perchlorate. For cation, go to Metal Analysis, Section 8.
   b. Yellow with brown precipitate: picric acid.
Warning! Metal perchlorates are shock-sensitive violent explosives!

Peroxide Test (EM Quant)

Reagents needed: peroxide test strip; distilled water.
1. Add the unknown liquid to form a pool the size of a dime or add ¼ of a pea-sized amount of the unknown solid to the watch glass.
2. Wet the peroxide test strip with 1 drop of distilled water.
3. Touch the wet peroxide test strip to the unknown.
4. **Results:**
   a. Blue: peroxide, also a possibility of weak chromic acid, which would be colored orange-green or brown: liquid unknown.

pH Test

Reagents needed: pH test strips; distilled water.
Interferences: Because pH test strips can change color during storage, dip a pH test strip into a test tube containing nothing but distilled water as a standard. Distilled water should have a pH of 7, so compare the pH strip result to the value of 7 illustrated on the box, and mentally adjust results as necessary to establish the pH of the unknown.

1. Dip the pH test strip into the unknown solution. Allow a few seconds for the colors to develop, and then compare them with the colors on the box of the pH test strips.
2. **Results:**
   a. pH of 2 or less: the unknown is an EPA acid corrosive.
   b. pH near 11: usually indicates the presence of nitrogen. Calcium, a white chalky material, may also have a pH of 11.
   c. pH of 12: may indicate cyanide, arsenic, strychnine, or amines; keep testing.
   d. pH of 12.5 or higher: the unknown is a basic corrosive according to the EPA.
   e. pH of 13 or higher: if inorganic, alkaline hydroxide; organic, piperidines, quanadines, or other multi-nitrogenated compounds are indicated.

Phenol Indicator Tube

Reagents needed: GasTec Phenol 60 indicator tube or a Draeger phenol 1/b indicator tube.
1. Break the tips off the indicator tube and insert with the arrow pointing toward the pump.
2. Add ¼ inch of the unknown liquid or a ½ a pea-sized amount of the solid unknown to a test tube.
3. Put the tip of the indicator tube into the headspace of the test tube and draw a sample through the indicator tube.
4. **Results:**
   a. A color change in the indicator tube from pale yellow to faint brownish gray: phenol

Phosphate Test (Hutton)

Reagents needed: ammonium hydroxide, silver nitrate, pH test strips, distilled water.
1. Add ¼ inch of the unknown liquid or ½ a pea-sized amount of the unknown solid to a test tube.
2. Do a pH test. If the pH is basic (pH > 7), go to step 3. If the pH is acidic (pH < 7), add 1 drop of the unknown to a test tube containing ½ inch of distilled water. Add ammonium hydroxide drop by drop until a precipitate appears or until the solution is basic.
3. Add ¼ inch of silver nitrate.
4. **Results:**
   a. Yellow precipitate (precipitate usually is a ring at the border of the solution and the ammonium hydroxide): phosphate.

Phosphide Test

Reagents needed: mercuric chloride in methanol, zinc metal, paper strips, 10 N hydrochloric acid.
Interferences: low levels of arsenic.
1. Add 1 drop of mercuric chloride/methanol to a paper strip. Allow drying.
2. Add ¼ inch of 10 N hydrochloric acid to a test tube containing ½ inch of the unknown solution in a test tube.
3. Sprinkle enough zinc metal into the test tube to cover the surface of the unknown.
Warning! Extremely toxic phosphine gas may be generated at this point.
4. Immediately insert an impregnated paper strip into the test tube.
5. Allow the test to stand for about 40 seconds before reading the results.

6. **Results:**
   a. Lemon yellow: phosphide.

Phosphorus Test (Kramer)

Reagents needed: 4-(4-nitrobenzyl)pyridine.
   1. Add a small scoop of 44NBP to a test tube.
   2. Use a swab to clean 2 inches squared of the suspected surface contamination.
   3. Forcefully push the end of the swab into the 44NBP in the test tube.
   4. **Results:**
      a. Cherry red color covers the 44NBP: indicates nerve agent.
      b. Deep purple liquid: indicates possible nerve agent.
      c. Bright yellow liquid: indicates possible phosphorus in a NMAP.
      d. Brown: indicates sulfur.

Picric Acid Test

Reagents needed: methylene blue in methanol.
   1. Add a few drops of methylene blue in methanol to ½ inch of the unknown solution in a test tube.
   2. **Results:**
      a. Brown-purple precipitate: picric acid.

Polytec IV Test (Gastec)

Reagents needed:. GasTec Polytec IV indicator tube No 27.
   1. Break the tips off the indicator tube and insert with the arrow pointing toward the pump.
   2. Add ¼ inch of the unknown liquid or ½ a pea-sized amount of the solid unknown to a test tube.
   3. Heat the test tube with a torch until vapors appear at the top of the test tube.
   4. Put the tip of the indicator tube into the headspace of the test tube and draw a sample through the indicator tube.
   5. **Results: see table next page.**

Protein Test

Reagents needed: ninhydrin in glycol
   1. Add about ½ inch of ninhydrin in glycol to a test tube.
   2. Unknown liquid, add several drops to the test tube.
   3. Solid; add a pinch of the solid to the test tube.
   4. Heat the test tube very gently; use a butane lighter.
   5. **Results:**
      a. Solution turns blue or purple; protein is present.

Psilocin Test

Reagents needed: p-dimethylaminobenzaldehyde in alcohol, hydrochloric acid, filter paper.
   1. Add p-dimethylaminobenzaldehyde in alcohol to the unknown.
   2. Add the solution to a filter paper.
   3. Add hydrochloric acid to the paper above the p-dimethylaminobenzaldehyde in alcohol.

GasTec Polytec IV Tube (No. 27)

Sampling time: 30 seconds for 1 pump stroke (100 mL)
Shelf life: 1 year
Reaction principle: See the table below

Detecting Layer	No. Name Original color	1 NH_3 Purple	2 HCl Yellow	3 H_2S White	4 SO_2 Blue	5 NO_2 Yellow	6 CO Yellow	7 CO_2 Blue
Substances		Measurement Results						
Ammonia Amines	Yellow Brown							
Hydrogen chloride		Red						
Hydrogen cyanide							Clear	
Hydrogen sulfide			Brown				Dark brown	
Chlorine					Yellow	Yellow		
Sulfur dioxide					Yellow			
Nitrogen dioxide					Purple	Orange		
Acetylene							Brown	
Carbon monoxide							Brown	
Ethylene							Brown	
Phosphine							Dark brown	
Hydrogen							Gray	
Methyl mercaptan							Yellowish orange	
Propylene							Gray	
Carbon dioxide								Brown

4. Results:

   a. Look for a color change.

Silver Test

Reagents needed: α,α′-dipyridyl ferrous sulfate solution.
   1. Add a few drops of α,α′-dipyridyl ferrous sulfate solution to the unknown solution in a test tube.
   2. **Results:**
      a. Cream-yellow precipitate: silver.

Sugar Test (Fehling)

Reagents needed: cupric sulfate solution, alkaline tartrate solution, 3 N hydrochloric acid.
1. Add 2 drops of 3 N hydrochloric acid to a test tube containing the unknown liquid solution.
2. In a second test tube, add equal amounts (about 1/4 inch) of cupric sulfate solution, and alkaline tartrate solution. Shake the mixture until it turns a rich clear blue color. If the mixture does not turn blue, add a little more alkaline tartrate solution and shake again.
3. Over torch flame, gently heat the acidified unknown solution until it is almost boiling.
4. Add several drops of the cupric sulfate–alkaline tartrate mixture prepared in step 2 to the heated unknown solution.
5. If a precipitation does not form immediately, gently heat the mixture in step 4 a little more over the torch flame.
6. **Results:**
   a. Yellowish orange-to-reddish brown precipitate: sugar.

Sulfate Test (Hahn)

Reagents needed: barium chloride solution.
1. Add a few drops of barium chloride to 1/2 inch of the clear or filtered solution of the unknown in a test tube.
2. **Results:**
   a. White precipitate: sulfate indicated.

Sulfide Test (ASTM method 4978-95)

Reagents needed: lead acetate paper, 3 N hydrochloric acid, distilled water.
1. Add the unknown liquid to form a pool the size of a dime or put 1/2 a pea-sized amount of the unknown solid on a watch glass.
2. Moisten a lead acetate paper with 1 or 2 drops of distilled water.
3. Add 5 drops of 3 N hydrochloric acid to the unknown on the watch glass.
4. Touch the moistened sulfide test strip to the effervescing unknown.
5. **Results:**
   a. A brown, black, or silver color: sulfides.

Sulfite Test (Paul and Smith)

Reagents needed: zinc metal, 3 N hydrochloric acid, lead acetate paper, distilled water, 10 N hydrochloric acid.
1. Add a pea-sized amount of zinc metal to 1/4 inch of the unknown solution in a test tube.
2. Add 1/4 inch of 3 N hydrochloric acid. There should be some effervescence. If there is no effervescence, add a few drops of 10 N hydrochloric acid.
3. Wet a lead acetate paper with 1 drop of distilled water.
4. Hold the wetted strip in the headspace of the test tube.
5. **Results:**
   a. Test strip becomes brownish black or silvery, and a rotten egg odor may be noted: sulfite (see interferences) To be certain, sulfites give off SO_3 when heated, which can be tested with the Gastec® PolyTec IV test.

Tetraethyllead Test (Fiegl)

Reagents needed: Whatman filter paper, dithizone in chloroform, an ultraviolet light such as that used for geology.
1. Place on the center of a filter paper a few drops of the unknown liquid that you suspect is an organometal liquid.
2. Radiate the filter paper using the UV light until the liquid has dried completely.
3. Add 1 or 2 drops of the dithizone/chloroform solution to the filter paper at the same spot at which the unknown liquid was added.
4. **Results:**
   a. Green-to-orange color change: lead indicated.

Thiocyanate Test

Test 1. Feigl
Reagents needed: 3 N hydrochloric acid, ferrous ammonium sulfate, distilled water.
1. a. **Liquid:** Add ½ inch of the unknown to a test tube.
   b. **Solid:** Add a pea-sized amount of the unknown to ½ inch of distilled water.
2. Add ¼ inch of 3 N hydrochloric acid.
3. Add ¼ inch of ferrous ammonium sulfate.
4. **Results:**
   a. Dark, blood red: thiocyanate.

Tin Test (Mazuir)

Reagents needed: potassium iodide, 2 N sulfuric acid, 10 N hydrochloric acid.
1. Add 2 or 3 drops of 2 N sulfuric acid to ½ inch of the unknown solution in a test tube.
2. Add a pinch of potassium iodide.
3. **Results:**
   a. Yellow and orange precipitate: tin indicated.

Titanium Test (Dobrolyubski)

Reagents needed: 3% hydrogen peroxide.
1. Add ¼ inch of 3% hydrogen peroxide to ½ inch of the unknown solution (suspension) in a test tube.
2. **Results:**
   a. The liquid becomes yellow: titanium.

Turpentine Test

Reagents needed: iodine, distilled water.
1. Add a spatula full of iodine crystals to the test tube containing the unknown.
2. **Results:**
   a. Boils vigorously, producing a cloud of purple smoke, and the tube becomes very hot: turpentine.

Urea Test (Kotney)

Reagents needed: methanol, copper wire.
1. Add ½ inch of the unknown liquid (pH of unknown liquid should be 7) or ½ a pea-sized amount of the unknown solid to a test tube. Urea is a waxy appearing solid, often in pellet or prill form. It is very soluble in water and is often in solution.

2. Clean the copper wire by heating it over the torch flame until it glows red. Continue to heat the copper wire until there is no green flame.
3. Plunge the hot copper wire into a test tube containing $\frac{1}{2}$ inch of methanol. Allow the wire to cool in the test tube until the alcohol stops boiling.
4. Reheat the copper wire, and place it in the unknown. Allow the unknown to melt or boil.
5. **Results:**
   a. Unknown turns blue: urea indicated.
   b. Unknown turns blue and then a dull purple: urea confirmed.

Urine Test (Moore)

Reagents needed: phosphotungstic acid, ammonium hydroxide.
1. Add $\frac{1}{2}$ a pea-sized amount of phosphotungstic acid to $\frac{1}{2}$ inch of the unknown liquid in a test tube.
2. Add $\frac{1}{4}$ inch of ammonium hydroxide.
3. Very gently heat the solution almost to a boil.
4. **Results:**
   a. Unknown turns dark blue: urine.

Vanadium Test: (Barreswil)

Reagents needed: concentrated sulfuric acid, 3% hydrogen peroxide.
1. Add several drops of concentrated sulfuric acid to $\frac{1}{2}$ inch of the unknown solution in a test tube.
2. Add $\frac{1}{4}$ inch of 3% hydrogen peroxide.
3. **Results:**
   a. Reddish-brown precipitate: vanadium.

Water Test

Reagents needed: Alka-Seltzer®.
1. Add $\frac{1}{2}$ a pea-sized amount of crumbled Alka-Seltzer into a test tube containing $\frac{1}{2}$ inch of the unknown liquid.
2. **Results:**
   a. Effervescence: water present in liquid.

Zinc Test (Deckert)

Reagents needed: sodium hydroxide, dithizone.
1. Add 1 pellet of sodium hydroxide to a test tube containing $\frac{1}{2}$ inch of unknown solution. Add a small clump of dithizone. Allow it to dissolve.
2. **Results:**
   a. Brilliant red pink precipitate: zinc.

GLOSSARY

The definitions offered in this glossary are for use in explaining, understanding, and using the book. Some definitions may stray from conventional chemistry definitions, but we do so in an effort to ease the use of this book.

acid A chemical with a pH below 7. An acid has the ability to release a hydrogen ion into water in the form of hydronium (H^+). The strength of the acid is a measure of an acid's ability to release a portion of its H^+ into water in a process called *disassociation*. The more H^+ dissociated, the stronger the acid. A strong acid ionizes (disassociates) its hydrogen completely, or almost completely; weak acids ionize to a lesser degree. The pH of water is 7 (neutral), or 1 part hydrogen ion in 10^7 (or 10,000,000) parts water. The pH range of acids is 0 to 7. An acid with a pH of 1 has 1 part hydrogen ion in 10 parts water. Substances with a pH less than or equal to 2 ($pH \leq 2$) are categorized as corrosives by the EPA and are highly irritating and corrosive to body tissues. Some organic acids may have a pH of 3 or 4. Acids react with bases and certain metals to form salts, water, and heat.

acidic Refers to a substance with a pH below 7 ($pH < 7$). *See also* acid.

ACGIH American Conference of Governmental Industrial Hygienists.

addition reaction A process in which two organic molecules are spliced together to make a new molecule. *See also* Grignard reagent.

alkali metal A metal from the first vertical row (1A) of the periodic table. Sodium (Na), potassium (K), and lithium (Li) are the most common. They form colorless or white binary salts that melt and recrystallize when heated. Their hydroxides are water soluble and form strong basic corrosives (pH 14). These metals are water reactive and burn with characteristic flame colors. The alkali metals are very strong reducing agents.

alkaline-earth metal Refers to magnesium (Mg), calcium (Ca), barium (Ba), and strontium (Sr), all of which are in the second vertical row (2A) of the periodic table. Their hydroxides are less soluble and therefore less corrosive than those of the alkali metals; calcium hydroxide is a skin irritant with a pH of 12.4. These metals burn with characteristic flame colors. These are also reducing agents.

amine An organic nitrogen compound with the functional group $-NH_2$. Amines are organic bases; they have a pH of 8 or higher and act very much like ammonia in the lungs, causing a delayed pulmonary edema if inhaled. The higher the pH, the more hazardous the amine.

Chemicals Used for Illegal Purposes: A Guide for First Responders to Identify Explosives, Recreational Drugs, and Poisons, By Robert Turkington
Copyright © 2010 John Wiley & Sons, Inc.

ammonium A cation that can replace any metal in inorganic salts. Written NH_4 or Amm.

ampoule A heat-sealed glass container often used to protect a pyrophoric material from igniting or to prevent sensitive compounds from being degraded by air.

analog A compound with similar electronic structure but different atoms.

anion An ion that has a negative charge. All electron-receiving elements form anions, and most elements (including nonmetals, metalloids, and metals) bond with oxygen to form polyatomic oxysalt anions. A cation is always associated with an anion; when a salt is named, the anion comes second (e.g., the salt formed by the sodium cation and the chloride anion is called *sodium chloride*).

anionic An organic molecule, usually an organic acid such as steric acid. Steric acid is ionized to become the large molecular anion in a sodium salt molecule that is called *soap*.

aromatic *See* BTEX.

Australian Group Group, now consisting of 39 countries, that has met regularly since 1985. When Iraq used chemical weapons in 1984, it became obvious that no one was keeping track of potential precursor chemicals for chemical weapons that were being sold legally to Iraq. Australia proposed a meeting of the countries exporting such chemicals to "harmonize" their procedures.

autoignition Ignition of an unknown or its vapors without contact with flame.

bait A food, sugar, or grain that is mixed with poison in order to kill animals or insects.

base A chemical with a pH above 7. A base has the ability to release its hydroxide into water in the form of an ion, OH^- or ties up a free hydrogen in water, increasing the relative number of hydroxyl anions in a solution. The strength (pH) of the base is a measure of a base's ability to release a portion of its OH^- into water in a process called *dissociation*. The more OH^- dissociated, the stronger the base. A strong base ionizes its hydroxide completely, or almost completely; weak bases ionize to a lesser degree. Substances with a pH of 12.5 or higher (pH 12.5 to 14) are categorized as corrosives by the EPA and are highly irritating and caustic to body tissues.

Note: Hhydroxide (OH^-) is ionically bonded and is not to be confused with the covalently bonded hydroxyl radical ($-OH$) used to form alcohols and other organic chemicals.

basic Refers to a substance with a pH above 7 (pH > 7). *See also* base.

blank A prepared solution containing no unknown, which is being tested concurrently with an unknown solution for purposes of comparison.

BTEX Acronym for a family of aromatic hydrocarbons consisting of benzene, toluene, ethylbenzene, and xylene. The carbon skeleton of these compounds comprises a very stable ring structure with alternating single and double bonds that move within the ring in resonance. Although unsaturated, their stability is similar to that of saturated compounds.

bulking Placing similar chemicals together to limit the number of containers required for shipping or disposal.

carcinogen A substance that causes cancer. Known carcinogens are those proven to cause cancer in general human or animal populations. Experimental, or suspected, carcinogens are those that have been found to cause cancer in laboratory animals.

catalyst A substance that accelerates or retards the rate of a chemical reaction without itself undergoing chemical change.

categorization Sorting compounds by Hazard Class only.

cation An ion that has a positive charge. A cation is always associated with an anion. All metals, most metalloids, and ammonium form cations. When a salt is named, the cation always comes first (e.g., the salt formed when the sodium cation bonds with the chloride anion is called *sodium chloride*). Cations always end in the suffix -ium.

caustic An alkaline material that has a corrosive or irritating effect on body tissues. Caustics have a pH above 12.5 and can damage body tissue without causing initial pain. *See* corrosive.

CDTA (Chemical Diversion and Trafficking Act) Lists chemicals that have an effect deemed to be intoxicating. Some schedules list chemicals that are intoxicating but serve no useful medical purpose. Some schedules list useful medical drugs that are also intoxicating. Other schedules list chemicals that are precursors for illegal drugs.

cheese A street mixture of tar heroin and Tylenol PM®, that has become a "kid-friendly" drug. Has become a major problem in several cities, causing a number of deaths among children from 11 to 19. This is snorted by the user.

clear Refers to a transparent liquid free of cloudiness, suspended solids, and particulates. It does *not* necessarily mean colorless. *See also* dissolves.

cloudy A liquid that is neither opaque nor clear, and which lacks the glowing refraction of opalescence. It can indicate a very weak emulsion or a tenuous suspension. Cloudy solutions usually clear when more water is added, whereas emulsions become more pronounced.

coating A layer of material (e.g., paint, shellac, sealant, varnish) applied to a solid surface for decoration and/or protection from the elements.

combustible Legally, a liquid that has a flash point above 141°F (60.5°C) and below 200°F (93°C). A liquid that quickly or slowly wicks through a match and burns with a small flame is considered combustible.

common Often found in commerce, industry, or nature and therefore is likely to be encountered. *See also* rare.

compound Denotes a homogeneous material made up of bonded elements.

concentration   The ratio of a given compound, such as an acid, to a second compound in which it is soluble. As a relative term, a concentrated solution would contain a high percentage of a compound.

conjugated   Alternating single and double bonds within a carbon skeleton.

controlled substances   Drug substances listed as of January 1 1988 under Schedules 1 through 5 of the U.S. Title 21 Code of Federal Regulations (CFR), Section 1300 to the end.

cookie dough   Slang for waste from the pill extraction process.

corrosive   Liquids or solids that cause full-thickness destruction of human skin at the site of contact within a specified period of time or exhibit a corrosion rate on steel or aluminum surfaces exceeding 0.25 inch (6.35 mm) a year at a test temperature of 130°F (55°C) (49 CFR 173.136).   Most corrosives can be identified by pH. According to the EPA, corrosives have a pH equal to or less than 2 (pH < 2); or a pH equal to or greater than 12.5 (pH > 12.5). Weak acids can also be corrosive; phenol, with a pH of 5, is typical.   The EPA definition groups acids (corrosives) and bases (caustics) together as *corrosives*.

covalent bonding   Bonding in which pairs of electrons are shared by atomic nuclei. Covalent bonding is the primary difference between organic and inorganic compounds and occurs significantly in organic chemistry.

crystals   A fuzzy deposit on the lid or around the top of a container, indicating the possible presence of a powerful shock- or friction-sensitive explosive such as peroxides or picric acid.   Crystals are also a solid with smooth faces and sharp edges, similar to quartz crystal. Ultimately, crystals refer to the structure of the molecule.

curdling   Formation of a thick white film or skin between the unknown and water.

decanting   Pouring off or pipetting a clear liquid from one test tube to another, leaving any heterogeneous solids or liquids in the first test tube.

deflagrates   Undergoes a rapid combustion of explosive particles typical of low explosives such as gunpowder. The shock wave is subsonic and may sound like a "thump." *See also* detonates *and* explosive.

deliquescent   Refers to a solid material that has a tendency to absorb water from the air and become liquid.

detection limit   The smallest level of a chemical that can be detected by a given method, often measured in parts per million.

detonates   Undergoes an extremely rapid explosive decomposition typical of high explosives such as nitroglycerin and picric acid. Detonation usually indicates the presence of nitrogen. The shock wave is supersonic and may sound like a sharp "crack." *See also* deflagrates *and* explosive.

dissociation   The breaking up of compounds into simpler units with the application of heat or a solvent.   For example, a salt in water will dissociate into cations and anions that move about in the water. In the case of acids and bases, water dissociates to hydronium cation and hydroxide anion.

dissolves Some or much of a liquid, not all, passes into water; however, a significant portion remains separate from the water. The compound becomes saturated in the water at some level less than 100%. This can range from trace amounts to percentages. At higher levels a compound that dissolves in water will form a flat meniscus.

DOT Department of Transportation. This U.S. government entity controls the transportation of hazardous materials on public transportation routes.

DTCC Part of the U.S. Department of Defense, which issues permits to people exporting chemical weapon–related chemicals.

EEC (European Economic Community) Free trade infrastructure used by several European nations.

effervescence Small, aggressive bubbles in acid or water, similar to the foam one sees in a glass of soda. This reaction can be cool (bicarbonate of soda) or hot (calcium metal), and may generate flammable and/or toxic gas.

element One of the 118 presently known substances that compose all matter at and above the atomic level. When an atom of an element gains an electron, it becomes an anion (–); when it loses an electron, it becomes a cation (+).

emulsion A uniform, milky, opaque formation created when certain liquids are added to water. Unlike a heterogeneous mixture such as salad dressing, a true emulsion does not separate.

EPA (U.S. Environmental Protection Agency) This U.S. government agency controls chemicals and actions which degrade the environment.

EPCRA (Emergency Preparedness and Community Right-to-Know Act) Mandates that public and local governments be provided with information about chemical hazards in their communities, and specifies requirements for emergency response. Under the provisions of this act, air, water, and land releases of hazardous chemicals are reported and entered into the national Toxic Release Inventory.

evaporation The change of a compound from a liquid to a vapor, in which the vapor has the same chemical structure as the liquid.

explosive A compound that detonates (high explosive) or deflagrates (low explosive) when subjected to heat, friction, and/or shock. Note that low explosives are not necessarily less hazardous than high explosives; the destructive potential of a given explosive depends on many other factors.

extremely flammable Flash point below 32°F.

flammable Flash point below 100°F OSHA; flash point below 140°F EPA.

flash point　We have not attempted to list how the flash points were determined. There are several methods, and each gives a different number. We feel that it is beyond the knowledge of our audience to understand the differences in a manner that would guide them in their chosen actions. Therefore, the flash points in this book are general and provide only a relative idea of how flammable a compound may be. We strongly recommend looking at the vapor pressure to determine the hazard from ignition of the material.

floats　Liquids are considered to float only if they form a distinct line of separation and remain on top of the water. Any liquid that floats is organic. A sharp line indicates a longer carbon chain; a more diffuse meniscus indicates a polar functional group.

flocculate　A clustering of suspended particles to form small clots or clumps.

foams　As an indicator of soap or detergent, foam should be substantial, such as one would see through the window of a washing machine. Foaming is an indicator of a surfactant.

fumigant　A compound that releases a toxic vapor, usually upon reaction with water or acid. Fumigants are used to kill rodents or insects in enclosed areas such as granaries or barns.

fuming　Refers to any vapor or smoke-like cloud visible above a compound. The true meaning is a recondensed metal over a molten metal. However, this word has more generally been used to describe the vapors over an acid.

functional group　An atom (other than hydrogen) or group of atoms covalently bonded to the carbon skeleton. The functional group defines the characteristics of an associated family of organic compounds.

gel　A viscous, jellylike substance, usually consisting mostly of water.

glass　A crystalline amorphous form of a compound.

gram-positive　When a sample is dyed with methyl violet or crystal violet, washed with iodine solution and ethanol, gram-positive bacteria turn blue. Anthrax is a gram-positive bacteria. Usually, gram-positive bacteria have a very thick cell wall, and teichoic acids are present, which act as chelating agents and for adherence.

granules　A free-flowing, chunky, amorphous material similar in size to coarse sand.

Grignard reagent　Adding magnesium metal to a halogenated hydrocarbon forms a Grignard reagent with the functional group R-MgCl. This compound reacts with other compounds in an addition reaction.

halogen　An element in group 7 of the periodic table. Listed in order of decreasing reactivity, the halogens are fluorine, chlorine, bromine, and iodine. These elements are aggressive electron receivers, and they form strongly oxidizing oxysalts.

halogenated hydrocarbon　A hydrocarbon in which one or more hydrogen atoms have been replaced by fluorine, bromine, chlorine, or (rarely) iodine. The vapors of many halogenated hydrocarbons are toxic, narcotic, and/or carcinogenic. Many fluorinated hydrocarbons, such as Freons®, are nontoxic but are volatile and their vapors are heavier than air. Brominated hydrocarbons, on the other hand, can have immediate effects, especially on the liver. Halogenated hydrocarbons can displace oxygen in confined areas and cause suffocation. Fluoro-, chloro-, and bromo hydrocarbons are used to put out fires. Mixtures of two halogens are common; these include Freons and chlorofluorocarbons (CFCs).

Hazard Class   DOT Hazard Classes are the numbers on the bottom of the plaquards.
  Explosives
      1.1   mass explosion hazard
      1.2   projection hazard
      1.3   predominately fire hazard
      1.4   no significant blast hazard
      1.5   very insensitive explosive with mass explosion hazard
      1.6   extremely insensitive articles
  Gases
      2.1   flammable gases
      2.2   nonflammable gases
      2.3   toxic gases
  Flammable liquids, 3
  Flammable solids
      4.1   flammable solids
      4.2   spontaneously combustible materials
      4.3   water-reactive substances/dangerous when wet
  Oxidizers
      5.1   oxidizing substances
      5.2   organic peroxides
  Toxic and infectious substances
      6.1   toxic substances
      6.2   infectious substances
  Radioactive materials, 7
  Corrosive substances, 8
  Miscellaneous hazardous materials, 9

headspace   The area above the compound inside a test tube.

heterogeneous   Any mixture or solution composed of two or more distinct substances. In heterogeneous mixtures it is usually possible to see the separate parts (e.g., sand and rocks, oil and vinegar).

homogeneous   Any mixture or solution composed of two or more distinct substances in which it is impossible to distinguish the components in the resulting mixture.

hydrocarbon   A gas, liquid, or solid composed exclusively of carbon and hydrogen covalently bonded. These are the basic building blocks of organic compounds.

hydrolyzing   Refers to an unknown that reacts with water to form two or more new compounds, often with the generation of heat. For example, dimethyl sulfate reacts with water to form methane and sulfuric acid. Such reactions can be very dangerous.

ICC (Interstate Commerce Commission)   Abolished in 1995. Its functions were transfered to The Surface Transportation Board.

identification   Knowledge of the specific name (e.g., acetonitrile) or family (e.g., single-bonded nitrile with a small carbon chain) of a previously unknown compound. Identification is necessary for accurate determination of hazard, especially in the case of poisons and certain oxidizers. *See also* categorization.

IDLH (immediately dangerous to life and health)   Any condition that poses an immediate or delayed threat to life, or that would cause irreversible adverse health effects, or that would interfere with a person's ability to escape unaided from a permit space (OSHA 1910.146).

inert   When applied to an unknown, this means that no visible activity is taking place; it is *not* meant to imply that the material is nonhazardous. When applied to a specific chemical or group of chemicals, such as titanium dioxide or silicates, it denotes the absence of any significant chemical reactivity.

inorganic   Inorganic chemistry deals with those substances (such as rocks, water, salts, and elements) that do not contain a covalent carbon–hydrogen bond and are not derived from a compound that contains a covalent carbon–hydrogen bond.

interference   A reaction that either obscures or prevents a positive result, causing a false negative or positive result.

intermediate   Usually, an organic chemical that represents a stage between a raw industrial compound and the manufactured end product. Often, these are very reactive and cannot be purchased off the shelf. They are more likely to be seen inside reaction vessels.

ion   An atom or group of atoms with an electric charge. *See also* cation *and* anion.

ionic bond   A bond formed by the electrostatic attraction of anions and cations.

kid-friendly mixes   Methamphetamine and heroin laced with candy and strawberry flavoring to appeal to youth. Since these are snorted and not injected, and sport names that do not include the primary drugs, they are more appealing to younger children.

laboratory bottle   Compressed gas cylinder, which is small enough to be used in the laboratory or the classroom, usually about 18 inches long.

LC_{50}   Lethal concentration to 50% of the animals exposed to an atmosphere.

LD_{50}   Lethal concentration to 50% of the animals exposed by ingestion or skin contact.

lecture bottle   *See* laboratory bottle.

MCA (Manufacturing Chemists' Association)   A private association that assigns levels of hazard to chemicals.

meniscus   Interface between a liquid and water in a test tube.

melts   The unknown becomes a liquid when heated and solidifies when cool, but does not change in any other way.

metallic   Refers to a substance that resembles a metal but is not necessarily a metal. The term *metal* is reserved for elemental metals. Thus, iodine (a nonmetal) is metallic, whereas a metal cyanide is a cyanide in which the cation is a metal but which may not have a metallic appearance (e.g., copper cyanide). *See also* nonmetallic.

metalloids   A group of nonmetallic elements so called because they are more similar to metals than to their fellow nonmetals. They are also known as *semiconductors*, because their electrical conductivity is between that of the nonmetals and the metals. The important metalloids are arsenic, antimony, boron, carbon, phosphorus, selenium, silicon, and sulfur. Chemistry books disagree as to which elements make up the melalloids. Most metalloids can form both anions and cations. *See also* metals *and* nonmetals.

metals   About two-thirds of the elements in the periodic table are metals. Most are crystalline solids with a metallic luster and high electrical conductivity. In salts, metals form cations.

metal salt.   An ionically bonded salt in which a metal is the cation.

milkshake   A fouled chemical reaction caused by the addition of guaifensin to a product which has been extracted.

miscible   A liquid, such as acetone, that mixes into water at any ratio. These are polar liquids with three or fewer carbons or which contain multiple functional groups. *See also* dissolves.

mordant   Any corroding substance used to etch lines into a surface. However, for dye processes a mordant is used to fix the dye to the material and is usually a metallic compound that combines with an organic dye to produce an insoluble colored compound.

MSDS (material safety data sheet)   This is a written description of a chemical concerned particularly with its hazardous characteristics.

multiphase   Describes a mixture with two or more distinct layers that are composed of one or more liquids and/or solids.

NaK   Sodium potassium–metal mixture. NaK is a liquid used in industrial thermostats, as a coolant for nuclear reactors, and for other high-temperature applications. NaK is extremely water reactive.

noble metals   The noble metals are gold, silver, platinum, and palladium. They are called "noble" because they react to a limited extent with other elements and tend to be found as metal as opposed to salt. They are often used in jewelry, as they do not tend to rust.

nonmetallic   Traditionally, this term refers to a group of about 25 elements with properties different from those of metals. It usually describes an element that does not have a metallic appearance, regardless of whether the unknown is actually a nonmetal.

nonmetallic acid precursors (NMAPs) A group of inorganic compounds made up of nonmetallic elements. These water-reactive solids and liquids generally hydrolyze in water, forming strong acids; chlorinated compounds such as boron trichloride and thionyl chloride release large amounts of hydrogen chloride gas when wet.

nonmetals Nonmetals are elements with markedly different properties from metals. They generally form anions, whereas metals generally form cations. About 25 elements are classed as nonmetals. Metalloids (semiconductors) separate the rest of the nonmetals from the metals. Many nonmetals are gases (e.g., hydrogen, nitrogen, oxygen, and the halogens), bromine is a liquid, and iodine, a solid. The solids include sulfur and phosphorus. *See also* metalloids *and* metals.

nonpolar Organic compounds that do not have polar functional groups, or derivatives that contain nonpolar elements. These include hydrocarbons and halogenated hydrocarbons. Inorganic binary elements, such as chlorine gas, and most elements in their base state. *See also* polar.

nonreactive *See* inert *and* reactive.

nonsoluble All things are soluble at some point. "Nonsoluble" indicates compounds where solubility is so slight, indication of solubility is not visible to the naked eye.

n.o.s. not otherwise specified.

oil A liquid with a viscosity of 3 to 5, derived from petroleum, vegetables, or animals. Oils are liquid at room temperature, 68 to 77°F or 20 to 25°C. Fats are solids.

opalescent Exhibits a play of colors similar to the refracted light of an opal. Opalescence is a translucent color that is more than white, as it shows flashes of all the colors throughout the surface. It is similar to usually not as spectacular as the inside of an oyster shell.

organic Describes a compound that contains covalently bonded carbon and hydrogen atoms, or is derived from such a compound. *See also* inorganic.

OSHA Occupational Safety and Health Agency.

overpack The combination of various packaged chemicals into a single container such as a drum.

oxidizer An element or compound that takes electrons in its reactions with other elements or compounds. This diverse group of chemicals includes chemically reactive compounds such as nitric acid, and thermally reactive compounds such as picric acid. Many oxidizers contain sufficient excess oxygen to initiate or promote the combustion with other materials.

oxysalt A salt in which oxygen bonds with another element to form a polyatomic anion. The reactivity of an oxysalt depends on the number of oxygen atoms present and the characteristics of the other elements in the salt.

PC Packing Class.

PEL (permissible exposure limit)   The permissible concentration of air contaminants to which nearly all workers may be exposed repeatedly eight hours a day, 40 hours a week, over a working lifetime of 30 years, without adverse health effects, per OSHA.

pellets   Materials similar to pills or prills, which appear to have been manufactured by a machine.

PG   Packing Groups. The Packing Group for a chemical indicates the degree of hazard associated with its transportation.

    Group 1 (Great danger)
    Group 2 (Medium danger)
    Group 3 (Minor danger)
    Group 4 (n.o.s.) with proper Hazard classification label.

pH   A scale of 0 to 14 that represents the acidity or alkalinity of an aqueous solution based on excess hydrogen or hydroxide ions present. Pure water has a pH of 7, with a balance of covalently and ionically bonded water at 1 part per 1×10^7 (or 10,000,000) parts ionic water. (*Note*: One H^+ added to an OH^- equals H_2O.) Bases, with and excess of OH^-, range from 7.1 to 14; acids, with an excess of H^+, range from 0 to 6.9. There is NO pH if there is no water. The technical definition of pH is the inverse logarithm of the hydronium concentration in water. The logarithm scale shows changes of concentration in powers of 10. A change from pH 6 to a pH 4 is 10 times more than a change from pH 6 to pH 5 since the changes are measured in powers of 10. *See also* acid *and* base.

polar   An organic compound with a functional group that allows it to mix with water, as opposed to the complete separation typical of nonpolar liquids. Polar functional groups include alcohols, ketones, amines, acids, and nitriles. Many salts are polar inorganic compounds. These tend to dissolve in water.

ppm (parts per million)   Represents the concentration by volume of a particular substance in a liquid or solid. 1% = 10,000 ppm.

precipitate   A clustering of small solid particles into an opaque mass. Precipitates form when a reagent is added to a liquid unknown as the result of an insoluble salt forming or a chelating agent pulling a metal out of solution. Many reagents work by reacting to form precipitates.

precursor   A chemical used to make another chemical. In this book it often refers to scheduled chemicals used to make illegal drugs or chemical weapons.

prills   Small manufactured balls. Ammonium nitrate and urea are often found in this form.

pyro-   A prefix indicating that a salt can burn.

pyrophoric   A substance that ignites spontaneously in air at 130°F (54°C) or below. In certain cases the reaction is caused by moisture in the air. Pyrophoric materials are often stored under inert gases such as argon, under solvents such as kerosene, or in glass ampoules.

qualitative   Quality or presence. Refers to detection of the presence of an unknown chemical. Although detection levels are significant, this usually refers to identification, not determination, of the amount of the material.

quantitative   Quantity or amount. Refers to detection of a specific concentration or amount of an unknown chemical. *See also* detection limit.

rare   Refers (in this book) to a chemical that is seldom used in industry and does not occur widely in nature. It is, therefore, not likely to be found.

RCRA (Resource Conservation and Recovery Act)   Principal federal law governing solid and hazardous waste.

reaction   A change or transformation caused by the interaction of a chemical compound with another chemical compound by heat or cooling or over time.

reactive   A chemical that reacts with a wide veriety of other types of chemicals. Does not play well with others.

reagent   A chemical that reacts in a predictable manner with some other chemical and is therefore used to identify that chemical.

recrystallize   After melting when heated, the unknown cools to a crystalline solid that is the same or resembles the compound before heating. This reaction is definitive for alkali metal salts.

reducer   An element or compound that gives up electrons in its reactions with other elements or compounds. Reducers are the opposite of oxidizers, which take electrons. The oxidizer/reducer scale is much like the pH scale, in that the farther the two compounds are from each other on the scale, the more violent the resulting reaction of the two will be. Organics are reducing, which is why they usually react violently with oxidizers. The alkali metals are the strongest reducers.

refractory   Describes a compound that is difficult to melt or work when heated. Refractories are used to line furnaces or as insulation.

regulated waste   A hazardous waste whose transport and disposal must be carried out according to federal regulations. Failure to comply with these regulations may lead to fine or imprisonment.

R-group   R is used to denote a part of an organic molecule that does not have a part in the reaction being considered. Since R can mean anything, it is simply a way to indicate that there is more to the molecule than is seen in a diagram.

room temperature   This is not a specified number, and can vary; however, it is considered to be between 20 and 25°C (68 to 77°F).

RTECS (Registry of Toxic Effects of Chemical Substances)   Database of toxicological information, maintained and updated by NIOSH.

SADT (self-accelerating decomposition temperature) In organic peroxides, the temperature at which irreversible molecular decomposition begins. The industry definition defines an SADT as the temperature at which after one week, 50% of the product will be degraded. Temperatures above the SADT will accelerate the decomposition. The reaction may result in fire or explosion.

salt The compound formed by an anion and a cation. Salts are formed when the hydrogen in an acid is replaced by ammonia or a metal and in ionically bonded compounds.

SARA (Superfund Amendments and Reauthorization Act) This act amended CERCLA. The purpose was to prevent future Superfund sites by controlling waste.

saturated A single-bonded carbon chain; *saturated* means that no more hydrogen atoms can be added to the chain.

Schedule 1 A list of substances that are controlled under U.S. federal laws, are deemed to have a high potential for abuse, and for which there is no accepted medical use.

Schedule 2 A list of substances that are controlled under U.S. federal laws, are deemed to have a high potential for abuse, and for which there is an accepted medical use.

screening test A test that confirms or denies the presence of a broad range, or family, of compounds, or ascertains a hazard category.

SCUBA (self contained breathing apparatus) Consists of a compressed air cylinder, a regulator, and a mask.

sinks Liquids are considered to sink when they form a distinct line of separation with the water above.

sludge A thick, viscous mixture of various compounds (especially oils, resins, solvents, and suspended solids with water).

soap bark Bark of a tree (*Quillaja saponaria*) that contains guillaic acid and guillaia-sapotoxin. It is very poisonous and contains marked foam-producing properties. It has been suggested to be medicinal, and is used in taxidermy to preserve furs.

solubility Refers to the tendency of an unknown to mix in water, although not necessarily completely, at ambient temperature. *See also* miscible, nonsoluble, *and* soluble.

soluble A compound that mixes partially when added to water. Solubility implies that a percent, as low as parts per billion or up to levels as high as 25% of the compound, has mixed with the water. Any mixing with the water will pollute the water, as biological systems are quite likely to be disrupted by foreign compounds even at the parts per billion level. More soluble compounds (5% or so) can be identified by a flat meniscus.

solution Any mixture of an unknown with water. It is more formally correct to say: any two compounds that have been mixed to form what looks like a homogeneous mixture.

solvent   A liquid capable of dissolving another substance. The most common solvent is water. Usually, the term describes a volatile liquid with the general appearance and viscosity of water. Most industrial and household solvents are flammable and/or toxic.

spiderwebs   Stringy, gritty fragments of partially burned organics or siloxanes which appear in smoke and remain in the air after the smoke is gone. Spiderwebs may be black, white, or gray.

stable   Refers to a compound that is not likely to react in contact with most other compounds. Note that such a compound is not necessarily inert. *See also* inert.

strawberry-quick   Street name for a mixture of methamphetamine and a strawberry-flavored powdered drink mix.

streamers   Long, solid strings that fly from a mouth of a test tube when a compound is heated in the tube. One end of the string remains attached to the rim of the tube.

sublimes   A solid that goes from the solid state to the gaseous state without going through the liquid state at a specific set of temperatures and pressures. At different pressures and temperatures, the compound may go through the liquid phase.

suspension   An opaque solution that differs from an emulsion in that it is a solid suspended in a liquid, whereas an emulsion is a liquid suspended in a liquid. All suspended solids will eventually settle.

tars   Forms a dark, sticky liquid similar to burned caramel when heated.

TCLP (toxicity characteristic leaching procedure)   Designed by the EPA to identify organic and inorganic toxins such as pesticides, heavy metals, and solvents that are likely to contaminate groundwater if mishandled. Any waste that contains these compounds at or above the toxicity characteristic (TC) regulatory level is assigned a specific EPA Hazardous Waste number.

TLV   Exposure guidelines intended for use in the practice of industrial hygiene for control of potential health hazards. Published by the ACGIH.

toxicity   The ability of a material to injure or kill living organisms through ingestion, inhalation, or absorption. The hazard of a toxic compound depends upon such factors as its volatility, solubility, rapidity of action, and mode of entry into the body.

transition metals   Elements in the center of the periodic table, the salts of which exhibit several oxidation states usually associated with bright coloration. The transition metals that are encountered most often are titanium, vanadium, chromium, manganese, iron, cobalt, nickel, copper, molybdenum, silver, tungsten, and gold.

TSDF (treatment, storage, and disposal facility)   Facility licensed to process hazardous waste.

unsaturated   Double- or triple-bonded carbon chains or rings. Unsaturated hydrocarbons turn red in the Iodine Crystal Test and generate black smoke and black spiderwebs when ignited.

unstable   Refers to a compound very likely to react, often violently, in contact with air, water, most other compounds, or when heated.

viscosity   A measurement of a particular liquid's resistance to flow relative to that of water. We have not used the unit centipoise because very few people understand that standard. In place of centipoise we have used words like *oily* or have assigned viscosity numbers from known materials for comparison.

Viscosity 1	methanol
Viscosity 2	ethylene glycol
Viscosity 3	corn oil
Viscosity 4	automatic transmission fluid
Viscosity 5	10W-30 motor oil
Viscosity 6	glycerin, USP

volatility   The tendency of a compound to evaporate. The more quickly a compound evaporates, the more potential it has to cause injury. *See also* evaporation.

w   Used to mark compounds that we could not find in reference books, on the Web, or in specialty chemical catalogs. These may be obscure synonyms for a chemical, a short-lived laboratory intermediate, chemicals that were banned by being listed, and or chemicals improperly named by amateur chemists making contraband materials.

water reactive   A material that explodes, burns, changes volume, changes temperature, effervesces, fumes, ignites, or evolves toxic or flammable gases after contact with water.

zone   Zones are assigned to certain chemicals by DOT and listed in the Emergency Response Guide. ZoneA: gases LC_{50} less than 200 ppm; ZoneB: gases LC_{50} greater than 200 but less than 1000 ppm; ZoneC: gases LC_{50} greater than 1000 but less than 3000 ppm; ZoneD: gases LC_{50} greater than 3000 but less than 5000 ppm.

BIBLIOGRAPHY

Aldrich Chemical Company, *Catalog Handbook of Fine Chemicals,* 1996–1997.

Amoore, John E., and Earl Hautala, *Odor as an Aid to Chemical Safety*, draft manuscript 4/27/81, Department of Human Services, State of California.

Arthur, Paul, and Otto Smith, *Semi-micro Qualitative Analysis*, McGraw-Hill, New York, 1942.

Brady, George S., Herry R. Clauser, and John A. Vaccari, *Materials Handbook,* 14th ed., McGraw-Hill, New York, 1997.

Brown, Robert E., *LSD: The Psychedelic Guide to Preparation,* Eucharist, 1967.

Burdick, Brett A., ed., *Hazardous Materials Training: Public Safety Response to Terrorism*, Student Manual, Virginia Department of Emergency Services, Richmond, VA, 1997.

Chandlee, Grover, C. Mack Pauline Beery, and Arnold J. Currier, *Experiments in General Chemistry: Qualitative Analysis*, 9th ed., The Pennsylvania State College, State College, PA, 1949.

Chewbacca Darth, *The Whole Drug Manufacturers' Catalog,* Prophet Press, Manhattan Beach, CA, 1977.

CIA Field Expedient Methods For Explosives Preparations, no author given, Desert Publications, El Dorado, 1977.

Cooper, Donald A., *Future Synthetic Drugs of Abuse,* Drug Enforcement Administration, McLean, VA.

Curtman, Louis J., *Introduction to Semi-micro Qualitative Chemical Analysis*, Macmillan, New York, 1950.

Emergency Management Institute, *Radiological Emergency Management Independent Study Course*, EMI, Emittsburg, MD, 1994.

Federal Emergency Management Agency, *The Federal Response Plan for Public Law 93-288*, FEMA, Washington, DC, as amended, 1992.

Federal Emergency Management Agency, *The Federal Response Plan for Public Law 93-288: Terrorism Incident Annex,* FEMA, Washington, DC, 1996.

Feigl, Fritz, *Qualitative Analysis by Spot Tests*, 3rd ed., Elsevier, New York, 1946.

Feigl, Fritz, *Spot Tests in Organic Analysis*, 7th ed., Elsevier, New York, 1966.

Feigl, Fritz, and Vinzenze Anger , *Spot Tests in Inorganic Analysis*, 6th ed., Elsevier, New York, 1972.

Gosselin, Robert E., Harold C. Hodge, Roger P. Smith, and Marion N. Gleason, Acute poisoning, in *Clinical Toxicology of Commercial Products,* 4th ed., Williams & Wilkins, Baltimore, 1976.

Gottlieb, Adam, *The Psilocybin Producers Guide,* 1976.

Hampel, Clifford A., *Rare Metals Handbook,* 2nd ed., Reinhold, London, 1961.

Handbook of Chemistry and Physics, 55th ed., CRC Press, Cleveland, OH, 1975.

Hawley, Gessner G., *The Condensed Chemical Dictionary,* 10th ed., Van Nostrand Reinhold Company, New York, 1981.

Hetayama, H. K., J. J. Chen, E. R. de Vera, R. D. Stephens, and D. L. Storm, *A Method for Determining the Compatibility of Hazardous Wastes,* U.S. Environmental Protection Agency, Grant R804692, Municipal Environmental Research Laboratory, Cincinnati, OH, 1980.

Hutton, Willbert, *General Chemistry Laboratory Text, with Qualitative Analysis,* Charles E. Merrill, Columbus, OH, 1968.

Internal Revenue Service, *Methods of Analysis for Alkaloids, Opiates, Marijuana,Barbiturates and Miscellaneous Drugs,* Publication 341, Rev. 12-66, IRS, Washington, DC, 1966.

Jacobs, Morris B., *The Analytical Chemistry of Industrial Poisons, Hazards and Solvents,* Interscience, New York, 1941.

Jungreis, Ervin, *Spot Test Analysis: Clinical, Environmental, Forensic, and Geochemical Applications,* Wiley, New York, 1985.

Kosanke, K.L. and B.J., Kosanke, lecture notes for pyrotechnic chemistry, parts 1 and 2, *Journal of Pyrotechnics,* 1997.

Laboratory Guide for Investigators and Chemists, Drug Enforcement Agency, Washington, DC, 1993.

Laboratory Methods for Preparing Promising Explosives, no author or publisher given.

Ledgard, Jared, *The Preparatory Manual of Explosives,* 2nd ed., Paranoid Publishing Group, La Port, 2002.

Ledgard, Jared, *The Preparatory Manual of Narcotics,* Vol 1, Paranoid Publishing Group, La Port, 2003.

Ledgard, Jared, *The Preparatory Manual of Chemical Warfare Agents,* Paranoid Publishing Group, La Port, 2004 .

Leichnitz, W.M., Air investigations and technical gas analysis with Drager tubes, *Detector Tube Handbook,* 7th ed., July 1998, Dräger Safety AG & Co., Lübeck, Germany, 2001.

Lewis, Grace, *1001 Chemicals in Everyday Products,* 2nd ed., Wiley, New York, 1999.

Meidl, James H., *Explosive and Toxic Hazardous Materials,* Macmillan, New York, 1970.

Merck, *Merck Index,* 5th ed., Merck, Rahway, NJ, 1940.

Merck, *Rapid Test Handbook,* Merck, Darmstadt, Germany, 1987.

Meyer, Eugene, *Chemistry of Hazardous Material,* Prentice-Hall, Englewood Cliffs, NJ, 1977.

Middlebrook, John L., Chapter 33, in *Botulinum Toxins,* Dugway, UT.

National Fire Protection Association, *Guide on Hazardous Materials,* 8th ed., National Fire Protection Association, Quincy, MA, 1984.

National Research Council, *Odors from Stationary and Mobile Sources,* NRC, Washington, DC, 1979.

New Jersey Department of Health and Senior Services, Auramine, in *Hazardous Substance Fact Sheet,* State of New Jersey, February 2004.

Oregon Lung Association, *Warning Properties of Industrial Chemicals,* Occupational Health Resource Center, OLA, Tigand, OR.

Patty, Frank A., *Industrial Hygiene and Toxicology, 2nd rev. ed.,* Interscience, New York, 1963.

Pickett, Mike, *Explosives: Identification Guide,* Delmar, Albany, NY, 1998.

Potential Military Chemical/Biological Agents and Compounds, NAVFAC P-467 AFR 355-7.

Powell, William, *The Anarchist Cookbook,* Barricade Books, Fort Lee, NJ, 1971.

Sax, Irving, *Dangerous Properties of Industrial Materials, 2nd ed.*, Reinhold, New York, 1963.

Saxon, Kurt, *Poor Man's James Bond,* Atlan Formularies, 1987.

Schilt, Alfred A., *Perchloric Acid and Perchlorates*, The G. Frederick Smith Chemical Company, Columbus, OH, 1979.

Shea, Dana A., *Small-Scale Terrorists Attacks Using Chemical and Biologicals: An Assessment Framework and Preliminary Comparisons,* Congressional Research Service, Washington, DC, May 2004.

Sorum, C. H., and J. J. Lagowski, *Introduction to Semi-micro Qualitative Analysis, 5th ed.*, Prentice-Hall, Englewood Cliffs, NJ, 1977.

Steiner, Bradley J., *Death Dealer's Manual,* Paladin Press, Boulder, CO, 1982.

Stevens, Deborah, *Deadly Doses: A Writer's Guide to Poisons,* Writer's Digest Books, Cincinnati, OH, 1990.

The Terrorist's Encyclopedia, no author or publisher given.

Turkington, R., *HazCat Chemical Identification Manual,* Haztech Systems, Mariposa, CA, 2001.

Turkington, R., *Methamphetamine Chemical Waste Identification System,* Haztech Systems, Mariposa, CA, 2001.

Turkington, R., *Weapons of Mass Murder,* Vanity, 2005.

Uncle Fester, *Secrets of Methamphetamine Manufacture,* 4th ed., Loompanics Unlimited, Port Townsend, MA, 1996.

U.S. Army, *Material Safety Data Sheets for: GA, GB, VX, GD, CS, Lewisite, and HD*, U.S. Army Edgewood Research, Development and Engineering Center. Class handout.

U.S. Army, *CBDCOM Emergency Response to Incidents Involving Chemical and Biological Warfare Agents.*

U.S. Army Medical Research Institute of Chemical Defense, *Medical Management of Chemical Casualties Handbook,* 3rd ed., Aberdeen Proving Ground, MD, 2000.

U.S. Army Medical Research Institute of Infectious Diseases, *Medical Management of Biological Casualties Handbook,* 4th ed., U.S. Army, Frederick, MD, 2001.

U.S. Department of Health Education and Welfare, National Institute for Occupational Safety and Health, *NIOSH Manual of Analytical Methods,* 2nd ed.,1977.

U.S. Department of Health Education and Welfare, National Institute for Occupational Safety and Health, *NIOSH Pocket Guide to Chemical Hazards*, June 1994.

U.S. Department of Health Education and Welfare, National Institute for Occupational Safety and Health, *Occupational Diseases: A Guide to Their Recognition*, June 1977.

U.S. Department of Transportation, *North American Emergency Response Guidebook*, U.S. DOT, Washington, DC, 1996.

U.S. Environmental Protection Agency, *Acute Hazardous Events Data Base,* EPA Contract 68-02-4055, December 1985.

Von Oettingen, W. F., *Poisoning*, Harper Brothers, New York, 1952.

White House, *United States Policy on Counterterrorism,* Presidential Decision Directive 39, Washington, DC;1995.

WEB SITES VISITED:

Airliquid.sa.france

akzomobel.com

alb2c3.com

albemarle.com

allaboutbeer.com

ansci.cornell.edu

answers.com

aps.anl.gov

aquacheck.com
arcadiaherbsandalternatives.com
aristatek.com
arkema.com
artcraftchemicals.com
australiagroup.net
bbc.co.uk
biotech.icmb.utexas.edu
botanical.com
brantleycenter.com
bt.cdc.gov
cannabis-seeds.biz
cbwinfo.com
cdnisotopes.com
ces.nesu.edu
cfsan.fda.gov
chembargains.com
chemblink.com/products
chemexper.com
chemicalland21.com
chemindustry.com
chip.med.nyu.edu
cmcsb.com
ctd.mdibl.org
designer-drug.com
dictidie.net
diseyes.lycaeum.org
dmt123.com
dosfan.uie.edu
drugfree.org
ecu.edu/oehs
environmentalchemistry.com
epa.gov
erowid.org
everything2.com
fasthealth.com
fishersci.com
gardenguides.com
globalsecurity.org
glymes.com
greydragon.org
healthed.msu.edu
health.enotes.com
health.howstuffworks.com
hummelcroton.com
iaspub.epa.gov
itcilo.it
jtbaker.com
kathryncramer.com

kuhj.com
leeners.com
lenntech.com
library.thinkquest.org
llnl.gov
mathesontrigas.com
medsafe.govt.nz
medterms.com
members.tripod.com
museum.gov v-serv.com
nanotech.wisc.edu
ncnatural.com
neis.com
neonjoint.com
newsearchch.chemexper.com
nextag.com
nicotinevictims.com
nj.gov/health
nti.org
optcorp.com
orgsyn.org
osha.gov
oxy.com
par-chem.com
pesticideinfo.org
physchem.ox.ac.uk
pmep.cce.cornell.edu
powerlabs.com
praxair.com
predatordefense.org
projectghb.org
ptcl.chem..ox.ac.uk
pubmedcentral.nih.gov
pubs.acs.org
purdue.edu
rhodium.ws
roguesci.org
rotteneggs.com
sciencelab.com
sciencestuff.com
sigmaaldrich.com
skepticfiles.org/new/007
skylighter.com
space-rockets.com
textfiles.com/destruction/nic.txt
  (Captain Hack)
thefreedictionary.com
theodoregray.com
toxprof.crcpress.com

truestarhealth.com
unblinkingeye.com
uncensored.citadel.org
usdoj.gov
users.lycaeum.org
vet.purdue.edu

webelements.com
wikipedia.org, the online encyclopedia
wired.com
yarchive.net
zoot2.com

REFERENCE DOCUMENTATION FOR CONFIRMATION TESTS

Acetate Test 1 Draeger, Gastec

Test 3 Ware, GasTec Corporation, 6431 Fukaya, Ayase-City, 252-1103, Japan; Draeger Safety, AG & Co. KgaA, Revalstrasse 1, 23560 Luebeck, Germany

Alcohol Test (1209) Fearon, Mitchell, *Analyst* 57, 372 (1932)

Aldehyde Test 1 (1213) Fehling, *Ann. Chem.* 72, 106 (1849)

Aluminum Test (997) Dubsky, Hrdlicka, *Mikrochemie* 22, 116 (1937)

Ammonia Gas Test Test 1 (These tests are a procedure based on the fact that ammonia has a pH of 11.)

Aniline Test (2446) Letheby, *Zeitschr. Anal. Chem.* 1, 375 (1862)

Antimony Test Test 1 (2974) Nilson, *Zeitschr. Anal. Chem.* 16, 417 (1877); Test 2 (1851) Hoffer, Drogenhandler 104 (1921)

Barium Test Feigl, *Qualitative Analysis by Spot Tests*, Elsevier, New York, p.165 (1946)

Benzoyl Peroxide Test EM Quant peroxide test papers, 609423 6300, manufactured by EM Science, 480 Democrat Rd., PO Box 70, Gibbstown, NJ 08027 (1960)

Bismuth Test (2288) Kubina, Plichta, *Zeitschr. Anal. Chem.* 72,12 (1927)

Boron Test (3431) Rosenblat, *Zeitschr. Anal. Chem.* 18 (1887) (modified)

Bromide Test (4213) Vitali, *Pharm. Zeitschr. Russland.* 216 (1888), Chem Zentr 810 (1888)

Cadmium Test Feigl, *Qualitative Analysis by Spot Tests*, Elsevier, New York, p. 74 (1946)

Calcium Test (1683) Kotney, State of California Air and Industrial Hygiene Laboratory, Berkeley, CA

Carbon Dioxide Test Gastec Corporation, 6431 Fukaya, Ayase-City, 252-1103, Japan; Draeger Safety, AG & Co. KgaA, Revalstrasse 1, 23560 Luebeck, Germany

Carbonyl Test No reference found

Chloropicrin Test (602) Thompson, Black, *Ind. Eng. Chem.*, 12, 1067 (1920)

Chloride Test Chandlee, Beery, Currier, *Experiments in General Chemistry: Qualitative Analysis*, 9th ed., The Pennsylvania State College, State College, PA, 1949

Chromium Test Cazeneuve, *Compt. Rend.* 131, 346 (1900)

Copper Test (3583) Sarzeau, *J. Pharm.* 516 (1830)

Copper Wire Test Beilstein, *Ber.* 620 (1872); Ruigh, *Ind. Eng. Chem. Anal. Ed.* 11, 250 (1939)

Cyanide Test (1942) Kotney, State of California Air and Industrial Hygiene Laboratory, Berkeley, CA

Cyanide Gas Test Gastec Corporation, 6431 Fukaya, Ayase-City, 252-1103, Japan; Draeger Safety, AG & Co. KgaA, Revalstrasse 1, 23560 Luebeck, Germany

Ephedrine Test Becton Dickinson Public Safety, 1720 Hurd Dr., Irving, TX 75062

Flame Test Orsino C., Smith, *Chemical Analysis of Minerals*, van Nostrand Company, Inc., New York (1948)

Flour Test Lugol, Bardach, *Zeitschr. Physiol. Chem.* 34, 355 (1829); Morikawa, *J. Pharm. Soc. Japan* 163 (1900)

Fluoride Test (775) DeBoer, *Rec. Trav. Chim* 44, 1071 (1925); *Glastech. Ber.* 13, 86 (1935).

Hydrochloric Acid Gas Test GasTec Corporation, 6431 Fukaya, Ayase-City, 252-1103, Japan; Draeger Safety, AG & Co. KgaA, Revalstrasse 1, 23560 Luebeck, Germany

Hydrazine Test (2867) Montignie, *Bull. Soc. Chim.* 51, 127 (1932)

Iodine Test (4382) *Sandlund Chem.* 128 (1894)

Iodide Test Courtoes, France (1811)

Iodine Crystal Test (4382) *Sandlund Chem.* 128 (1894)

Iron Test 1,10-phenanthroline, Kim, Carter, Kupel, *Test for Screening Asbestos*, National Institute for Occupational Safety and Health, Cincinnati, OH (1979)

Lead in Gasoline Feigl, *Qualitative Analysis by Spot Tests*, 3rd ed., Elsevier, New York (1946)

Lead Test Kotney, State of California Air and Industrial Hygiene Laboratory, Berkeley, CA

Lewisite Test Test 1 (1934) Ilosvay, *Ber.* 32, 2697 (1899); Czako, *Zeitschr. Angew. Chem.* 44, 388 (1931)

Magnesium Test Kim, Carter, Kupel, *Test for Screening Asbestos*, National Institute for Occupational Safety and Health, Cincinnati, OH 1979; see also (370)

Magnet Test Sweetzer (This test is based on the magnetic characteristics of iron, manganese, and cobalt).

Manganese Test HazTech Systems, Inc.

Mercury Test Test 1 (1851) Hoffer, *Drogenhandler* 104 (1921); Test 2 (1237) Feigl, Neuber, *Zeitschr. Anal. Chem* 62, 369 (1923)

Methamphetamine Test Test Q, for methamphetamine, Becton Dickinson Public Safety, 1720 Hurd Dr., Irving, TX 75062

Nickel Test (4100) Tschugaev, *Ber.* 38, 2520 (1905); detection level 2.5-ppm, Merck, 5th ed.

Organic Acids Gas Test Gastec Corporation, 6431 Fukaya, Ayase-City, 252-1103, Japan; Draeger Safety, AG & Co. KgaA, Revalstrasse 1, 23560 Luebeck, Germany

Organophosphate Quick Test Suggested by Arizona State Police

Oxalate Test (3059) Patschowsky, *Ber.* 35, 542 (1913)

Oxidizer Test Runge (1839) (No published test found; however, this test is referenced as ASTM method D.)

Perchlorate Test Schilt (1585) Alfred, *Perchloric Acid and Perchlorates*, The G. Frederick Smith Chemical Company, Columbus, OH (1979)

Peroxide Test EM Quant peroxide test papers, 609423 6300, manufactured by EM Science, 480 Democrat Rd., PO Box 70, Gibbstown, NJ 08027 (1960)

Phenol Gas Test GasTec Corporation, 6431 Fukaya, Ayase-City, 252-1103, Japan; Draeger Safety, AG & Co. KgaA, Revalstrasse 1, 23560 Luebeck, Germany

Phosphate Test Hutton, *General Chemistry Laboratory Text, with Qualitative Analysis*, Charles E. Merrill Columbus, OH (1968)

Phosphide Test Feigl, *Qualitative Analysis by Spot Tests*, Elsevier, New York, p. 74 (1946)

pH Test Sorensen, Denmark (1909) [Test is suggested in EPA SW 846 as a field method. We recommend EM Quant pH paper, EM Quant peroxide test papers, manufactured by EM Science, 480 Democrat Rd., PO Box 70, Gibbstown, NJ 08027 609423 6300, (1960).]

Picric Acid Test (3486) Rozier, *Bull. Sci. Pharmacol.* 81 (1917)

Polytec IV Test Gastec Corporation, 6431 Fukaya, Ayase-City, 252-1103, Japan

Protein Test (883) Deniges, *Bull. Trav. Soc. Pharm. Bordeaux* 54, 49 (1016) (1915)

Silver Test (2558) Lombardo, *Arch. Farmacol. Sper.* 7, 400

Sugar Test (1213) Fehling, *Ann Chem.* 72, 106 (1849)

Sulfate Test (1687) Hahn, *Ber.* 5613, 1733 (1926)

Sulfide Test Hutton, *General Chemistry Laboratory Text, with Qualitative Analysis*, Charles E. Merrill, Columbus, OH, (1968)

Sulfite Test Arthur, Smith, *Semi-micro Qualitative Analysis*, McGraw-Hill, New York (1942)

Sulfur Mustard Test Jacobs, *The Analytical Chemistry of Industrial Poisons*, Interscience, New York, p. 959 (1944)

Tabun Test Scandia Laboratories, Livermore CA, (1967)

Thiocyanates Arthur, Smith, *Semi-micro Qualitative Analysis*, McGraw-Hill, New York (1942)

Tin (2525) Mazuir, *Ann. Chim. Anal. Chim. Appl.*, 2, 9 (1920)

Titanium Test 1 (1948) Jackson, *Chem. News* 47, 157 (1883)

Titanium Test 2 (939) Dobrolyubski, *J. Appl. Chem.* 11, 123 (1938); detection level .00005 mg

Turpentine Test Haztech Systems, Inc.

Urine Test (2871) Moore, *Biochem. Zeitschr.* 149, 575 (1924)

Urea Test Kotney, State of California Air and Industrial Hygiene Laboratory, Berkeley, CA

Vanadium Test (221) Barreswil, *Ann. Chim. Phys.* 20, 364 (1847)

Water Test Haztech Systems, Inc.

Zinc Test (780) Deckert, *Zeitschr. Anal. Chem.* 100, 386 (1935)

Analyst	The Analyst (England)
Ann. Chem.	Annalen der Chemie, Justus Liebig's (Germany)
Ann Chim. Anal. Chim. Appl.	Annales de Chimie Analytique et de Chimie Applique et Revue de Chimie Analytique Reunies (France)
Ann. Chim. Phys.	Annales de Chimie et de Physique (France)
Ber.	Berichte der Deutschen Chemischen Gelsellshaft (Germany)
Biochem. Zeitschr.	Biochemische Zeitschrift (Germany)
Bull. Soc. Chim.	Bulletin de la Societe Chimique de France (France)
Bull. Sci. Pharmacol.	Bulletin des Sciences Pharmacologiques (France)
Bull Trav. Soc. Pharm. Bordeaux	Bulletin des Travaux de la Societe de Pharmacie de Bordeaux (France)
Chem. News	Chemical News (England)
Chem.Zentr.	Chemisches Zentralblatt (Germany)
Compt. Rend.	Comptes Rendus Hebdomadaires des Seances de l'Academie des Sciences (France)
Drogenhandler	Drogenhandler (Germany)
Glastech. Ber.	Glastech Nische Berichte (Germany)
Ind. Eng. Chem.	Industrial and Engineering Chemistry (USA)
J. Pharm.	Journal de Pharmacie et des Sciences Accessoires (France)
J. Pharm. Soc. Japan	Journal of the Pharmaceutical Society of Japan (Japan)
Mikrochemie	Mikrochemie: Internationales Archiv fur Deren Gesamtgebiet (Austria)

Pharm. Zeitschr. Russland   Pharmazeutische Zeitschrift Russland (USSR)
Rec Trav. Chim.   Recuiel des Travaux Chimiques des Pays-Bas (Holland)
Zeitschr. Anal. Chem   Zeitschrift fur Analytische Chemie (Germany)
Zeitschr. Angew. Chem.   Zeitschrift fur Angewandte Chemie (Germany)
Zeitschr. Physoil. Chem.   Zeitschrift fur Physiologische Chemie (Germany)

CHEMICAL ABSTRACTS SERVICE REGISTRY NUMBERS

CAS	Chemical	CAS	Chemical
00100-07-0	Hexamine	10042-76-9	Strontium nitrate
00100-07-0	Hexamethylenetetramine	10043-01-3	Aluminum sulfate
00100-07-0	Methenamine	10043-35-3	Boric acid
00100-07-7	Portable cooker	10043-52-4	Calcium chloride
0315-04-4	Antimony pentasulfide	10045-94-0	Mercuric nitrate
1-01-0	Hydrochloric acid	101-68-8	Diphenylmethane-4-
1-01-0	Muriatic acid		diisocyanate
100-15-2	*N*-Methyl-*p*-nitroaniline	10102-43-9	Nitrogen oxide
100-37-8	Diethylaminoethanol	10102-44-0	Nitrogen dioxide
100-42-5	Styrene	10102-45-1	Thallium nitrate
100-44-7	Benzyl chloride	10112-91-1	Calomel
100-52-7	Benzaldehyde	10124-56-9	Calgon
100-54-3	Caproyl hydride	10125-13-0	Copper(II) chloride
100-54-3	Hexane	102-07-8	Diphenyl urea
100-61-0	Phenylhydrazine	102-71-6	Triethanol amine
100-62-7	Benzoic aldehyde	10294-33-4	Boron tribromide
100-66-3	Anisole	10294-34-5	Boron trichloride
100-66-3	Methoxybenzene	10294-38-4	Barium chlorate
100-81-1	Methyl styrene	10294-40-3	Barium chromate
100-86-1	Pyridine	103-32-2	Benzyl formic acid
10024-97-2	Nitrous oxide	103-79-7	1-Phenyl-2-propanone
10025-67-9	Sulfur chloride	103-79-7	P-2-P
10025-67-9	Sulfur monochloride	103-82-2	Phenylacetic acid
10025-69-1	Stannous chloride	103-84-4	Acetanilide
10025-69-1	Tin chloride	10361-44-1	Bismuth nitrate
10025-70-4	Strontium chloride	104-15-4	*p*-Toluenesulfonic acid
10025-87-3	Phosphorus oxychloride	10545-99-0	Sulfur dichloride
10026-13-8	Phosphorus pentachloride	10588-01-9	Sodium dichromate
10028-15-6	Ozone	106-46-7	Dichlorobenzene
10028-22-5	Ferric sulfate (anhydrous)	106-46-7	Paradichlorobenzene
10034-85-2	Hydriodic acid	106-46-7	Urinal cakes
10035-10-6	Hydrobromic acid	106-95-6	Allyl bromide
10035-10-6	Hydrogen bromide	106-99-8	Butane

Chemicals Used for Illegal Purposes: A Guide for First Responders to Identify Explosives, Recreational Drugs, and Poisons, By Robert Turkington
Copyright © 2010 John Wiley & Sons, Inc.

CAS	Chemical	CAS	Chemical
1066-50-8	Ethylphosphonyl dichloride	110-71-4	Dimethoxyethane
		110-71-4	Dimethyl Cellosolve®
107-02-3	Ethylene chlorohydrin	110-71-4	Dimethyl glycol
107-02-8	Acrolein	110-71-4	Ethylene glycol dimethyl ether
107-05-1	Allyl chloride		
107-06-2	1,2-Dichloroethane	110-71-4	Monoglyme
107-06-2	Ethylene chloride	110-77-0	2-Ethylthioethanol
107-07-3	Chloroethanol	110-80-5	Ethyl Cellosolve®
107-07-3	2-Chloroethanol	110-82-7	Cyclohexane
107-12-0	Ethyl cyanide	110-83-8	Cyclohexene
107-12-0	Propionitrile	110-89-4	Piperidine
107-21-1	Ethylene glycol	110-91-8	Morpholine
107-22-2	Biformyl	111-42-2	Diethanolamine
107-22-2	Glyoxal solution	111-46-6	Diethylene glycol
107-23-1	Dioctyladipate	111-48-8	Thiodiethanol
107-23-1	DOA	111-48-8	Thiodiglycol
107-45-3	Ethylenediamine	112-80-1	Oleic acid
108-10-1	Methyl isobutyl ketone	112945-52-5	Silica
108-10-1	MIBK	113-45-1	Methylphenidate
108-13-4	Malonamide	115-10-6	Dimethyl ether
108-18-9	Diisopropylamine	115-10-6	Methyl ether
108-20-3	Diisopropyl ether	115-77-5	Pentaerythritol
108-20-3	Isopropyl ether	1184-64-1	Copper carbonate
108-24-7	Acetic anhydride	120-12-7	Anthracene
108-24-7	Acetyl oxide	120-57-0	Piperonal
108-39-4	m-Cresol	120-72-9	Indole
108-45-2	2-Chlorohydrin	120-80-9	Catechol
108-46-3	Resorcinol	120-80-9	Pyrocatechol
108-58-7	Resorcinol diacetate	12027-06-4	Ammonium iodide
108-59-2	Carbolic acid	12054-85-2	Ammonium molybdate
108-59-2	Phenol	121-44-8	Triethylamine
108-67-8	Mesitylene	121-45-9	Trimethyl phosphite
108-67-8	Trimethylbenzene	121-51-3	Isoamyl alcohol
108-86-1	Bromobenzene	121-51-3	Potato spirit oil
108-88-3	Toluene	12124-97-9	Ammonium bromide
108-90-7	Chlorobenzene	12125-01-8	Ammonium fluoride
108-93-0	Cyclohexanol	12125-12-9	Ammonium chloride
108-94-1	Cyclohexanone	1213-57-61	Ammonium sulfide
109-66-0	Pentane	12135-76-1	Ammonium bisulfide
109-69-3	n-Butyl chloride	12167-74-4	Calcium phosphate (tribasic)
109-77-3	Malonitrile		
109-87-5	Dimethoxymethane	12167-74-4	Tricalcium phosphate
109-95-5	Ethyl nitrite	122-39-4	Diphenyl amine
109-99-9	Tetrahydrofuran	122-52-1	Triethyl phosphite
109-99-9	THF	123-31-9	Hydroquinone
110-00-9	Furan	123-57-3	Fusel
110-21-4	Glyme	123-73-9	Crotonaldehyde

CAS	Chemical	CAS	Chemical
123-75-1	Pyrrolidine	13268-42-3	Ferric ammonium oxalate
123-91-1	Dioxane	1330-20-7	Xylene
124-38-9	Carbon dioxide	1330-96-4	Borax
124-38-9	Dry ice	1332-37-2	Ferrous oxide
124-40-3	Dimethylamine	1332-37-2	Iron oxide (red)
12403-82-1	Styphnic acid	1332-40-7	Copper(II) oxychloride
12612-12-4	Arsenic sulfide	1333-74-0	Hydrogen
127-00-4	Propylenechlorohydrin	1333-83-1	Sodium bifluoride
127-65-1	Chloramine	1336-21-1	Ammonium hydroxide
127213-77-8	4,5-Dichloro-1,3-dioxolan-	1336-21-6	Nitrosil
	2-one	134-03-2	Sodium ascorbate
129-51-1	Ergot	134-49-6	Phenmetrazine
13011-54-6	Ammonium phosphate	1341-49-7	Ammonium bifluoride
1302-09-6	Silver†(I) selenide	1344-09-8	Sodium silicate
1304-29-6	Barium peroxide	1344-28-1	Alumina
1304-76-3	Bismuth trioxide	13463-67-7	Titanium dioxide
1305-62-0	Calcium hydroxide	13472-35-0	Sodium phosphate
1305-62-0	Slaked lime	13560-89-9	Decachlorane
1309-37-1	Ferric oxide	13823-29-5	Thorium nitrate
1309-37-1	Iron oxide (black)	13840-33-0	Lithium hypochlorite
1310-58-3	Potassium hydroxide	13840-33-0	Shock
1310-66-3	Lithium hydroxide	139-87-7	Ethyl diethanolamine
1310-73-2	Caustic soda	13943-58-3	Potassium ferrocyanide
1310-73-2	Red Devil® cleaner	140-29-4	Benzyl cyanide
1310-73-2	Sodium hydroxide	1401-55-4	Tannic acid
1312-73-8	Potassium bisulfide	141-43-5	Ethanol amine
1313-13-9	Manganese dioxide	141-78-6	Ethyl acetate
1313-60-6	Oxone	142-84-5	Heptane
1313-60-6	Sodium peroxide	143-33-0	Sodium cyanide
1313-82-2	Salt cake	144-55-3	Sodium bicarbonate
1313-82-2	Sodium sulfide	144-55-8	Baking soda
1314-11-0	Strontium oxide	144-55-8	Bicarbonate of soda
1314-20-1	Thorium oxide	144-62-7	Oxalic acid
1314-28-1	Aluminum oxide	1450-14-2	Hexamethyl disilane
1314-41-6	Lead oxide	14635-75-7	Nitrosonium
1314-41-6	Litharge		tetrafluoroborate
1314-56-3	Phosphorus pentoxide	14807-96-6	Talc
1314-62-1	Vanadium pentoxide	149-91-7	Gallic acid
1314-80-3	Phosphorus pentasulfide	149-91-7	3,4,5-Trihydroxybenzoic
1314-85-8	Phosphorus sesquisulfide		acid
1314-88-5	Palladium black	1498-48-4	Ethylphosphinyl dichloride
1317-37-9	Ferrous sulfide	151-36-4	Aziridine
1317-37-9	Iron sulfide	151-50-8	Potassium cyanide
1317-39-1	Copper oxide	15293-77-3	Sodium metaborate
1317-65-3	Calcium carbonate	156-62-7	Cadmium
1319-46-6	Lead(II) hydroxide	15715-41-0	Diethylmethyl phosphonite
1319-46-6	Lead nitrate	160-08-0	Carbon monoxide

CAS	*Chemical*	CAS	*Chemical*
167-63-0	Iso-HEET	281-23-2	Adamantane
167-63-0	2-Propanol	2811-04-9	Solidox
167-63-0	Rubbing alcohol	299-42-3	Ephedrine
16731-55-8	Potassium metabisulfite		(pseudoephedrine)
16731-55-8	Potassium pyrosulfite	300-57-2	Allyl benzene
16853-85-3	Lithium aluminum hydride	3012-65-5	Ammonium citrate
16893-85-9	Sodium hexafluorosilicate	302-01-2	Hydrazine
16940-66-2	Borohydride	303-130-8	Penethyl bromide
16940-66-2	Sodium borohydride	30525-89-4	Paraformaldehyde
16961-25-4	Gold chloride	311-28-4	*n*-Butylammonium
17194-00-2	Barium hydroxide		iodide
1723-14-6	Blue death	3189-13-7	2-Methoxyindole
1762-95-4	Ammonium thiocyanate	3251-23-8	Copper(II) nitrate
1782-42-5	Graphite	33719-74-3	3,5-Dichloroanisole
18480-07-4	Strontium hydroxide	3554-74-3	3-Hydroxy-1-
1910-42-5	Gramoxine		methylpiperidine
1910-42-5	Paraquat	3568-94-3	4-Methylaminorex
19168-23-1	Ammonium	3646-85-0	Zinc chloride
	hexachloropalladate(IV)	3672-37-1	1,3,5-Trifluorobenzene
1918-02-1	Picric acid	3811-04-9	Potassium chlorate
19513-05-4	Manganese(III) acetate	40-63-4	1,4-Butanediol
202-589-0	Eugenol	407-25-0	Trifluoroacetic (anhydride)
2036-96-6	1-Chlorobutane	4261-68-1	*N,N*-Diisopropyl-2-
20667-12-3	Silver oxide		aminoethyl chloride·HCl
21908-53-2	Mercuric oxide	4418-61-5	5-Aminotetrazole
22708-90-3	Aluminum silicate		(monohydrate)
2319-78-9	Gun bluing	46-48-0	Dihydrofuran-2(3*H*)-one
2319-78-9	Hydrogen selenide	461-58-5	Cyanoguanidine
2319-78-9	Selenious acid	461-58-5	Dicyanodiamide
23389-93-5	Magnesium carbonate	463-04-7	Amyl nitrite
2404-03-7	Diethyl *N,N*-	463-04-7	Isoamyl nitrite
	dimethylosphoramidate	463-04-7	Rush
2421-59-0	Tin dioxide	464-07-3	Penacoyl alcohol
245-59-7	Safrol	464-07-3	Pinacoyl alcohol
2465-65-8	*o,o*-Diethylphosphoro-	465-04-7	Snappers
	dithioate	48-78-4	Isopentane
2465-65-8	*o,o*-Diethylphosphoro-	492-41-1	Norephedrine
	thioate	495-76-1	Piperonyl alcohol
25-64-9	Isobutyl propane	497-18-7	Carbodihydrazide
25265-76-3	*m*-Phenyleneamine	497-19-8	Sodium carbonate
25322-69-4	Polypropylene glycol	5-07-0	Acetaldehyde
257-07-8	Dibenz[*b,f*]-1,4-	5-07-0	Ethanol
	oxazephine	5-07-0	Formaldehyde
25895-60-7	Sodium cyanoborohydrate	5-07-0	Formaline
2647-10-1	Palladium chloride	5-07-0	Methanol
26628-22-8	Sodium azide	50-70-4	Sorbitol
27-09-3	Sodium acetate	500-66-3	Olivetol
	(anhydrous)	502-87-0	Lead thiocyanate

CAS	Chemical	CAS	Chemical
506-59-1	Dimethylamine·HCl	590-29-3	Potassium formate
506-68-3	Cyanogen bromide	598-94-7	Dimethylurea
506-77-4	Cyanogen chloride	60-00-4	EDTA
506-87-6	Ammonium carbonate	60-12-8	Phenylethyl alcohol
506-93-4	Guanidine nitrate	60-29-7	Ethyl ether
509-14-8	Tetranitromethane	60-29-7	Diethyl ether
509-34-2	Rhodamine B	60-29-7	Engine-starting fluid
51-05-8	Procaine	60-29-7	Ether
516-02-2	Barium oxalate	606-20-2	2,6-Dinitrotoluene
53125-86-3	Sodium tungstate	6099-90-7	Phloroglucinol
532-27-4	Chloroacetopherone	611-71-2	Mandelic acid
532-27-4	Phenylacyl chloride	611-73-4	Benzoylformic acid
532-32-1	Sodium benzoate	6163-75-3	Dimethylethyl phosphonate
5329-14-6	Sulfamic acid	62-53-3	Aniline
5378-49-4	5-Aminotetrazole	62-56-6	Thiourea
540-51-2	Bromoethanol	62-74-8	Compound 1080
540-72-7	Sodium thiocyanate	62-76-0	Sodium oxalate
540-81-2	Ethylene bromohydrin	622-24-2	Phenylethyl chloride
541-41-3	Ethyl chloroformate	6238-13-7	3-Quinuclidinol
541-42-3	Ethyl chlorocarbonate	624-92-0	Methyl disulfide
5414-24-1	Isopropyl nitrite	629-06-1	2-Chloroheptane
542-88-1	Bis(2-chloroethyl ether)	63-23-1	Diethyladipate
544-92-3	Copper cyanide	63-42-3	Lactose
547-66-0	Magnesium oxalate	6303-21-5	Hypophosphorous acid
5470-11-1	Hydroxylamine·HCl	631-61-8	Ammonium acetate
553-91-3	Lithium oxalate		(anhydrous)
554-13-2	Lithium carbonate	633-08-9	Diethylmethyl phosphonate
556-89-8	Nitrourea	6381-59-5	Potassium sodium tartrate
56-03-1	Baquacil	6381-59-5	Rochelle salts
56-03-1	Biguanidine diperchlorate	64-10-8	Phenylurea
56-23-5	Carbon tetrachloride	64-12-5	Everclear®
56-38-2	Pentaborane	64-17-5	Ethanol
56-81-5	Glycerin	64-17-5	Ethyl alcohol
563-97-3	1-Bromo-1-nitroethane	64-18-6	Bilorin
564-00-1	Erythritol	64-18-6	Formic acid
5647-15-6	Sodium bromide	64-19-2	Glacial acetic acid
5677-24-9	Cyanotrimethylsilane	642-38-6	Quebrachitol
57-11-4	Steric acid	64315-11-3	Charcoal (activated)
57-24-9	Strychnine	646-06-0	1,3-Dioxolane
57-27-2	Morphine	6484-52-2	Ammonium nitrate
57-50-1	Sucrose	6487-48-5	Potassium oxalate
57-55-6	Propylene glycol	65-85-0	Benzoid acid
5794-28-5	Calcium oxalate	67-56-1	HEET
5798-79-8	Bromobenzyl cyanide	67-56-1	Methanol
58-88-9	Lindane	67-56-1	Wood alcohol
58-89-9	Benzene hexachloride	67-63-0	Isopropyl alcohol
58-89-9	Hexachlorocyclohexane	67-64-1	Acetone
584-08-7	Potassium carbonate	67-66-3	Chloroform

CAS	Chemical	CAS	Chemical
67-68-5	Dimethyl sulfoxide	7440-22-4	Silver
67-68-5	DMSO	7440-23-5	Sodium
67-71-0	Dimethyl sulfone	7440-31-5	Tin (powdered)
67-72-1	Hexachloroethane	7440-36-0	Antimony
676-85-5	Methylphosphinyl dichloride	7440-38-2	Arsenic
		7440-42-8	Boron (anhydrous)
676-97-1	Methylphosphonic dichloride	7440-50-8	Copper
		7440-66-6	Zinc
676-98-2	Methylphosphonothio dichloride	7440-67-7	Zirconium
		7446-08-4	Selenium dioxide
676-99-3	Methylphosphonic difluoride	7446-09-5	Sulfur dioxide
		7446-11-9	Sulfur trioxide (anhydrous)
68-12-2	Dimethylformamide	7446-18-6	Thallium sulfate
68-12-2	DMF	7446-70-0	Aluminum chloride
6823-67-4	Silica gel	7446-70-0	Aluminum trichloride
69-65-8	Mannitol	7446-793-0	Zinc sulfate (hydrated)
69-72-7	Salicyclic acid	7447-40-7	Potassium chloride
69-74-8	Sodium monofluoroacetate	7487-94-7	Chalchlor
69-79-4	Maltose(anhydrous)	7487-94-7	Mercuric chloride
71-36-3	*n*-Butylamine	75-00-3	1-Chloroethane
71-41-0	Amyl alcohol	75-00-3	Ethyl chloride
71-43-2	Benzene	75-01-4	Vinyl chloride (gas)
74-82-8	Methane	75-05-8	Acetonitrile
74-82-8	Natural gas	75-05-8	Methyl cyanide
74-83-5	Methyl bromide	75-09-2	Dichloromethane
74-84-0	Ethane	75-09-2	Ethylene dichloride
74-85-1	Ethylene (ethene)	75-09-2	Methylene chloride
74-86-2	Acetylene	75-11-6	Methylene iodide
74-87-3	Chloromethane	75-15-0	Carbon disulfide
74-87-3	Methyl chloride	75-20-7	Bangsite
74-88-4	Methyl iodide	75-20-7	Calcium carbide
74-89-5	Monomethylamine	75-21-8	Ethylene oxide
74-90-8	Hydrogen cyanide	75-29-6	Isopropyl chloride
74-98-6	Propane	75-31-0	Isopropyl amine
7429-90-5	Aluminum	75-36-5	Acetylchloride
7439-87-6	Iron	75-44-5	Phosgene
7439-87-6	Iron filings	75-52-5	Nitrocarbol
7439-92-1	Lead acetate	75-52-5	Nitromethane
7439-92-1	Sugar of lead	75-64-9	2-Amino-2-methylpropane
7439-93-2	Lithium	75-64-9	*tert*-Butylamine
7439-95-4	Magnesium	75-74-1	Tetramethyllead
7439-97-6	Mercury	75-77-8	Pinacolone
7440-28-0	Thallium	755-56-2	Iodine
7440-02-0	Nickel	7550-45-0	Titanium tetrachloride
7440-02-3	Raney nickel	756-79-6	Dimethylmethyl phosphonate
7440-05-3	Palladium		
7440-06-4	Platinum	7587-88-9	Bath salts
7440-09-7	Potassium	7587-88-9	Epson salts

CAS	Chemical	CAS	Chemical
7587-88-9	Magnesium sulfate	7727-37-9	Nitrogen (liquid)
76-93-7	Benzilic acid	7727-37-9	Nitrogen (gas)
760-06-2	Chloropicrin	7727-54-0	Ammonium persulfate
7601-89-0	Sodium perchlorate	7732-18-5	Ice
7601-90-3	Perchloric acid	7732-18-5	Water
7616-94-6	Perchlorofluoride	7740-32-6	Titanium
762-08-9	Diethyl phosphite	7740-33-7	Tungsten
7631-86-9	Cab-o-Sil	775-08-0	Ferric chloride (anhydrous)
7631-89-4	Tree stump remover	775-14-6	Sodium hydrosulfite
7631-90-5	Sodium bisulfite	7757-79-1	Potassium nitrate
7631-99-4	Sodium nitrate	7757-79-1	Saltpeter
7632-00-0	Sodium nitrite	7757-82-6	Disodium sulfate
7632-100-0	Nitrox	7757-82-6	Sodium sulfate
7646-78-8	Stannic chloride	7757-83-7	Sodium sulfite
7647-01-0	Hydrogen chloride	7758-08-0	Potassium nitrite
7647-14-5	Salt	7758-62-3	Potassium bromide
7647-14-5	Sodium chloride	7758-99-8	Copper sulfate
7647-14-5	Table salt	7759-02-6	Strontium sulfate
7664-33-2	Phosphoric acid	7761-88-8	Silver nitrate
7664-37-3	Hydrogen fluoride	7772-98-7	Antichlor
7664-39-3	Potassium fluoride	7772-98-7	Sodium thiosulfate
7664-39-4	Hydrofluoric acid	7775-09-9	Sodium chlorate
7664-41-7	Ammonia gas	7778-18-9	Calcium sulfate
7664-93-9	Sulfuric acid	7778-18-9	Gypsum
7681-25-9	Sodium hypochlorite	7778-53-2	Potassium phosphate
7681-11-0	Potassium iodide	7778-54-3	Calcium hypochlorite
7681-38-1	Sodium bisulfate	7778-74-7	Potassium perchlorate
7681-49-4	Sodium fluoride	778-50-9	Potassium dichromate
7681-52-9	Bleach	778-80-5	Potassium sulfate
7681-57-4	Sodium metabisulfite	7782-41-4	Fluorine
7682-07-2	Boron trifluoride	7782-44-7	Oxygen
7697-37-2	Nitric acid	7782-49-2	Selenium
7699-41-4	Silicic acid	7782-50-5	Chlorine
77-09-8	Phenolphthalein	7782-83-0	Ferrrous sulfate
77-13-6	Urea		(pentahydrate)
77-78-1	Dimethyl sulfate	7783-06-4	Hydrogen sulfide
77-92-9	Citric acid	7783-13-8	Ammonium thiosulfate
7704-34-9	Sulfur	7783-20-2	Ammonium sulfate
7711-64-7	Potassium permanganate	7783-82-6	Tungsten hexafluoride
7715-09-9	Weed killers	7783-93-9	Silver perchlorate
7719-09-7	Thionyl chloride	7783-96-21	Silver iodide
7719-12-2	Phosphorus trichloride	7784-01-2	Silver chromate
7722-84-1	Hydrogen peroxide	7784-34-1	Arsenic trichloride
7722-84-1	Wood bleach	7784-46-5	Sodium arsenite
7723-14-6	Phosphorus (red)	7787-71-5	Bromine trifluoride
7723-14-6	Rat nip	7789-00-6	Potassium chromate
7725-14-6	Red phosphorus	7789-09-5	Ammonium dichromate
7726-95-6	Bromine	7789-29-9	Potassium bifluoride

CAS	Chemical	CAS	Chemical
7789-60-8	Phosphorus tribromide	89-69-4	Tartaric acid
7789-69-7	Phosphorus pentabromide	89-83-8	Thymol crystals
7790-94-5	Chlorosulfonic acid	892-80-8	Auramine
7790-98-9	Ammonium perchlorate	9000-01-5	Gum arabic
7791-03-9	Lithium perchlorate	9002-18-0	Agar
7791-25-5	Sulfuryl chloride	9002-86-2	Poly(vinyl chloride)
7798-38-0	Sodium bromate	9002-86-2	PVC
78-00-2	Tetraethyllead	9003-53-6	Polystyrene
78-38-6	Diethylethyl phosphate	9008-8-0	Starch
78-83-1	Isobutanol	9009-70-0	Nitrocellulose
78-89-1	Methyl benzilate	9011-05-6	Nitroform
78-93-3	Butanone	91-20-3	Naphthalene
78-93-3	MEK	91-43-2	Benzol
78-93-3	Methyl ethyl ketone	91-66-7	N,N-Diethanolethylene-
78-95-5	Chloroacetone		diamine
7803-51-2	Phosphine	93-14-1	Guaifenesin
7803-55-8	Ammonium metavanadate	93-97-0	Benzoic anhydride
7803-55-8	Ammonium vanadate	94-36-0	Benzoyl peroxide
784-84-9	Toluene-2,4-diisocyanate	94-89-5	Methylamine
79-03-8	Propionyl chloride	9440-50-8	Copper wire
79-04-9	2-Chloroacetyl chloride	945-59-7	4-Allyl-1,2-methylene
79-11-8	Chloroacetic acid		dioxybenzene
79-11-8	Monochloroacetic acid	95-14-7	Benzotriazole
79-24-3	Nitroethane	95-92-1	Ethyl oxalate
79-37-8	Oxalylchloride	953-98-0	Ethylphosphinyl
800-66-19	Gasoline		difluoride
8001-25-0	Olive oil	96-23-1	1,3-Dichloro-2-propanol
8001-26-1	Linseed oil	96-48-0	γ-Butyrolactone
8006-64-2	Turpentine	96-48-0	Paint stripper
8008-20-6	Kerosene	96-48-125	Floor stripper
8014-95-7	Oleum	96-79-7	N,N-Diisopropyl-2-
8014-95-7	Pyrosulfuric acid		aminoethane chloride
8015-86-9	Carnuba wax	96-80-0	N,N-Diisopropyl-2-
8032-32-4	Ligroin		aminoethanol
8036-14-7	Marijuana	97-56-1	Methyl alcohol
82-45-1	Aminoanthroquinone	98-07-7	Benzoic trichloride
830-79-5	Trimethoxybenzaldehyde	98-07-7	Benzo trichloride
831-52-0	Sodium picramate	98-88-4	Benzoyl chloride
84-74-2	Dibutylphthalate	98-95-3	Nitrobenzene
85-98-3	Diethyldiphenylurea	99-09-2	m-Nitroaniline
868-85-9	Dimethyl phosphite	99-33-2	3,5-Dinitrobenzoyl
87-66-1	Pyrogallic acid		chloride
88-72-2	Nitrotoluene	99-34-3	3,5-Dinitrobenzoic acid
886-84-2	Potassium citrate	993-13-6	Methylphosphonic acid

INDEX

Printed in the United States
By Bookmasters